THEORY OF PHYSICAL AND TECHNICAL MEASUREMENT

FUNDAMENTAL STUDIES IN ENGINEERING

Vol. 1 Coelho/Physics of Dielectrics for the Engineer
Vol. 2 Szücs/Similitude and Modelling
Vol. 3 Scharf, Lisieski/Amplitude Distribution Spectrometers
Vol. 4 Mohilla, Ferenz/Chemical Process Dynamics
Vol. 5 Baker et al./Explosion Hazards and Evaluation
Vol. 6 Sobczyk/Stochastic Wave Propagation
Vol. 7 König/Shakedown of Elastic-Plastic Structures
Vol. 8 Čačko, Bílý, Bukoveczky/Random Processes
Vol. 9 Kubáček/Foundations of Estimation Theory
Vol. 10 Cătuneanu, Mihalache/Reliability Fundamentals
Vol. 11 Boros/Measurement Evaluation
Vol. 12 Baker, Westine, Dodge/Similarity Methods in Engineering Dynamics
Vol. 13 Baker, Tang/Gas, Dust and Hybrid Explosions
Vol. 14 Piotrowski/Theory of Physical and Technical Measurement

FUNDAMENTAL STUDIES IN ENGINEERING 14

THEORY OF PHYSICAL AND TECHNICAL MEASUREMENT

JANUSZ PIOTROWSKI
Silesian Technical University
Gliwice, Poland

ELSEVIER
Amsterdam-London-New York-Tokyo
PWN — POLISH SCIENTIFIC PUBLISHERS
Warszawa
1992

Revised translation from the Polish original *Teoria Pomiarów*, published in 1986 by Państwowe Wydawnictwo Naukowe, Warszawa

Translated by Przemysław Siuda

Distribution of this book is being handled by the following publishers:

For the USA and Canada
ELSEVIER SCIENCE PUBLISHING CO., INC.
655 Avenue of the Americas, New York, NY 10010

For Albania, Bulgaria, Cuba, Czecho-Slovakia, Hungary, Korean People's Democratic Republic, Mongolia, People's Republic of China, Poland, Romania, the USSR, Vietnam and Yugoslavia
POLISH SCIENTIFIC PUBLISHERS PWN Ltd.
Miodowa 10, 00-251 Warszawa, Poland

For all remaining areas
ELSEVIER SCIENCE PUBLISHERS B.V.
Sara Burgerhartstraat 25
P.O. Box 211, 1000 AE Amsterdam, The Netherlands

Library of Congress Cataloging-in-Publication Data
Piotrowski Janusz, doc. dr inż.
[Teoria pomiarów. English]
Theory of physical and technical measurement / Janusz Piotrowski
[translated by Przemysław Siuda]
p. cm. — (Fundamental studies in engineering: 14)
"Rev. translation from the Polish original Teoria pomiarów" — T.p. verso.
Includes bibliographical references and index.
ISBN 0-444-98693-6
1. Mensuration. I. Title. II. Series.
T50. P56 1991
620'. 0044 — dc20 91-15429
CIP

ISBN 0-444-98693-6 (Vol. 14)
ISBN 0-444-41756-7 (Series)

Printed in Poland by D.N.T.

Preface

Developments in mathematical statistics, theories of automatic control and computer science have transformed outlooks on measurement problems. Applications of statistical methods to experiment design, pattern recognition processes and object identification methods developed within the framework of automatic control theory, all include many interesting mathematical procedures which seem to have anticipated the development of metrology.

The growing complexity of the measurement problems and the increasing accuracy and complexity of measuring instruments have made it necessary to apply more advanced methods of analysis and synthesis. The fundamental tool is the theory of measurement.

So far, the development of measurement theory within metrology has lagged far behind the development of measurement methods and techniques. The theory has only interpreted known metrological phenomena, and this situation remains unchanged.

The secondary role of theory, in relation to the methods applied in different kinds of research (especially in physics), is specific to the theory of measurement. The ingenuity and subtlety of the research methods and techniques invented are really astonishing. However, metrology starts at the moment a research technique becomes widely or repeatedly applied, or expands into other disciplines.

The variety of problems addressed by modern science results in a diversity of models of physical and metrological phenomena, and leads to the view that there is not now, and may never be, a uniform theory of measurement. Indeed, there is no reason for creating a uniform model of the existing variety of objects. There is a need, however, to establish a foundation common to all measurements and studies. This would actually itself constitute a theory and is the author's main aim, allied to attempts to create a universal theory of measurement. A concept fundamental to the theory is to examine the purpose of measurement. The aim of every measurement is to formulate a declarative statement about the state of a measured part of reality. In other words, a true and objective image of this reality is to be produced. The theory of measurement should provide the means of proving that the image obtained is true and objective.

At the beginning of the theory of measurement, the following four axioms are accepted.

1. Isomorphic relations occur between the states of a given quantity and the values of that quantity.

2. The mapping of a state of a given quality into an image of the state is ambiguous—a point is mapped into a set.

3. The ambiguity of the mapping of a state into an image, produced with a measuring instrument, may be determined from the mathematical model which describes the metrological properties of the instrument.

4. The declared image of reality is related to some agreed reference states.

The above axioms have been known and used for some time, although they have never been explicitly formulated together.

The model of error is an essential part of every theory of measurement. Here, error is assumed to be a multidimensional, non-stationary random process. The process may be reduced to a stationary form by means of appropriate transformations. According to the author, such a model well describes the metrological properties of a number of measuring instruments and also allows the methods of mathematical statistics to be more fully applied.

As in every theory, here also the reader may find different interpretations of some of the notions and methods from others found in the literature on this subject. For example, a different interpretation is given of the regression and least square methods. The distinction—if noticed at all—is based on the model class. The deterministic model involves the least square method, and the stochastic model the regression method. In the author's opinion, this separation has a historical origin, but the division itself is not essential. It is the primary assumptions which are really important, as well as the way in which they are applied. In the second kind of the regression method, the basic assumptions for the form of the model and constant variance of the stochastic component are sharply defined. The reason is that the assumptions influence the estimates of errors of the model coefficients and the estimation process of the random component. Further assumptions are therefore less significant.

Some details of the models, for example the interpretation of the accuracy class, may be disputed. However, unsatisfactory models must be accepted unless better ones are proposed, since they do allow interpretation of some phenomena.

Due to the time and space limitations, the author has not attempted to present all the applicable methods. His intention is to trace the entire process and show that a certain problem may be solved by a given method. Analysis of other methods applicable to solving the same problem may extend the treatment given in this book, and will no doubt enrich the theory.

The reader should be familiar with mathematical statistics and the theory of probability, as well as with metrology. Thus no mathematical proofs will be given, as they may be easily found in the bibliography.

Janusz Piotrowski

Contents

PREFACE V
Chapter 1. Establishing algorithms for the cognition process 1
1.1 The logical and formal rules for producing images of reality 1
1.2 Physical scales and the ambiguity of images of reality 18
1.3 Multidimensional scales 28

Chapter 2. Standards and the propagation of standards 37
2.1 Establishing the standards of units of measurement 37
2.2 Propagation of standards 42
2.2.1 Single comparison 46
2.2.2 n-fold comparison with the same primary standard 49
2.2.3 m-fold comparison with different standards 50
2.2.4 n-fold comparison with m different primary standards 51
2.2.5 m times larger measure of a secondary standard 52
2.2.6 m times smaller measure of a secondary standard 54
2.3 The certified reference materials 60

Chapter 3. Modelling of measurement systems 67
3.1 The general principles of modelling 67
3.2 Modelling of the characteristics of measurement systems 72
3.2.1 The static model 73
3.2.2 Dynamic models 76
3.2.3 Discrete-time models of measurement systems 86
3.3 The modelling of error sources. Probabilistic models 89
3.3.1 Nonlinear elements 95
3.3.2 Dynamic elements 98
3.3.3 Non-stationary error sources 100
3.4 The general model of error 104
3.4.1 The continuous-time model 104
3.4.2 The discrete-time model 111
3.4.3 The correlation of errors 113
3.5 Identification of models of measurement systems 117
3.5.1 Determination of the variables of a model 118
3.5.2 Correlation analysis 121
3.5.3 Regression analysis 130
3.5.4 The least squares method 140
3.5.5 Other methods 144

Chapter 4. The theory of error 146
4.1 Images of reality in measurement systems. Principles of image production . . 146

4.2 Optimization of calibration procedures 150
4.3 Definitions and classification of errors 153
4.4 The metrological characteristics of measurement systems and standards 164
4.5 Metrological characteristics. Methodology of their determination and interpretation 174
4.5.1 Determination of class 1 characteristics 174
4.5.2 Theoretical determination of the limiting distributions of errors 185
4.5.3 Experimental determination of the limiting distributions of errors . . . 193
4.5.4 Interpretation of the concept of accuracy class 199
4.6 The basic concepts of inference in the theory of error 207

Chapter 5. Measurement procedures. 220
5.1 Definition of measurement 220
5.2 Procedures describing measurement methods. 225
5.3 Calibration procedures 230
5.4 Reading checking and adjustment of measurement systems 242
5.5 Measurement procedures for constant quantities 247
5.5.1 Single measurement 248
5.5.2 Multiple measurement—the measurement series 250
5.6 Measurement procedures for time-dependent quantities 256
5.6.1 Single measurement 257
5.6.2 Multiple measurement 265

Chapter 6. The design of measurements and experiments 266
6.1 Selection of standards for calibration of measuring instruments 266
6.1.1 Transfer of a measure of a standard onto measuring instruments 266
6.1.2 Selection of the calibration points 272
6.1.3 The boundary problems of G-optimum designs 279
6.1.4 The design principles of calibration processes 282
6.2 Selection of the error sampling interval 285
6.2.1 Error estimation in a correlated-error measurement series 285
6.2.2 Total error analysis for a measuring instrument 289
Appendix. The Lebesgue measure of a set 292
References 294
Index . . . . .

Chapter 1

Establishing Algorithms for the Cognition Process

1.1 The Logical and Formal Rules for Producing Images of Reality

The concept of measurement is understood differently in different sciences. In philosophy, measurement is presented as a tool of the cognition process.

In terms of philosophy, the theory of cognition concerns the forms of acquiring information about the surrounding reality. It also deals with the essence and the results of the cognitive act. It is essential to determine the sources of knowledge, the limits of cognition and the veracity of the knowledge, so as to know whether the knowledge is objective, verifiable and valuable.

There are two forms of cognition: sensory and rational. In *sensory cognition*, the subject (man) acquires sensory impressions which are an image of reality. These impressions may be expressed in the form of declarative statements. *Rational cognition* includes analysis, synthesis, translation and inference of the images of reality, and is thus performed in the domain of abstraction. Questions and hypotheses are formulated in the process of rational cognition, and relationships between the phenomena and the objects are deduced.

Observations and experiments are the tools of sensory and rational cognition. When the impressions are to determine quantitative relations, the observations are called *measurement*. Thus, measurement is an act of cognition of reality. The result of measurement must meet the condition of objective truth. In order an image of reality produced in the domain of abstraction in an act of measurement to be made objective, certain rules must be obeyed. The *object of measurement* is a selected *quality* of a given object or phenomenon that is an element of reality. The quality is revealed with different levels of intensity in the subjects (or one subject) investigated. The levels of intensity are expressed as different *states* of the quality, as shown in Fig. 1.1. The cognition of individual objects is the distinction of the state of a given quality. The number of these states is hard to determine a priori—this is an effect of the uniqueness of the quality. The number of states also depends on the set of objects having the same quality, as well as on some different factors, in particular on the subject of cognition. The *subject of cognition*

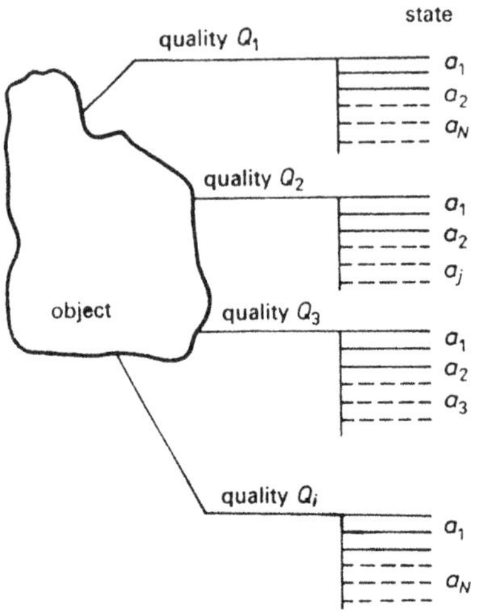

Fig. 1.1. Basic notions in an act of cognition of an object.

is man, and in the act of cognition he has to decide how profound the cognition should be and what is its aim.

The given quality may be specific to an object when the properties of this object are being examined. The number of states may then be smaller than the number of states of a set of objects with the same quality.

The image of the state of a selected quality of the object investigated is a symbol which maps the state of the quality in the domain of the image, i.e., the abstraction. Most frequently, the domain of the image is a set of real numbers or selected subsets of them. The element which maps the state of the quality is called the *measure of the state* or a *value of the quantity*. Unambiguity of the assignment is the necessary condition of unambiguity of the mapping of the states of a given quality in the domain of symbols (expressing the values of the quantity, i.e., the values of the states of the quality).

The following sets exist in the real domain:

— the set of objects M;

— the set of qualities Q as a property of one or more objects;

— the *set of states* $A = (a_1, a_2, \ldots, a_N, \ldots)$ assumed by each of the qualities.

The following sets are created in the domain of abstraction:

— the set of the objects' names M';

— the set of quantities Q' expressing the qualities distinguished;

— the *set of values* of the quantity $B = (b_1, \ldots, b_N, \ldots)$.

An illustration of these is presented in Fig. 1.2.

The *image of reality* is a *mappings*:

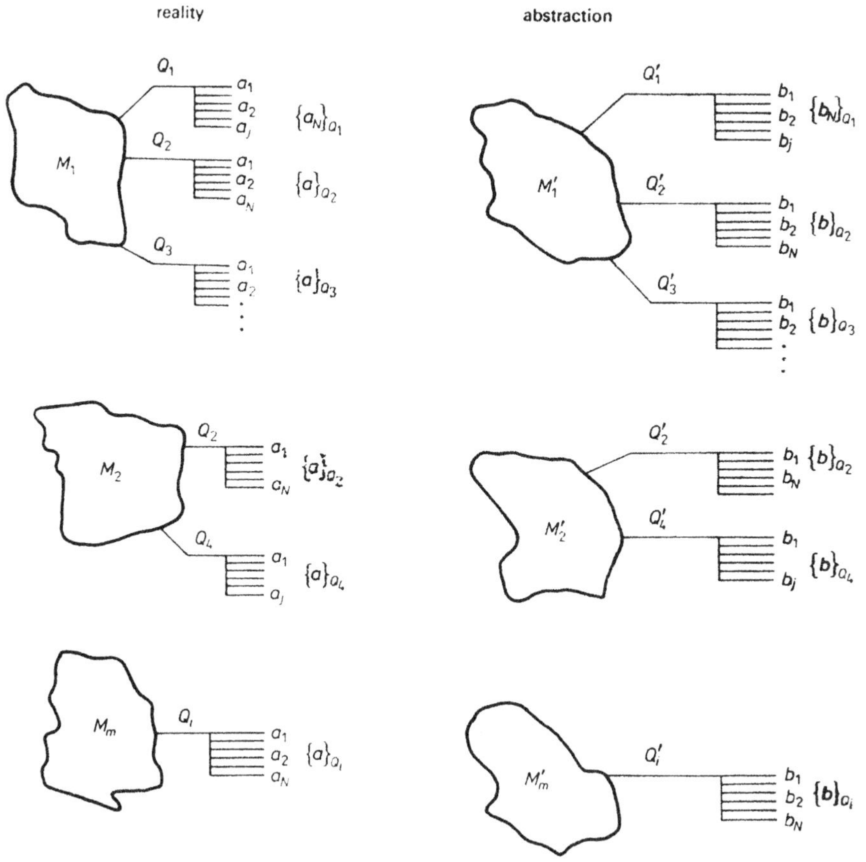

Fig. 1.2. A part of reality and its image in the domain of abstraction.

$$\{M'\} \equiv \{M\}, \qquad \{Q'\} \equiv \{Q\}, \qquad \{A\} \rightarrow \{B\}, \tag{1.1}$$

$$b = f(a), \tag{1.2}$$

such that for all the elements of the set (A) an inverse mapping

$$a^* = \varphi(b) = f^{-1}(b) \tag{1.3}$$

exists and satisfies the condition

$$a^* \equiv a \tag{1.4}$$

for all the elements of $\{A\}$.

The symbol f^{-1} denotes the mapping inverse to f. This means that in the domain of the set $\{A\}$ the mapping f is one-to-one mapping. There is no requirement that the powers of the sets $\{A\}$ and $\{B\}$ be equal. Elements of the set $\{B\}$ which are the mapping of no element of set $\{A\}$ may also exist.

The mapping f_Q is called the *scale of the quatity Q*. Such a scale is further called the *empirical scale*. The empirical scale allows the creation of images of reality in the form of declarative statements, by stating the equivalence of the states $\{A\} = (a_1, \ldots, a_N)$ considered the reference with the examined states $\{X\} = (x_1, \ldots, x_N)$ of the same quality. Thus, when a state x_i has an image b_j attributed under the scale f_x:

$$b_x = f_x(x_i), \tag{1.5}$$

then because of the existing relation defining the scale of the quantity x:

$$b_j = f_x(a_q), \tag{1.6}$$

and because of the identity of b_j based on the inverse mapping

$$x_i = f_x^{-1}(b_j), \quad f_x^{-1}(b_j) = a_q, \tag{1.7}$$

we obtain

$$x_i \Leftrightarrow a_q. \tag{1.8}$$

As a consequence of the one-to-one mapping, the identity of values means the equivalence of states of the quantity.

The image of reality produced in this way must be very inadequate, as it does not determine the interrelations of the states of a given quality distinguished. The relations between the real numbers, which produce the images of reality, are well known and need no explanation. The basic aim of cognition of reality is to distinguish and determine the qualities which express the properties of different objects and phenomena. Another aim is to recognize the relations occurring between the states of a given quality.

Mass and length relations were the first to be recognized, and knowledge of them was the most profound. The length relations of intervals were the subject of geometry and related empirical studies have been performed as well. The introduction of the metric system by the end of the eighteenth century (in 1795)[1] codified the knowledge of those times. The *image domain* is a set of positive real numbers marked on the real axis. The relations of the values of length are valid also for the states of the quantity, and hence they are isomorphic. This property is easily verifiable empirically. The following relations are of basic significance:

---

[1] The standard was established in 1799, the Gaussian system of units was created in 1830, and the Metre Convention was internationally created in 1875. The starting point were the prototype definitions of the metre and the kilogram in 1889.

— *equivalence*—every two elements of a set B are related in one of two possible ways, either

$$b_1 = b_2 \quad \text{or} \quad b_1 \neq b_2, \tag{1.9}$$

— *strict order*—every element of the set B has its proper position on the real axis; every two elements are related in one of two ways:

$$b_1 < b_2 \quad \text{or} \quad b_1 > b_2, \tag{1.10}$$

— *equivalence and strict order of the intervals*—every difference of two elements is related to the difference of two others as follows:

$$b_1 - b_2 = b_3 - b_4 \quad \text{or} \quad b_1 - b_2 < b_3 - b_4, \quad \text{or} \quad b_1 - b_2 > b_3 - b_4, \tag{1.11}$$

— *equivalence and strict order of quotients*—every quotient of two elements is related to a quotient of two other elements in one of the following ways:

$$\frac{b_1}{b_2} = \frac{b_3}{b_4} \quad \text{or} \quad \frac{b_1}{b_2} < \frac{b_3}{b_4}, \quad \text{or} \quad \frac{b_1}{b_2} > \frac{b_3}{b_4} \tag{1.12}$$

(the elements b_2, b_4 must not be zero).

The corresponding elements of the set A are related identically. Geometry shows how to verify the above relations. Identical relations apply to the states of mass. The states and the differences of states may be verified using an equal-arm balance, while for quotients a balance with a variable ratio of arm lengths must be applied.

As other qualities and relations of states were recognized, they were represented by real numbers. This implied transfer of these relations onto other quantities. The experimental verification of these relations was performed by formulation of physical laws. It was assumed implicitly at the time that the images of reality are expressed in the domain of real numbers, i.e., on *metric scales*.

The relations of the states of two dependent quantities are strictly determined, but they are not necessarily isomorphic. For example, the kinetic energy of a uniformly moving body is

$$E = \tfrac{1}{2}mv^2. \tag{1.13}$$

Doubling of the velocity of a body having a mass m causes its energy to increase four times. The situation is easy to imagine when the velocity is expressed by the values

$$v^2 = v', \quad v' \in R, \tag{1.14}$$

i.e., A would be the set of the states v' and B the set of the values R^+. Then doubling of the velocity v' causes the energy to be doubled as well. The law describing the relationship of the velocity v', the mass m and the kinetic energy is now

$$E = mv'. \tag{1.15}$$

A similar situaton arises for a three-argument relationship expressing some physical law, e.g.

$$P = UI. \tag{1.16}$$

By introducing one-to-one mapping, for example $f(x) = e^x$, while

$$P = e^z, \quad U = e^x, \quad I = e^y, \tag{1.17}$$

the relationship

$$z = x+y \tag{1.18}$$

is obtained. This means that the choice of relation among the values of a quantity influences the form of the relationships between different states of the qualities of objects and phenomena. The relations of equivalence and of strict order of the states, intervals and quotients are valid for the elements of the sets P, U, I. However, not all of the relations apply to the states of the sets X, Y, Z.

If $U_1 \lesseqgtr U_2$, then $x_1 \lesseqgtr x_2$,

but if $U_1 - U_2 \lesseqgtr U_3 - U_4$, then $x_1 - x_2 \lesseqgtr x_3 - x_4$ does not hold,

and if $\dfrac{U_1}{U_2} \lesseqgtr \dfrac{U_3}{U_4}$, then $\dfrac{x_1}{x_3} = \dfrac{x_3}{x_4}$ does not hold,

but if $\dfrac{U_1}{U_2} \lesseqgtr \dfrac{U_3}{U_4}$, then $x_1 - x_2 \lesseqgtr x_3 - x_4$.

Therefore, a representation of the *values* by positive real numbers has been assumed when defining the *states of qualities* of objects. Thus, a basis has been created for producing the elementary *images of reality*. This is the basic principle for expressing and establishing the physical laws determining the relations of the states of different qualities. If some other mappings were applied, the image of reality produced by physics would be different and would be described by different laws. The physical laws in turn allow determining the proportions of the intervals and the quotients of the states of quantities describing more complex phenomena of reality by means of rational cognition.

Let us consider once again the *mapping function* which maps the states of quantities onto the values (1.2), together with the properties of the set $B = R^+$. When the relations of equivalence and of strict order of the states, intervals and quotients are all valid, the same relations are valid for

$$b = kf(a), \tag{1.19}$$

where k = constant. In substance, k is unknown, but by applying different values of k, different sets B are obtained as a linear mapping of a primary set (for $k = 1$). For the base quantities, it is set convention that $k = 1$, giving the value of 1 to the distinguished state. The values of k for other quantities (derived quantities) are also conventional. When $k = 1$, the derived quantities are included in a coherent system of units. At present, the International System of Units (SI) is generally accepted by convention.

Formulation of the images of reality is performed in two groups of physical quantities:

(i) quantities called the *metric quantities*; their units are included in the SI and their images are the real numbers;

(ii) quantities measured under *empirical scales* and quantities represented in the domain of real numbers, but not included in the SI.

Let us review the state of knowledge for each of these groups, deciding if the state is stable and what problems still remain to be solved by metrology.

One more truism has to be mentioned here. The problem of measurability of the phenomena, and therefore also the problem of the units of measurement, is not only a basic problem of metrology, but of all the cognitive sciences. A field of knowledge becomes an exact science when the character of the study of relations becomes not only qualitative but also quantitative [1, 137]. The measurability of phenomena, being the subject of enquiry of a given science, is a measure of the exactness and progress in the science.

In respect of the measurable quantities, the historic development of systems of units is a reflection of the development of knowledge and science. The *metric system* included two quantities: length and mass. The Gaussian (CGS) system extended the range of coherent phenomena. The units for electric quantities (described by Giorgi), thermodynamic and light quantities were further included in the CGS and MKS systems, and thus the coherent system comprised nearly all domains of the material world. Possibly other branches of science will also be included in the SI.

The quantities measured under empirical scales are characteristic for the second group. The well known scales are:

— the Beaufort Scale—a 12-degree scale of wind force, introduced in 1806. Part of the scale is presented in Table 1.1;

— Mohs' Scale of Hardness, based on ten separate minerals, introduced in 1812;

— the Sörensen Scale of reaction determination by the value of pH, 1909;[2]

— Engler's Scale of viscosity;

— the scale of water colour, hardness, turbidity, etc.

These scales were created for practical purposes, based on the level of knowledge of their times.

---

[2] The pH scale has the following theoretical interpretation: it expresses the concentration of hydrogen ions as a logartihmic function. The concentration is a metric quantity.

TABLE 1.1
An excerpt from the Beaufort scale (degrees 9 to 12 omitted).

Degree	Wind	Symptoms on the open sea	Symptoms in land areas	Wind velocity m/s
0	calm	mirror-smooth sea	smoke goes up vertically	0–0.2
1	light air	scale-like ripples	smoke column slightly deflected	0.3–1.5
2	light breeze	small, short not-breaking waves	a light wind may be felt on face	1.6–3.3
3	gentle breeze	small, short waves with breaking crests; glossy looking foam	moving leaves	3.4–5.4
4	moder. breeze	small waves starting to elongate, a few white crests	small branches moving	5.5–7.9
5	fresh breeze	medium, elongated waves, a lot of white crests	larger branches moving	8.0–10.7
6	strong breeze	big waves with white foamy crests	thick branches moving	10.8–13.8
7	moder. gale	swell sea, the wind blows out foam from breaking waves	thin trunks moving	13.9–17.1
8	fresh gale	medium high, long waves, the wind blows off the breaking wave tops in form of whirling spray	walking against the wind is difficult; thick trunks pushed down	17.2–20.7

The logical rules of formulation of *additive* (*metric*) *scales* were published by Helmholtz in 1887. The axiomatic properties of measurable quantities were discussed by Hölder [36], based on the axiomatic Euclid geometry and on the works on irrational numbers by J. W. R. Dedekind. The properties of different scales were later generalized by Campbell [10], Stevens [72], and others. The contemporary logical and formal *scale theory* and the theory of measurable quantities were presented by Pfanzagl [85]. These problems are considered in detail in Refs. [54], [120]. Works [1], [131] present a popular approach to the theory. The types of scales, depending on the relations between the states of a given quality, are shown in Table 1.2.

When a quantity is expressed by a quotient (metric) scale, the relations between the states are better determined than in the case of the Beaufort scale. The Beaufort scale representation within the metric scale is given in the last column of Table 1.1.

TABLE 1.2
Types of measurement scales.

Item	Scale type	Relations of the states	Number of reference states	Examples
1	Nominal	Equivalence of the states	N	Colour patt., image shape
2	Ordinal	Equivalence and strict order of the states	N	Mohs', Beaufort scales
3	Interval scale	Equivalence and strict order of the states and intervals of the states	2	Calendar time scale, Celsius temperature scale
4	Metric (quotient)	Equivalence and strict order of the states, intervals and quotients of the states	1	Scales of the quantities covered by the CGS and SI systems
5	Natural	Equivalence and strict order of the states, intervals and quotients of the states	0	Counting of a number of elements of a set

The Engler viscosity scale is represented in the metric scale of a coherent system of units (E—Engler viscosity, ν—kinematic viscosity) by

$$\nu = \frac{E}{100} \times 7.6^{(1-E^{-3})} \times 10^{-4}\ \mathrm{m^2/s}, \tag{1.20a}$$

or

$$\nu = \left(7.31 \times E - \frac{6.31}{E}\right) \times 10^{-6}\ \mathrm{m^2/s}. \tag{1.20b}$$

An empirical scale is used when the level of knowledge about a phenomenon is insufficient to determine relations between the states of a given quality. Also, when the empirical scale is convenient and its application is sufficient for practical purposes, it may be employed. Here are some examples.

Until recently colour was measured on a nominal scale. Atlases and colour patterns with several distinguished and named colours, or with several hundreds of named and numbered colours, are available. The fact of matching or not matching may be stated after a sensory comparison of the colours of investigated objects against the pattern. The result of the comparison makes a proper declarative statement possible, producing an image of reality on a nominal scale. The pattern states are the reference in this case. Colour, as a quality subjected to sensory investigation, appears very complex. The sensory impressions include three independent quantities: shade, saturation and brightness. Colour is a property of an object, concerning its surface. However, it is also a property of radiation transmitted or scattered by the object. In an ordinary visual field, the visual sensations depend on the shape of an object, on its dimensions, the distance to the object, etc., as well as on the properties of the eye and on the mental state of the observer. Thus, the cognition of the quality has two aspects—physical and psychophysical. In the first aspect, the problem was solved when the wave nature of light and its spectral properties were recognized. The second aspect concerns human sensations.

Colour mixing experiments have been undertaken, based on Grassman's laws formulated in 1853. In 1931, the International Commission on Illumination (CIE[3]) developed the trichromatic colour scale. The quality of an object is its chromatic colour, consisting of three primary colours: red (R), green (G) and blue (B). The spectral characteristics of the primary colours are shown in Fig. 1.3.

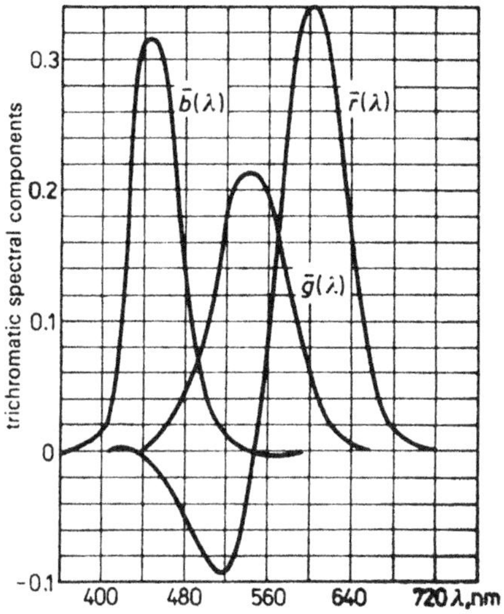

Fig. 1.3. Spectral characteristics of the primary components of a chromatic colour, according to CIE (*Proc. 8th Session*, Cambridge, 1931).

The numerical value of a measure of colour is a vector with three components $\bar{r}(\lambda)$, $\bar{g}(\lambda)$, $\bar{b}(\lambda)$, in a three-dimensional space. The principles of operations of vector addition and vector multiplication by a number are different here from the case of a vector space in the general sense. The coordinates of the space are the real axes. All the vectors lying on the line passing through the origin of the coordinates have the same colour. The value of luminance increases with the distance of the point from the origin of coordinates.

[3] CIE—Commision International de l'Eclairage.

To characterize a colour, it is only necessary to know the vector components of the length, meeting the condition:

$$C = \frac{R}{R+G+B}(R) + \frac{G}{R+G+B}(G) + \frac{B}{R+G+B}(B) = 1. \tag{1.21}$$

The symbols in parentheses denote a primary component and the letters R, G, B the amount of this component. The fraction determines the relative part of a given component.

The addition of two colours described by the vectors:

$$C_1 = r_1(R)+g_1(G)+b_1(B), \tag{1.22a}$$

and

$$C_2 = r_2(R)+g_2(G)+b_2(B), \tag{1.22b}$$

gives a colour

$$C = C_1+C_2 = r(R)+g(G)+b(B), \tag{1.23a}$$

where

$$r = \frac{r_1+r_2}{r+g+b}, \quad g = \frac{g_1+g_2}{r+g+b}, \quad b = \frac{b_1+b_2}{r+g+b}. \tag{1.23b}$$

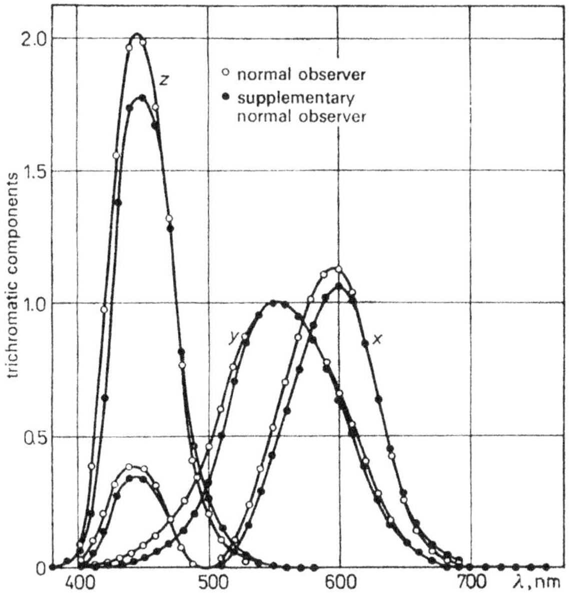

Fig. 1.4. The spectral characteristics of the imaginary trichromatic components x, y, z for a visual field of 2° (normal observer), according to CIE 1931, and for a visual field of 10° (supplementary normal observer), according to CIE 1964 (*Colorimetry, Official CIE Recommendations*, Publ. No 15, 1971).

By means of proper transformations, new coordinates X, Y, Z of the component imaginary colours have been introduced. The spectral characteristics of the new colours are shown in Fig. 1.4. The specific character of the addition operation makes it convenient to present the vector of colours on the plane as a right-angled triangle. For this triangle, the coordinate $z = 0$, and the distribution of colours is more uniform. The rules of transformation and the colourimetric calculations are too complex to be presented here [20].

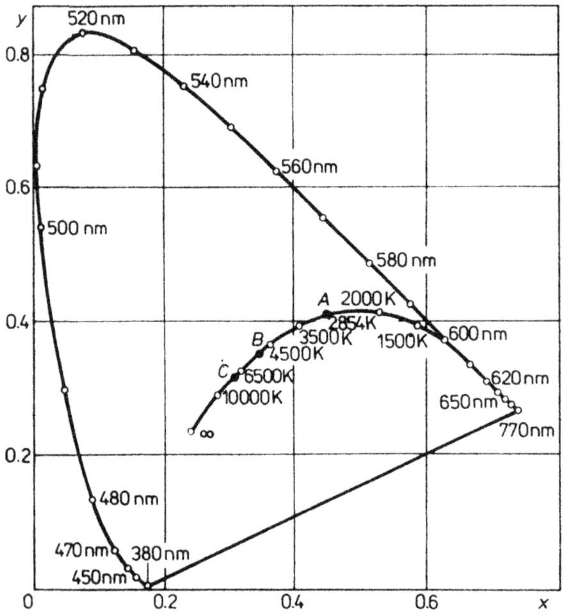

Fig. 1.5. Colour region and colour curve of a black body, presented in a chromaticity diagram, according to CIE 1931. *A*, *B*, *C* denote the standardized stimulae.

Figure 1.5 shows the all-colours region together with the colours of a black body at different temperatures, in the coordinates used for analyses and calculations. A measure of the colour is provided by the point coordinates (x, y). The neighbourhood of the point (x, y) is the region with barely discriminated colours. As shown in Figure 1.6, the region dimensions vary, depending on the colour.

To complete the definition of chromatic colour, it is necessary to evaluate the visual sensations of man and to include the influence of stimulae. It has been accepted by convention that the eye characteristic of the normal observer is as shown in Fig. 1.4. The CIE normal stimulae, i.e., the conditions in which the colour is measured, were also established. This last convention made it possible to assess the sensations when different light sources are involved. Also the anom-

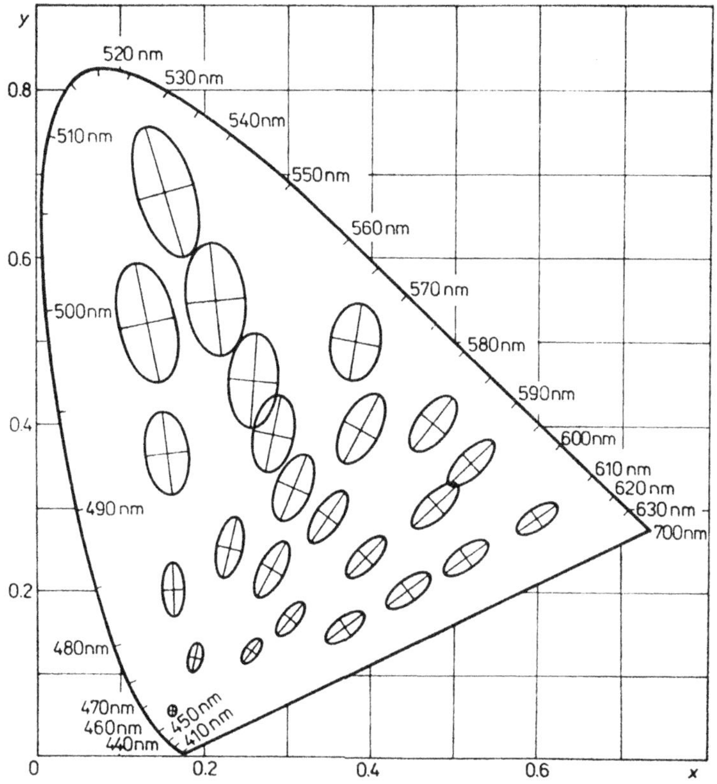

Fig. 1.6. The Stiles diagram illustrating colour discrimintation. The ellipses correspond to three threshold values of the differences of stimulae. The centre to the ellipse determines the chromatic coordinates (x, y) of a colour [20].

alies of the eye have been reported, which makes it possible to evaluate and compare the sensations.

Because of the conventions listed above, numerical factors have been introduced. The factors determine unambiguously the psychophysical colour of many stimulae, thus making colour a measurable quantity. The relations between the states of this quality are not determined, because a measure of this quality is a vector of two or three dimensions. The equivalence relation may then occur, generating a nominal scale. Relations may be determined for the vector components, with the measure of each of the components being the luminance.

The luminances of unit amounts of the primary components R, G and B of a trichromatic colour are related as follows:

$$L_r : L_g : L_b = 1.0000 : 4.5907 : 0.0601. \tag{1.24}$$

Thus the luminance of vector C may be determined from the formula

$$L_C = rL_r + gL_g + bL_b, \tag{1.25}$$

or, after introduction of proportion (1.24),

$$kL_C = 1.0000r + 4.5907g + 0.0601b. \tag{1.26}$$

The unit of measure of luminance L is $\mathrm{cd \cdot m^{-2}}$. The radiation flux is described by the 4-dimensional vector $C = (R, G, B, L)$.

The metrology of colour is the subject of trichromatic colourimetry. The images of reality are objective, as referred to the accepted reference states.

Another example of a quantity measured under an empirical scale is water turbidity. The quality of the object being investigated is the concentration of the suspended solids in water. The most common type of measurement is the nephelometric one. In this method, the relationship between the concentration of suspended solids and the scattered light intensity is applied, the intensity of the incident light beam being constant. The principle of measurement is shown in Fig. 1.7.

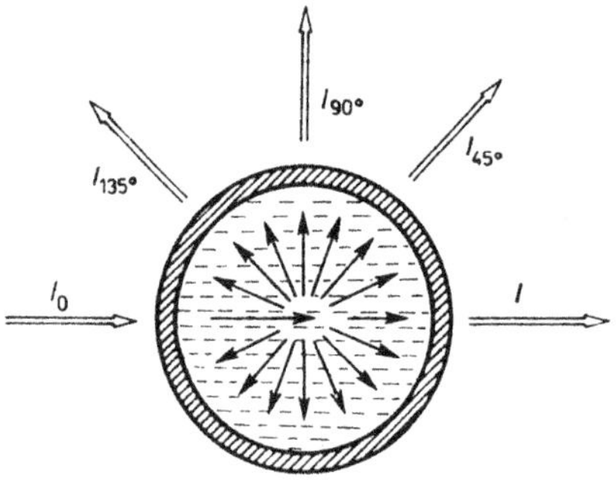

Fig. 1.7. The principle of nephelometric water turbidity measurement.

The reference states are the concentrations of diatomaceous earth or diatomite aqueous suspensions. The measured quality is the ability to scatter light. In this case, the suspension concentration is expressed on the metric scale, like the light intensity. The relationship between the two quantities, called *Wells' formula*,[4] is

$$\log \frac{I}{I_0} = k \frac{lcd^3}{d^4 + \alpha\lambda^4}, \tag{1.27}$$

where c is the concentration of particles, l the layer thickness, d the mean particle diameter, λ the wavelength of the incident light, and k, α are constant coefficients.

The complexity and ambiguity of relations caused by the variety of shapes and dimensions of suspensions, and of chemical constituents and physicochemical

[4] Wells' formula applies to the light scattered at right angles. Some other empirical formulae have been derived as well.

properties, are the reason why the measurement result is only a fragmentary image of reality. The turbidity of the investigated liquid and that of a standard are equivalent in the sense of an identical ability to scatter light. The relations between the states of a standard are established on a conventional empirical scale which expresses the relations of equivalence and strict order of states, intervals and quotients. With respect to the ability to scatter light, the relations of the states of the measured quantity and the states of the reference are isomorphic. However, they are not so when the physicochemical properties or the contents of the suspension are concerned.

Application of the nephelometric method for measuring water turbidity is not a consequence of ignorance. It is possible either to measure the quantities in formula (1.27), and thus to limit the ambiguity of the measured result, or to apply some other method of measurement, such as counting the particles in unit volume of water and then determining the spectrum of their dimensions, shapes, compositions, etc. However, such measurements are very laborious and time-consuming, and the time delay in obtaining the results makes the method less useful. The reasons for using the nephelometric method are entirely practical. A simplified image of the phenomenon is quite sufficient in this case, and though the measurement result is less valuable in terms of cognition, its usefulness is greater to man's purposeful activity.

A interesting separate problem in the realm of scientific research is that of the so-called *unmeasurable quantities*. This problem, however, should be regarded differently in the arts from the natural sciences, where the subject of interest is the natural world. In the natural sciences, the problem is to extend the limits of cognition of reality, or to formulate images of different quantities. In the images, the complexity of the phenomena is reduced, but the simplified image is sufficient and convenient for the practical activities of man.

Each of the natural sciences is concerned with some fragment of reality. Assessment of the cognitive results of these sciences is certainly not the subject of metrology, which has its own fragment of reality to recognize: the phenomena applied for the construction of standards and measuring tools. In this author's opinion, metrology never has, and does not now, create primary images of reality [91].

Very symptomatic, relatively new, and thus easy to follow, is the research on the *Josephson effect*. Its history starts in 1911, when the effect of superconductivity was discovered by Kamerlingh-Onnes. The theory of superconductivity developed in the 1950s led to determining the properties of so-called Cooper electron pairs. The Cooper pairs form the oscillator described by the equation

$$\frac{d\varphi}{dt} = \frac{2\pi}{\Phi_0} V, \tag{1.28}$$

where φ is the phase difference of the Schrödinger wave function, V the superconductive junction voltage, $\Phi_0 = h/2e$ is the magnetic flux quantum, h the Planck constant and e the electron charge. The properties of this oscillator became the subject of interest of electronics engineers. The effect was intended to be applied in the construction of high-frequency diodes, triodes, generators, modulators and detectors, as well as in cryotrons (logic elements). The subject of interest for metrology was the applicability for magnetic flux measurements (fluximeters) and current and voltage measurements at low temperatures, as well as the construction of a voltage standard.

As research proceeded, the Josephson effect, which sets up a new fundamental physical constant $h/2e$, became a key problem of physics. Studies showed that reproducibility of this constant shows a much better accuracy than obtained previously. Thus the possibility arose for the whole of science, and for metrology in particular, of defining a reference state for the images of electrical and electromagnetic phenomena. The Comité International des Poids et Mesures (CIPM) recommends the conventional value of the Josephson constant

$$K_{J-90} = \frac{2e}{h} = 483\ 597.9\ \frac{\text{GHz}}{\text{V}} \tag{1.29}$$

as a definition of the volt reproduced by frequency measurement applying the Josephson effect, according to formula (1.28). This definition is used from 1st January 1990.[5]

The problem of physical unmeasurable quantities involves also a need for the creation of simplified images of reality, which is connected with the possibility of the construction of some measurement tools with good metrological and practical properties. This is a subject of interest to metrology.

The problem of measurability of the phenomena which are the subject of the arts[6] is complex, as it concerns the impressions of man as an individual as well as the human community. The scope of impressions is enormous: from identification of colours and evaluation of properties of different images, through the sensations of hearing, smell, taste and touch, ending with the evaluation of art products, morality, usefulness, etc.

The characteristic feature of measurement within the arts is the function of man as a measurement tool, and the lack of material reference states. The reference states are created by man in his imagination. At the same time, the human being is subjected to learning, adaptation and fatigue, he experiences different timulae, remembers and forgets things. Therefore, his properties as a measures-

[5] The first temporary definition $K_J = 483\ 594\ 000\ \text{MHz}/\mu\text{V}_{\text{BIPM}}$ declared in 1972 by the Comité Consultatif d'Electricité (CCE).

[6] The arts, as understood here, include also sociology, psychology, economics, etc.

ment tool may not be, and are not, constant. For example, Schumacher[7] questions the objectiveness of economics (including econometry). He asks the question: What does "uneconomic" mean? and responds:

"And the answer to this question cannot be in doubt: something is uneconomic when it fails to earn an adequate profit in terms of money. The method of economics does not, and cannot, produce any other meaning. Numerous attempts have been made to obscure this fact, and they have caused a very great deal of confusion, but the fact remains.

Society, or a group, or an individual within society, may decide to hang on to an activity or asset for non-economic reasons—social, aesthetic, moral or political—but this in no way alters its uneconomic character. The judgement of economics, in other words, is an extremely fragmentary judgement; out of a large number of aspects which in real life have to be seen and judged together before a decision can be taken, economics supplies only one—whether a thing yields a money profit to those who undertake it or not." And further on: "Economics, moreover, deals with goods in accordance with their market value and not in accordance with what they really are. The same rules and criteria are applied to primary goods, which man has to win from nature, and secondary goods, which presuppose the existence of primary goods and are manufactured from them. All goods are treated the same, because the point of view is fundamentally that of private profit-making, and this means that it is inherent in the methodology of economics to ignore man's dependence on the natural world."

In economics, the measurement and its result play an important role in the formulation of declarative statements concerning reality. And how is it in other, "less measurable", sciences?

The arts also tend to achieve a qualitative cognition of the phenomena of interest to a given science. To make a result of cognition objective, the measurement scales must be defined. Attempts to do so were, and still are, being undertaken [85, 54]. In this author's opinion, the interpretation of human sensations and of the properties of the senses and mentality is a subject of investigation for physiology, psychology and other sciences but not for metrology. If the properties of a particular fragment of reality are being recognized by quantity, the definitions of the scales of new quantities should be included in metrology. An example is the trichromatic colour scale [20]. Thanks to this scale, colour may be objectively measured and the recognized relations of states are richer than in the case of the nominal scale.

When recognizing phenomena which still remain unmeasurable, the general

---

[7] E. F. Schumacher: *Small is Beautiful. A Study of Economics as if People Mattered,* Abacus—Sphere Books Ltd., London, 1974, pp. 35, 36.

philosophical and metrological problem arises whether a measurement is possible at all if the parameters of the measurement tool are unknown, its properties change and there are no material reference states (standards). The answer of the classical theory of measurement is that it is not. In fact, measurements like that are performed. These include test measurements of intelligence products, usefulness measurements, etc. What are the properties of the images produced in these measurements? What is their cognitive and practical value? These questions are till to be answered.

1.2 Physical Scales and the Ambiguity of Images of Reality

Measurement scales were presented above according to their mathematical or logical interpretation. Physical measurements do not provide such conditions as those of mathematical experiments. Because of the properties of human senses and of the tools used by man, there is no physical possibility of an unambiguous proof of equivalence of two states of a given quality.

The images of reality produced in the process of measurement are ambiguous. The state a of a measured quantity x is represented in the domain of abstraction as a *set of values*, namely

$$a \Rightarrow B_a \equiv \{b\}. \tag{1.30}$$

Applying the language of mathematics, one may say that the *mapping* (1.30) transforms state a into a random variable and B_a is the set of values of this variable. The random variable, which is an image of state $a \in A$, has the distribution $p_a(b)$. The assignment of a number of elements of the set B (when generally $B \in \mathbf{R}$) to one state a is not equally preferred. It may be supposed that the probability of the mapping $b = f(a)$ is either greater or smaller than the probability of the mapping $b+\varepsilon = f(a)$ when $\varepsilon > 0$. That is why the selection of the elements of the set B which maps state a is described by a measure $p \in \langle 0, 1 \rangle$, for which

$$\sum_{b \in B_a} p_a(b) = 1, \tag{1.31a}$$

or

$$\int_{(B_a)} p_a(b)\,\mathrm{d}b = 1. \tag{1.31b}$$

The mapping takes the form

$$a \Rightarrow \{b, p_a(b)\} \tag{1.32}$$

instead of (1.30), while the powers of the sets B and P are equal. Function $p_a(b)$ shown in Fig. 1.8 is an example of mapping under relation (1.32).

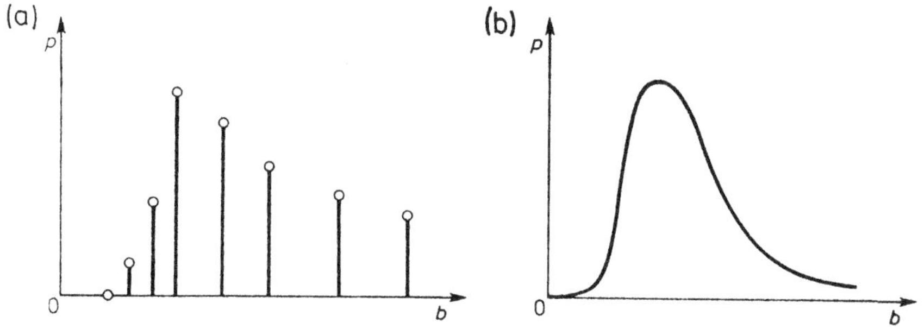

Fig. 1.8. The probability of mapping of a state a in the domain of abstraction: (a) for a discrete set of values, (b) for a continuous set of values.

A property of the assignment relation (1.32) is that the complete set of events described by condition (1.31a) or (1.31b) is assigned to state a, realized with the probability $P = 1$. Therefore, to define a measurement scale, only one element b^* being the best mapping of the state a should be selected from the set B_a. The selection is based on the decision D:

$$b^* = D(p_a) \tag{1.33}$$

Different *decision rules* may be applied but the decision rule must be known. For example, D_1 states that the best mapping of state a is the most probable value b^*, that is the modal value

$$D_1(p_a) = b_1^* \quad \text{such that} \quad p_a(b_1^*) = \max_{(b \in B_a)} p_a(b). \tag{1.34a}$$

Another rule is the choice of a mean value, i.e.,

$$D_2(p_a) = b_2^* \quad \text{such that} \quad \int b p_a(b) \mathrm{d}b = b_2^* \quad \text{or} \quad \sum_i b_i p_{ai} = b_2^*, \tag{1.34b}$$

or a median

$$D_3(p_a) = b_3^*, \quad \text{while} \quad P_a(b_3^*) = \tfrac{1}{2} \quad \text{or} \quad P_a(b \leqslant b_3^*) \geqslant \tfrac{1}{2}, \quad P_a(b \geqslant b_3^*) \geqslant \tfrac{1}{2}. \tag{1.34c}$$

The point $b^* \in B$ need not be a point of set B_a. The relations between the values $b_1^*, b_2^*, \ldots$ may be determined for the known decision rules $D_1, D_2, \ldots$ if the measure $p_a(b)$ is known. The values of $b_1^*, b_2^*, \ldots$ are generally different, though in some cases they may be identical. For example, for a function $p_a(b)$ symmetrical at a point $b_3^*, b_2^* = b_3^*$. Selection of the value b^* as an image of the state a is essential, because the assignment

$$b^* = f^*(a) \tag{1.35}$$

should be valid for these elements, as well as the inverse mapping

$$a^* = \varphi^*(b^*), \tag{1.36}$$

where $\varphi^* = f^{*-1}$ is the *operator (function) of the inverse mapping*. To distinguish from the postulated mathematical scales (1.2), we shall denote the mappings generating physical scales by an asterisk.

Another important property of the mapping operations f^* and φ^* is the requirement for them to be definite, not only for state a or value b^*, but also in the neighbourhood of this state or value. Denoting by u the deflection of state a, the extension of the mapping region may be written as

$$b^*+v = f^*(a+u), \tag{1.37}$$

where v is the deflection from the point b^*. The form of the inverse mapping is

$$a^*+u^* = \varphi^*(b^*+v) = f^{*-1}(b^*+v). \tag{1.38}$$

This property, as shown by Ajdukiewicz [1] and others, is fulfilled for additive quantities measurable under a metric scale. The sum $a+u$ may be realized physically by fulfilling the requirements of isomorphism of the relations for the states of a quality (in a physical reality) and between the relations of real numbers which are images of these states. The consequence is the mapping [1]:

$$b^*+v = f^*(a+u) = f^*(a)+f^*(u), \tag{1.39}$$

and then, considering (1.35),

$$v = f^*(u) \tag{1.40a}$$

and

$$u = \varphi^*(v). \tag{1.40b}$$

Property (1.39) means that the measurement function f^* [1] (the mapping operator) is a linear operator in the case of metric scales.

Considering again the mapping (1.30), when the set B_a is found to be in the neighbourhood of the point b^*, then, from (1.40), it should be possible to map the set $\{b_a\}$ element by element into a set of states denoted as

$$\{a^*\} = A_a^* = \varphi^*(B_a). \tag{1.41}$$

The interpretation of the result obtained from the inverse mapping operation may be as follows. The image of the state a is the set B_a. This image may be expressed, either in the domain of images or in the domain of states of a specific quality, as a set A_a^*. The meaning of this fact is that several states $a_i \in A_a^*$, $i = 1, 2, \ldots$, are interpreted as state a. Mapping (1.30) written in the form

$$a \Rightarrow B_a \Leftrightarrow A_a^* \tag{1.42}$$

is the postulate arising from the limited ability of man to distinguish, either directly or with measurement tools, between the states comprising the neighbourhood of a. In physical measurements, a measurement scale is generated by relation

(1.35) arising from the properties of a given quality. If relation (1.35) is identical with (1.2), the measurement scale will express the relations of the states that result from properties of scale (1.2).

An example of how the function f^* was reformed due to studies of the relations between states of a certain quality is provided by the subsequent definitions of the International Practical Temperature Scale (IPTS), formulated in 1948 and 1968.[8] For example, the fixed freezing point of gold was designated as 1063°C and 1064.43°C, respectively.

After mappings (1.35) and (1.36) are introduced, the question arises whether the states a, a^* are equivalent. Generally, they are not. The equivalence

$$a^* = \varphi^*[f^*(a)] = a, \tag{1.43}$$

i.e.,

$$a \Leftrightarrow a^*, \tag{1.44}$$

takes place within the limits of discrimination expressed by (1.42). Since the choice of a decision rule affects the value of b^*, as shown in Fig. 1.9, and thus

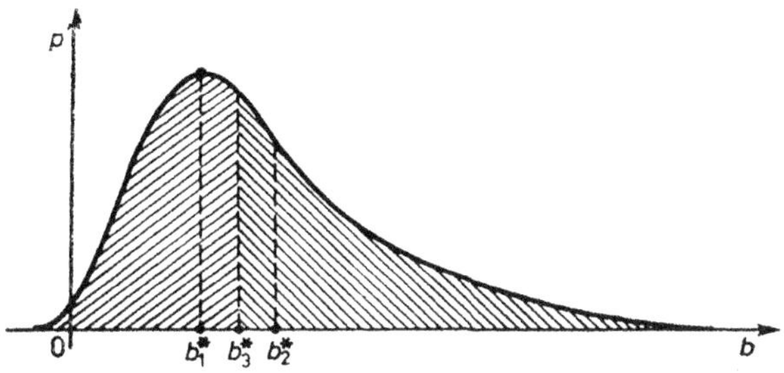

Fig. 1.9. The images of the state a for decision rules D_1, D_2, D_3.

also the state a^* (by relation (1.36)), a number of states a^* may be equivalent to the primary state a. Hence, the difference

$$a^* - a = \text{constant} \tag{1.45}$$

for every decision rule possible.

Point b^* selected in a decision D (1.33) does not describe the uncertainty of the image of state a obtained from measurement.

Use of the set $\{b, p_a(b)\}$ is inconvenient and is not always necessary. First, let us consider the case $p_a(b)$ = constant. The condition for the events to be complete implies

[8] The first International Temperature Scale was introduced in 1927. The freezing point of gold was attributed a value of 1063°C.

$$p_a(b) = \frac{1}{b_u - b_l}, \tag{1.46}$$

where b_u, b_l are the supremum and infimum of the set B_a, respectively. After a metric is defined in the space in which the events $b \in B_a$ are represented, the ambiguity of the image may be characterized by the distance of the elements of the set B_a from the point b^*,

$$\varrho(b^*, b). \tag{1.47}$$

As an effect of the previously accepted assumption $B \in \mathbf{R}$, the space of the events b is normalized.[9] The distances ϱ with respect to the applied *norms of the space* are

$$||x|| = \sup x, \qquad \varrho_1 = \sup(b - b^*), \tag{1.48}$$

$$||x|| = \sum_{i=1}^{n} |x_i|, \qquad \varrho_2 = \frac{1}{n}\sum_{i=1}^{n} |b_i - b^*|, \tag{1.49}$$

$$||x|| = \sqrt{\sum_{i=1}^{n} x_i^2}, \qquad \varrho_3 = \sqrt{\frac{1}{n}\sum_{i=1}^{n} (b_i - b^*)^2}, \tag{1.50}$$

$$||x|| = \left(\sum_{i=1}^{n} x_i^q\right)^{1/q} \qquad \varrho_4 = \left[\frac{1}{n}\sum_{i=1}^{n} (b_i - b^*)^q\right]^{1/q}. \tag{1.51}$$

For infinite sets B_a, by application of the parametric notation $x(t)$, $t \in T$, the distance may be expressed as the integral, e.g.,

$$||x|| = \left\{\int_{(T)} [x(t)]^q \mathrm{d}t\right\}^{1/q}, \qquad \varrho_5 = \left\{\frac{1}{T_u - T_l}\int_{T_l}^{T_u} [b(t) - b^*]^q \mathrm{d}t\right\}^{1/q}, \tag{1.52}$$

while $q > 1$, and T_u, T_l are the supremum and infimum of T, respectively. An important property of the distance is that the domain of ϱ is the domain of the set B.

The set B is divided by the point b^* into two disjoint subsets B_l and B_u for which

$$B = B_l \cup B_u. \tag{1.53}$$

The distances in formulae (1.48)–(1.52) may be determined separately for the sets B_l and B_u as ϱ_l, ϱ_u, respectively.

A different way of expressing the *ambiguity of the mapping* of a state a into a value $b \in B_a$ (for case (1.46) under consideration) is to apply a *measure of the*

[9] It is assumed that the set B is measurable under a Lebesgue measure.

set, e.g. the Lebesgue measure, Radon measure [115], etc. The values of the *Lebesgue measures* of sets B, B_u, B_l, as in (1.53), are[10]

$$\mu(B_l) = \mu_l = b^* - b_l > 0, \tag{1.54a}$$

$$\mu(B_u) = \mu_u = b_u - b^* > 0, \tag{1.54b}$$

$$\mu(B) = \mu(B_l \cup b^* \cup B_u) = \mu_l + \mu_u = 2\mu, \tag{1.54c}$$

whereas

$$\mu(B_a) = b_u - b_l. \tag{1.54d}$$

Usually $p_a(b) \neq$ constant and finding a simple measure of ambiguity of mapping (1.32) becomes more difficult. Such a measure should consist of a small number of parameters and impress the human imagination. Most important, however, is whether it is necessary to know the distribution $p_a(b)$ in order to determine the measure. The problem is that some of the probability distributions (Poisson, exponential, geometric, Student's, etc) are described by a single parameter, and others (normal, uniform and binomial distributions) by two parameters, i.e.,

$$p_a(b) = p(b, \theta_1, \theta_2, \theta_3, \ldots), \tag{1.55}$$

where θ_1, θ_2 are the parameters of a distribution function.

The ambiguity of mapping (1.32) and the value of b^* is completely determined by stating the kind of probability distribution, one (or two) parameters and a decision rule (1.33) of selection of the value b^*. It is then essential to be able to determine the measure of ambiguity of the mapping while not knowing the probability distribution $p_a(b)$; the measure should have the same sense for different distributions. One may say that there is no measure which could describe diversity in a less diverse manner, as the lost information on diversity may not be restored without additional information. For this reason, a variety of incomplete (as compared with $p_a(b)$) measures are applied to describe the ambiguity of the mapping, each of them expressing the ambiguity in some broad sense.

The first group includes measures similar to the measure of distance (1.48)–(1.52), while every element of the set B_a is attributed a weight $p_a(b_i)$ and condition (1.31a) is fulfilled for $i = 1, 2, \ldots, n$. Thus, by analogy to ϱ_2 (1.49), the mean weighted distance of the elements of the set B_a from the point b^* is

---

[10] The Lebesgue measure of a set is presented in the Appendix. As may be easily shown, exclusion of the point b^* from the set B does not change the measure of the set. In particular, the property of this measure

$$\mu\left(\bigcup_{i=1}^{N} B_i\right) = \sum_{i=1}^{N} \mu(B_i) \quad \text{and} \quad \mu(b^*) = 0$$

implies $\mu(B) = \mu_u + \mu_l$.

$$d_m = \sum_{i=1}^{n} [|b_i - \bar{b}| p_a(b_i)], \tag{1.56a}$$

or

$$\bar{d} = \int_{B_a} [|b - b^*| p_a(b)] \, db. \tag{1.56b}$$

The mean square deviation under the norm (1.50) is called the *standard deviation* of the probability distribution and is calculated from the formulae

$$\sigma = \sqrt{\sum_{i=1}^{n} [(b_i - \bar{b})^2 p_a(b_i)]}, \tag{1.57a}$$

or

$$\sigma = \sqrt{\int_{B_a} (b - \bar{b})^2 p_a(b) \, db}, \tag{1.57b}$$

where the value $\bar{b} = b_2^*$, according to decision D_2 (1.35).

For distributions $p_a(b)$ symmetric about point b^*, the condition $b_2^* = b_3^* = b^*$ is met. For $\bar{b} = b_2^*$, the qth order ($q = 2, 3, 4, \ldots$) central moments of the probability distribution are defined like the distance ϱ_4 (1.51):

$$m_q = \sum_{i=1}^{n} [(b_i - \bar{b})^q p_a(b_i)], \tag{1.58a}$$

or

$$m_q = \int_{B_a} (b - \bar{b})^q p_a(b) \, db. \tag{1.58b}$$

The second group of measures of the ambiguity of the mapping $a \to B_a$ are the interval measures which are similar to the Lebesgue measure when a subset is extracted from the set B_a under an additional condition. For example, the condition may be that the probability of realization of the element b_i is large enough,

$$p_a(b_i) > p_{\min} > 0. \tag{1.59}$$

Condition (1.59) is fulfilled in the set

$$B_p = \{b \in B_a : p(b) > p_{\min}\}, \tag{1.60}$$

as shown in Fig. 1.10a. The set B_p may be characterized by a Lebesgue measure (1.54) of a set

$$\mu(B_p) = \mu_1 + \mu_u = 2\mu > 0. \tag{1.61}$$

The advantage of condition (1.59) is that in the case of a theoretically infinite set B_a (e.g. a set with a normal distribution), the set of images is limited. A normal distribution choice of $p_{\min}$ such that

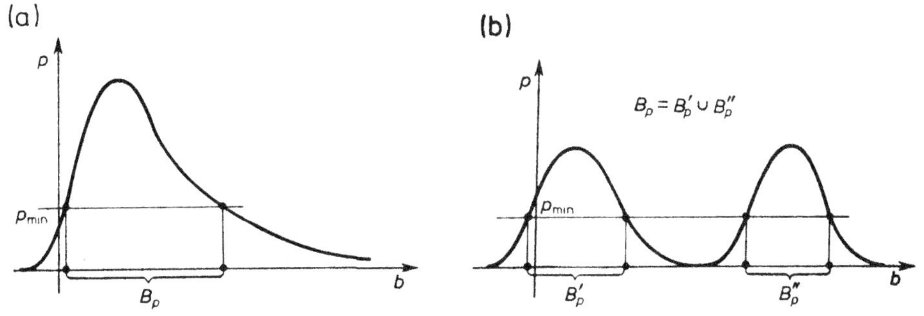

Fig. 1.10. The measure of ambiguity of an image when $p_a(b) > p_{min}$: (a) a graphical interpretation of the measure, (b) a case when the measure does not fully characterize the ambiguity of the image.

$$p_{min}/p_{max} = 0.1 \quad \text{gives} \quad \mu(B_p) = 2\times 2.15\sigma = 4.3\sigma, \tag{1.62}$$

$$p_{min}/p_{max} = 0.01 \quad \text{gives} \quad \mu(B_p) = 2\times 3.04\sigma = 6.08\sigma. \tag{1.63}$$

The Lebesgue measure of the set B_p is a given multiplication factor of the standard deviation. For some of the distributions, however, e.g. bimodal and incoherent ones, a measure like that is inconvenient, as shown in Fig. 1.10b.

The set B_a may also be limited by introducing a condition that the neighbourhood of the point b^* should include the α part of the elements of the whole set B_a. By selecting values $\varepsilon_l, \varepsilon_u > 0$, fulfilling the condition

$$\int_{b^*-\varepsilon_l}^{b^*+\varepsilon_u} p_a(b)\,db \geqslant \alpha < 1, \tag{1.64}$$

or two conditions

$$\int_{b^*-\varepsilon_l}^{b^*} p_a(b)\,db \geqslant \tfrac{1}{2}\alpha, \qquad \int_{b^*}^{b^*+\varepsilon_u} p_a(b)\,db \geqslant \tfrac{1}{2}\alpha, \tag{1.65}$$

a set $B_\alpha \in B_a$ is obtained for which the value of the Lebesgue measure is

$$\mu(B_\alpha) = \varepsilon_l + \varepsilon_u = 2\varepsilon > 0. \tag{1.66}$$

In this case the interval 2ε is called the *confidence interval* and α is called the *confidence level* or *confidence*.[11] A graphical interpretation of confidence intervals is shown in Fig. 1.11. For a normal distribution, the interval $2\times 2\sigma$ contains 95% of the elements of set B and the interval $2\times 3\sigma$ contains 99.7% of the elements.

[11] The confidence level is commonly denoted as $1-\alpha$, where α is the part of a set lying outside the interval 2ε.

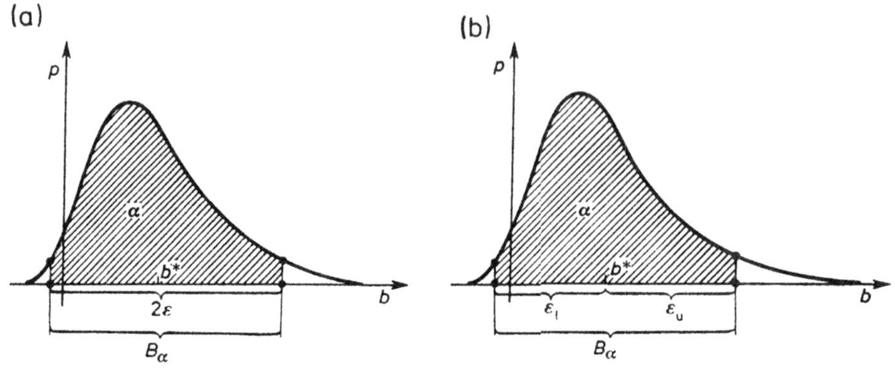

Fig. 1.11. The methods of extracting the set B_α: (a) according to condition (1.64), (b) according to condition (1.65).

One more interval measure used sometimes is the *range*. When the infimum and supremum of the set B_a are respectively b_l, b_u, the range W is

$$W = b_u - b_l > 0. \tag{1.67}$$

The range in this case is equal to the Lebesgue measure of the set B_a (1.54c). The sample range is of practical significance. The properties of this measure are listed on page 196 and its relationship with the standard deviation on page 198.

The third kind of measure of the ambiguity of the mapping $a \to B_a$ is *entropy* (or *negentropy*), expressed as

$$H = -\sum_{i=1}^{n} [p_a(b_i) \ln p_a(b_i)] > 0, \tag{1.68a}$$

or

$$H = -\int_{-\infty}^{\infty} [p_a(b) \ln p_a(b)] \mathrm{d}b > 0. \tag{1.68b}$$

It is characteristic of entropy that $H \notin B$, although it is represented in the real domain. Entropy is a logarithmic measure of an interval; the interval length depends on the probability distribution $p_a(b)$ and the dispersion parameters of the distribution. For a set with uniform distribution (1.46), the value of the entropy is

$$H = \ln(b_u - b_l), \tag{1.69}$$

although decimal logarithms and logarithms with base 2 are used as well. The entropy of a set with a normal distribution is

$$H = \ln(\sqrt{2\pi e}\,\sigma = \tfrac{1}{2}\ln(2\pi e) + \ln(\sigma). \tag{1.70}$$

The entropy is the measure of disorder of indeterminacy applied in thermodynamics. A set with uniform distribution has the greatest indeterminacy. Entropy should be regarded as a measure allowing the comparison of two sets with different probability distributions. When the distributions of two compared sets are the same, the sets may be compared by the ratio of the probability density function parameters, for example σ for a normal distribution or Δ for a uniform distribution.

Two sets having different distributions will be equivalent (as far as entropy is concerned) if their entropies are equal. The equivalence or order of a set in the context of entropy has no other meaning.

Up to now, we have been considering a single cognition act for a state a of a given quality. The image of the state a is the set of values $B_a = \{b\}$ and the set of probabilities $p_a(b)$ conforming to condition (1.31) of realization of the value b. The value of b^* is chosen here as the best image, according to one of the decision rules (1.33). The ambiguity of the mapping of a state a into the value b^* is characterized by a set of values

$$\{b\} \neq b^*, \tag{1.71}$$

comprising the neighbourhood of the point b^*, or by the set of states according to (1.42)

$$A^* \neq a. \tag{1.72}$$

Recognition of the states of matter and energy is realized by the multiple cognition acts. A number of images is obtained, as for the supposition of the same state a of a given quality, or a number of images is obtained, interpreted as different sets A^*. Thus it is necessary to compare the images, assuming that the state of the quality is the same. The values b^* and the probability distributions $p_a(b)$ or other measures of ambiguity of the images should be compared. The complexity of these procedures cannot be doubted; this is a problem in mathematical statistics, and some aspects important for metrology will be discussed in the sections on the theory of error and on measurement procedures.

When the ideas of producing images by physical measurements were presented, i.e., when the ambiguity of an image of the state of a quality was accepted as an axiom, the problem whether a mapping like (1.32) or (1.42) fulfils the formal assumptions of preservation of isomorphic relations of the values and of the states was not taken into consideration. In other words, the problem is whether mapping (1.32) is identical with (1.2). A similar problem was discussed by Wójcicki [137], who analysed the so-called approximation measurement system. Wójcicki discussed the case of a quantized set of states A, due to which the relation $b^* = f^*(a)$ becomes reproducible for the states a_i, $i = 1, 2, \ldots, n$. The ambiguity of the mapping operation concerns the interval $\langle a_i, a_{i+1} \rangle$ of the states. With additional conditions fulfilled (e.g. the states a_i, a_{i+1} should have a common reference state defining a common unit of measure) and considering the order of the states $a_{i+1} > a_i$, Wójcicki showed that such measurement is equivalent to the measurement according to equation (1.2), to the accuracy of the value of quantum $q = a_{i+1} - a_i$. Introduction of the ambiguity measure permits the set A (or B) to be quantized, and Wójcicki's reasoning may be followed for this case. Because of the different notional set, the problem is not presented again.

The present author [91] had previously assumed intuitively the quantization of a set of reference states to be a postulate arising from the physical limitations of the unambiguity of the mapping.

1.3 Multidimensional Scales

In studying reality there is a need for recognizing more complex phenomena, characterized by more than one quality. Our interest lies not in the relation of states of different qualities within the same phenomenon but in how to represent the image of this fragment of reality.

If a quality is regarded as an isolated property, the image of several quantities is the Cartesian product of the sets of states of each of these qualities. A phenomenon A is thus expressed as

$$A = P \times R \times S \times \ldots = P, R, S, \ldots \qquad (1.73)$$

The states of the phenomenon form the set $A = (a_1, a_2, \ldots, a_i, \ldots)$, where each state a_i is a realization of one state of each of the qualities. This means that

$$a_1 = (p_1, r_1, s_1, \ldots). \qquad (1.74)$$

Each quantity is mapped into a numerical axis. Therefore, the image of phenomenon A is represented in the space E^n with coordinates $U, V, W, \ldots$ While producing the images based on measurements and physical scales, the state a_i (a point in the space $P \times R \times S \times \ldots$) of the phenomenon is mapped into a set of values B_a, where

$$B = U \times V \times W \times \ldots = U, V, W, \ldots \qquad (1.75)$$

Since the mapping of a state of every quality takes the form of (1.32)

$$r \Rightarrow \{V_r, p_r(v)\}, \qquad (1.76)$$

in this case the mapping is

$$a \Rightarrow \{b(u, v, w, \ldots), p_a(b)\}. \qquad (1.77)$$

As in the case of a single quality (1.32), the best image $b^*(u^*, v^*, w^*, \ldots)$ must be selected. In this case the probability distribution

$$p_a(b) = p(u, v, w, \ldots) \qquad (1.78)$$

is multidimensional and thus the decision rule states that

$$D\{p(u, v, w, \ldots)\} = b^*. \tag{1.79}$$

The image of the state a of a phenomenon is b^* and the ambiguity of the image is described by the multidimensional probability distribution (1.78):

$$a \Rightarrow \{b^*, p_a(u, v, w, \ldots)\}. \tag{1.80}$$

Image ambiguity may be characterized by different measures.

1. The distance in space E^n between the best image and the set of images, according to one of norms (1.48)–(1.52):

$$\varrho[b^*(u^*, v^*, w^*, \ldots), b(u, v, w, \ldots)]. \tag{1.81}$$

2. The above distance with weighted set elements, the weights being expressed by $p_a(b)$ as in (1.56):

$$d = \int \ldots \int_{B_a} \int [(b-b^*)p_a(b)]\,\mathrm{d}u\,\mathrm{d}v\,\mathrm{d}w \ldots \tag{1.82}$$

3. The Lebesgue measure of a set

$$\mu(B_a) = (u_u-u_l)(v_u-v_l)(w_u-w_l) \ldots \tag{1.83}$$

4. A set B_p which is a subset of the set B extracted by rule (1.59)

$$p_a(u, v, w, \ldots) \geqslant p_{\min}. \tag{1.84}$$

The measure of the set B_p may be the Lebesgue measure (1.83) or, for example, the volume of the set. In the case of a normal distribution, the shape of B_p is ellipsoidal and its volume is[12]

$$V = \frac{\sqrt{[\pi(n+2)]^n|\Sigma|}}{\Gamma(n/2+1)}, \tag{1.85}$$

where $|\Sigma|$ is the determinant of the covariance matrix and Γ is the gamma function. For $n < 3$ the values obtained correspond to the length of interval (1.61) and to the area of the ellipse, respectively.

5. A set $B_\alpha \in B_a$ such that

$$\int_{B_a} \int \ldots \int p_a(u, v, w)\,\mathrm{d}u\,\mathrm{d}v\,\mathrm{d}w \ldots \geqslant \alpha. \tag{1.86}$$

The set B_a may be characterized by the Lebesgue measure, the set volume, etc. In the case of a normal distribution, the set B_α is an ellipsoid whose volume depends

---

[12] Note that (1.85) is the volume of an ellipsoid for

$$p_{\min} = \frac{\Gamma(n/2+1)}{\sqrt{[(n+2)\pi]^n|\Sigma|}} = \frac{1}{V}.$$

on the significance level, as for a unidimensional image (1.64). When the individual variables are independent, then

$$p_a(u, v, w, \ldots) = p_p(u)p_r(v)p_s(w) \ldots \tag{1.87}$$

and condition (1.86) takes the form

$$\alpha = \alpha_u \alpha_v \alpha_w \ldots \tag{1.88}$$

The above case of phenomena with multiple qualities mapped as multidimensional ones in a normalized vector space leads to the problem of *vector scales*. The vectors may be considered comparable when they are mapped in the same domain E^n. Let us first determine the relations between the images. Two vectors (1.75) are equal when their components are equal, i.e.,

$$b_1 = b_2, \tag{1.89}$$

when $u_1 = u_2$, $v_1 = v_2$, $w_1 = w_2$, ...

The relation of strict order of two vectors

$$b_1 < b_2 \tag{1.90a}$$

takes place when

$$u_1 < u_2, \quad v_1 < v_2, \quad w_1 < w_2, \ldots \tag{1.90b}$$

for all the components.

More complex relations such as equality and strict order of intervals may be established if the vectors are linearly dependent, i.e.,

$$b_2 = cb_1, \tag{1.91a}$$

and the relative ratio is expressed by the number c, i.e.,

$$b_1 < b_2 \quad \text{if} \quad c > 1, \tag{1.91b}$$

$$b_1 > b_2 \quad \text{if} \quad c < 1. \tag{1.91c}$$

For the scalar quantity c, relations of strict order of intervals and quotients of values (1.10)–(1.12) are defined. The same relations are valid for the vectors b_1, b_2 when the equality condition (1.91a) is fulfilled. Consider colour as an example. Multiplication of a colour vector by a constant $c > 0$ does not change the colour, but increases its luminance c times.

A specific phenomenon is the *time variability of a quantity* which is the image of time changes of the states of a quantity. The specific nature of this case lies in the fact that the abstract notion of time is introduced to the definition of a quantity. Time is represented in the real domain as an interrelation of different phenomena. An elementary phenomenon is the state of two qualities X, T, namely $a = (t, x)$. The image of these states is the point $b = (\tau, y)$ in a two-dimensional metric space with coordinates Y, τ. Making allowance for the ambiguity of the mapping (1.32), we obtain

$$(t, x) \Rightarrow \{[\tau^*, p_t(\tau)], [y^*, p_x(y)]\}. \tag{1.92}$$

When the relations

$$t^* = \varphi_\tau^*(\tau^*) \quad \text{and} \quad x^* = \varphi_y^*(y^*), \tag{1.93}$$

are valid, the mapping

$$(t, x) \Rightarrow \{t^*, x^*, p_{tx}(t^*, x^*)\} \tag{1.94}$$

takes place and

$$p_{tx}(t^*, x^*) = p_t(\tau)p_x(y), \tag{1.95}$$

because the coordinates t, x (and their images) are independent of one another. The set B_p (1.60) or B_α (1.64), as shown in Fig. 1.12, is the image of the state a in the case when the ambiguity of mapping is described by a two-dimensional normal distribution. The set is an ellipse whose axes overlap with the coordinates t, x.

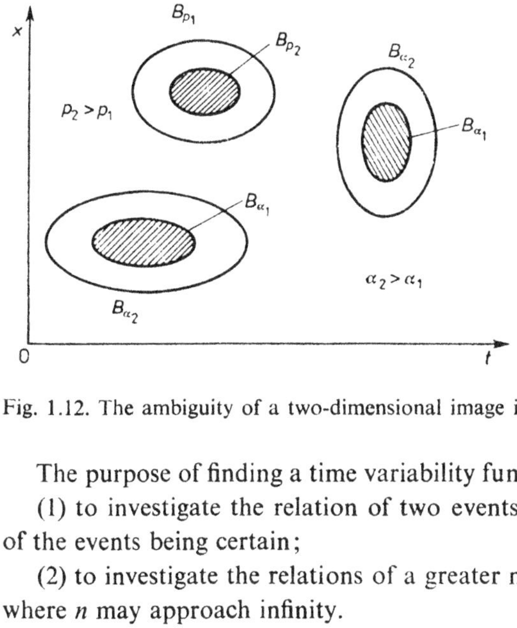

Fig. 1.12. The ambiguity of a two-dimensional image in the normal distribution case.

The purpose of finding a time variability function may be:

(1) to investigate the relation of two events $a_1 = (t_1, x_1)$, $a_2 = (t_2, x_2)$, each of the events being certain;

(2) to investigate the relations of a greater number of states $a_1, a_2, \ldots, a_n, \ldots$, where n may approach infinity.

Let us consider case (1). The distance of the states a_1 and a_2 is mapped as the *distance of two sets* $\{t_1^*, x_1^*\}$ and $\{t_2^*, x_2^*\}$, i.e.,

$$\varrho(a_1, a_2) \Rightarrow \varrho[\{t_1^*, x_1^*\}, \{t_2^*, x_2^*\}] = \varrho[\{t_2^* - t_1^*\}, \{x_2^* - x_1^*\}], \tag{1.96}$$

where the set $\{t_2^* - t_1^*\}$ is described by the distributions

$$p(t_2^* - t_1^*) = p_\tau(t_2^*) * p_\tau(t_1^*), \tag{1.97a}$$

$$p(x_2^* - x_1^*) = p_y(x_1^*) * p_y(x_2^*). \tag{1.97b}$$

The ambiguity of the image is expressed by the distribution which is a convolution of distributions (1.97). The distance is now two-dimensional and differs from that for the one-dimensional image. In practice, a two-dimensional image is usually reduced to a one-dimensional image in the domain of x, with the assumption that the ambiguity of the image in the time domain is negligible. In effect, the approximations

$$t^* \equiv t,\ p_t(\tau) = p_t(t) = \delta(t), \tag{1.98}$$

$$(t, x) \Rightarrow \{[t^*, y^*], p_x(y)\} \tag{1.99}$$

are being applied instead of mapping (1.92) [26]. This author does not accept such an interpretation, since the existing recorders of time variability function $x(t)$ do not provide suitable uniformity and accuracy of coordinate τ of the image, despite the potential abilities of these devices. To reduce the image ambiguity to the one-dimensional set, the image of the second coordinate should be corrected. When Δ expresses the ambiguity of the image of time, we obtain

$$y(t) \approx y(\tau) - \Delta_\tau \frac{\partial y}{\partial \tau}, \tag{1.100}$$

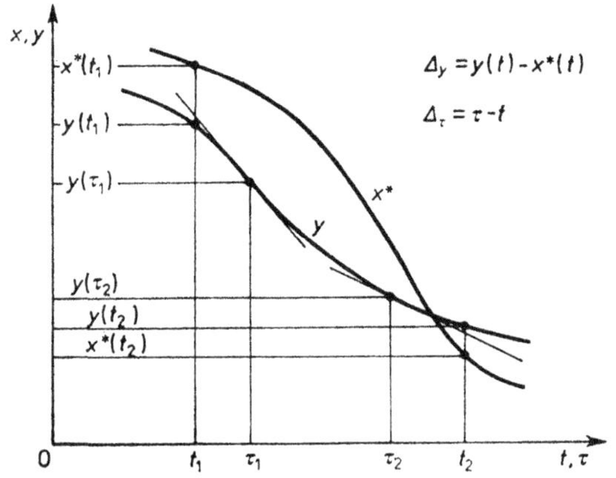

Fig. 1.13. A two-dimensional image reduced to one dimension.

as shown in Fig. 1.13. Since

$$x^*(t) = y(t) - \Delta_y, \tag{1.101}$$

so introducing the notation

$$\Delta = \Delta_y + \frac{\partial y}{\partial \tau} \Delta_\tau = \Delta_y + \Delta'_\tau, \tag{1.102}$$

we obtain

$$x^*(t) \approx y(\tau)-\Delta_y-\Delta_\tau\frac{\partial y}{\partial \tau} = y(\tau)-\Delta. \tag{1.103}$$

Since the image of the coordinate x is characterized by the set $\{y^*, p_x(y)\}$, we find on considering (1.102) that

$$p(\Delta) = p_y(\Delta_y)*p_\tau\left(\frac{\partial y}{\partial \tau}\Delta_\tau\right) = p_y(\Delta_y)*p(\Delta'_\tau). \tag{1.104}$$

Instead of (1.92), we may write

$$(t, x) \Rightarrow \{t^*, y^*, p_x(\Delta)\}. \tag{1.105}$$

The factor $\partial y/\partial \tau$ in formula (1.104) may be derived from the image or may be considered a random variable, depending on the specific case.

In case (2), the sequence of images $b_1, b_2, \ldots, b_n$ mapping the sequence of states $a_1, a_2, \ldots, a_n$ is expressed by [26, 129]:

(i) a class of images,

(ii) coefficients of the images of a given class.

The *class of images* may be characterized:

— by a function, e.g.

$$y = b_0+b_1\tau+ \ldots +b_q\tau^q, \tag{1.106}$$

$$y = A\sin(\omega\tau+\phi), \tag{1.107}$$

$$y = Be^{a\tau}, \tag{1.108}$$

where the domain of the coefficients $b_0, b_1, \ldots, b_q, A, \omega, \phi, B, a$ is a continuous domain of real numbers and y, τ are the image coordinates;

— by vectors, e.g.

$$\boldsymbol{y}_1 = \begin{bmatrix} y_{11} \\ y_{12} \\ \cdots \\ y_{1n} \end{bmatrix}, \quad \boldsymbol{y}_2 = \begin{bmatrix} y_{21} \\ y_{22} \\ \cdots \\ y_{2n} \end{bmatrix}, \quad \ldots, \quad \boldsymbol{y}_n = \begin{bmatrix} y_{N1} \\ y_{N2} \\ \cdots \\ y_{Nn} \end{bmatrix}; \tag{1.109}$$

— in other ways, for example as shown in Fig. 1.14.

The class of images is selected arbitrarily, while the image coefficient or the image number are determined by applying the methods of optimization, according to a chosen figure of merit. The problems of identification will be presented in Section 3.4.

For example, the images from Fig. 1.14 have been divided into 4 classes. The images (waveform $y(t)$) were sampled with constant sampling interval Δt. In effect, vector $\boldsymbol{y}_i$ was obtained. The form of the image which is the subject of classification is

$$\boldsymbol{y}_i^T = [y_{i1}, y_{i2}, \ldots, y_{iN}], \tag{1.110}$$

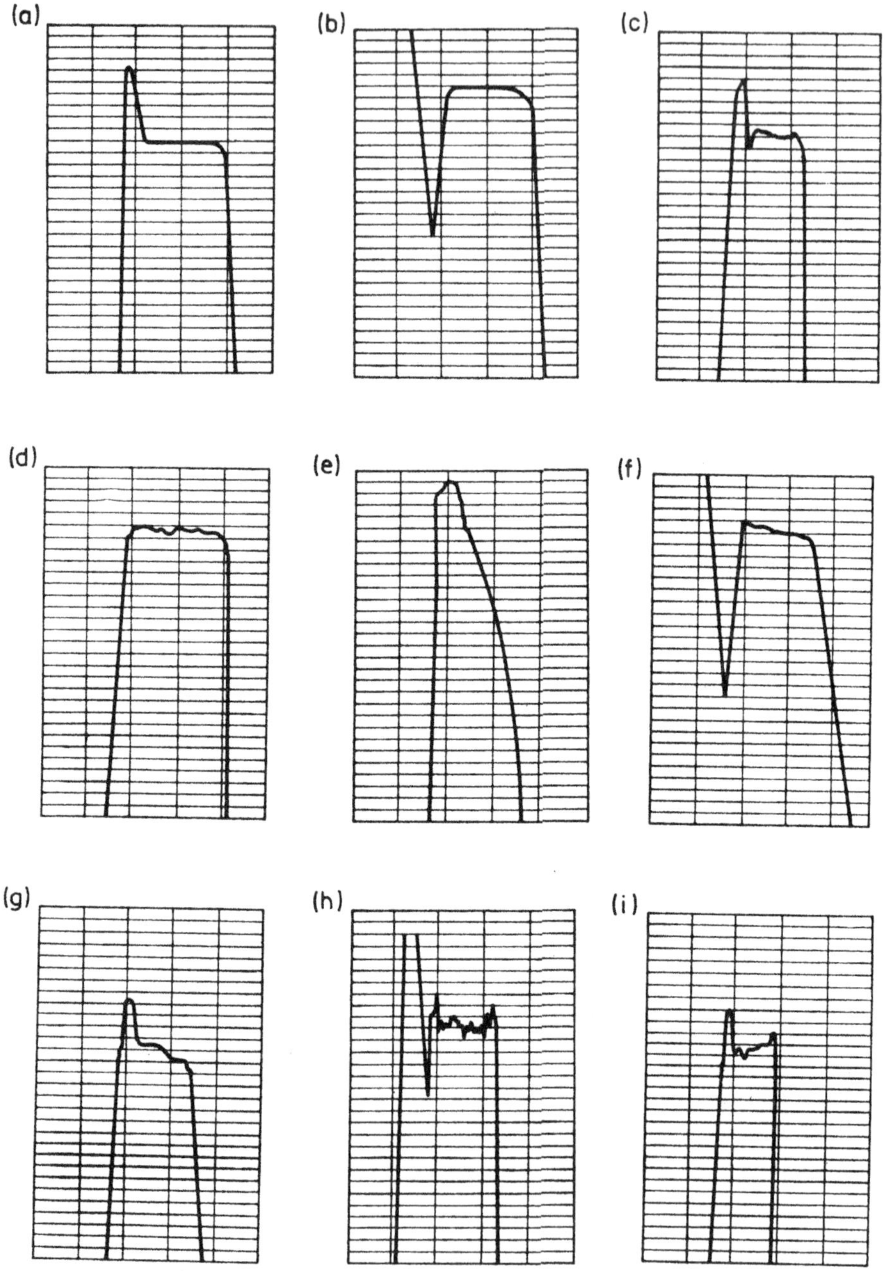
(a)
(b)
(c)
(d)
(e)
(f)
(g)
(h)
(i)

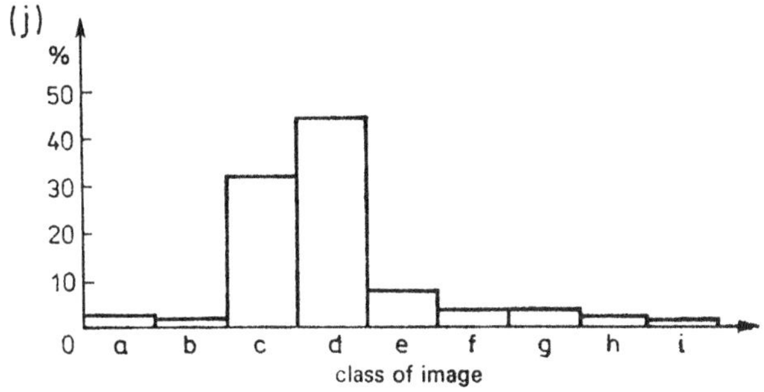

Fig. 1.14. An example of image classification. The temperature of molten steel has been measured with an expendable submerged thermocouple [57]: (a), (b) correct measurements; (c), (d), (e), (f), (g) disturbed measurements, but steel temperature still determinable; (h), (i) incorrect measurements; (j) the frequency of occurrence of an image of a given class.

where the distinctive features are the values y and N

$$|y_{i(j+1)}-y_{ij}| \leqslant \Delta y, \tag{1.111a}$$

$$|y_{iN}-y_{i1}| \leqslant \Delta y. \tag{1.111b}$$

When

$$\Delta y < \Delta y_A, \quad N > N_A, \quad \text{then the class is } A, \tag{1.112a}$$

$$\Delta y < \Delta y_B, \quad N > N_B, \quad \text{then the class is } B, \tag{1.112b}$$

$$\Delta y < \Delta y_C, \quad N > N_C, \quad \text{then the class is } C. \tag{1.112c}$$

The remaining images constitute the fourth class, the class of incorrect images. For each of the classes A, B and C, a different algorithm is applied to determine $y_{i\,\mathrm{av}}$, which is the measure of the measured temperature.

Generally, for images in the form of vectors, a proper allowance interval is established for every vector component. If a given image is included in an allowance field, it is qualified for a given class. For images in the form of a function of one or several arguments, the class of images is described by a vector of coefficients and a vector of allowance intervals. The result is qualified for a given class by estimating the values of the coefficients.

The assessment of the relations between the classes of images is seldom performed by a specialist in measurement science, and usually it depends on the purpose of measurement. Here is an example.

Let $y(t)$ be an output signal from a measuring instrument excited with the signal $x(t) = \sin\omega_1 t$. The following images may be realized in this case:

(i) $A\sin\omega_1 t$—class: a measuring instrument with linear characteristic;
subclasses: $A > 1$—amplifier,
$A < 1$—attenuator,
$A = 1$—biterminal;

(ii) $y \neq A\sin\omega_1 t$—class: the instrument is a nonlinear element;
subclasses: $y(t)$—terminated sine curve,
$y(t)$—series of harmonic components,
etc.

The relations of classes (i) and (ii) may be unknown. It may be known why the image obtained belongs to class (ii). The properties of the image may possibly be controlled, for example for small excitations the image may fall into class (i), and for excitations greater than U_1 into class (ii).

The casual connections may be more complex, and hard or even impossible to express. Some auxiliary tools (test signal generators) are used to measure some of the images, and are the conventional reference states. The signal generators are not standards, as they do not reproduce the states which define units of measurement. The specific states which are the quality of a test signal are defined through mathematical relations (in the abstract domain), and the relations are also valid in the real domain. This, however, is not sufficient for a test signal generator to be identified with a standard, and in any case there is no such need.

If material reference states (such as the generators of sine waves, square waves, white noise or other measurable quantities) are created, then the purpose is not to define reference states but to make measuring more convenient. It should be remembered that elementary waveforms (time variability functions), such as (1.107) and (1.108), are not the products of imagination but an imitation of phenomena occurring naturally in reality, such as pendulum deflection, mechanical vibrations, electrical oscillations, diffusion processes, chemical reactions, etc. One-dimensional scales are quite sufficient for the formulation of both two-dimensional images, e.g. $x(t)$, and multidimensional ones.

Chapter 2

Standards and the Propagation of Standards

2.1 Establishing the Standards of Units of Measurement

International standards are established by convention, in the form of resolutions of the General Conference of Weights and Measures. The basis of the convention is the current state of knowledge, and the resulting conviction that a new settlement is better than the previous one. To illustrate this, a historical outline of the most important settlements on the units of measurement of electric quantities will be presented. A number of the changes made in this field are quite significant.

The units of measurement of electric quantities were initially defined by Gauss in the CGS system via mechanical quantities. To combine the electric and magnetic measurement units, it was necessary to introduce an electric base quantity. In 1881, 1 ampere was accepted as the base unit of measurement of electric quantities. The ampere was defined in various ways:

— through the volt and the ohm,

— by Faraday's law, based on the mass of silver produced in the electrolysis of silver nitrate,

— more recently, by the force of interaction between two conductors which carry flowing current. This refers to the MKSA system of units.

The search for standard units of electric quantities continued for years, and different standards were in use.

1860—Standard resistance, constructed by Siemens. This was a mercury column 1 metre long with a cross-sectional area of 1 mm^2 at 0°C. This was used until 1911, and in Germany untill 1948. (In 1898, the column length was increased to 106.3 cm).

1863—The BAU standard—a metal alloy, applied in Great Britain.

1873—The Clark cell, accepted in 1893 as the standard element.

1881 —The ohm and the volt accepted as the base units, with the ampere defined through the ohm and volt; empirical units (standards) of ohm and ampere were defined—a current of 1 A yields 1.11810 mg of silver during 1 second of silver nitrate electrolysis.

1892—The Weston cell introduced as the new standard element.

1910—Application of the ampere definition based on electrolytic phenomena; the EMF of the standard cell was determined as 1.01830 V.

1927—An electrical laboratory created within the framework of BIPM;[1] beginning of studies on standard cells and standard resistors.

1948—The General Conference of Weights and Measures introduced the MKSA system of units.

1960—Introduction of the SI system of units.

1931, 1969—Corrections of the value of EMF of cells in the bureaus of measurement in different countries.

1962—The discovery of the Josephson effect, established in 1968 as the voltage standard.

1972—Constant $\Phi_0^{-1} = h/2e$ accepted by the CIPM.[2]

1977—Discovery of the quantum Hall effect [166].

1980—Construction of silicon MOSFET elements as the prototype resistance standard, applying the quantum Hall effect [187].

1990—Planned acceptance of the Josephson junction as the voltage standard.

Current state:

— base unit—ampere, reproduced with an accuracy of 6×10^{-6} using the current balance;

— best reproduced standard—Weston standard cell (group standard), accuracy better than 10^{-6};

— new standard being investigated—Josephson junction, accuracy 10^{-7}; the nominal value established arbitrarily by CIPM is [109]:

$$\Phi_0^{-1} = \frac{h}{2e} = 483\,597.9 \ \frac{\text{MHz}}{\mu\text{V}_{\text{BIPM}}}.$$

The accuracy obtained initially was 10^{-7} to 10^{-8}, it has now reached 10^{-9} [194] and is predicted to reach 10^{-12} [162]. The forecasts of Kose and Wöger [162] are presented in Fig. 2.1.

The new resistance standard based on the quantum Hall effect is the silicon MOSFET element or heterostructure of GaAs, AlGaAs, InGaAs, InP, whose resistance is

$$R_{Hi} = \frac{1}{i}\frac{h}{e^2}, \quad i = 1, 2, 3, \ldots$$

For $i = 1$, a value of

$$R_K = 25\,812.807$$

is called Klitzing constant. Its experimentally verified accuracy is 10^{-7}–10^{-8}

[1] BIPM—Bureau International des Poids et Mesures (in Sevres, Paris).

[2] CIPM—Comité International des Poids et Mesures.

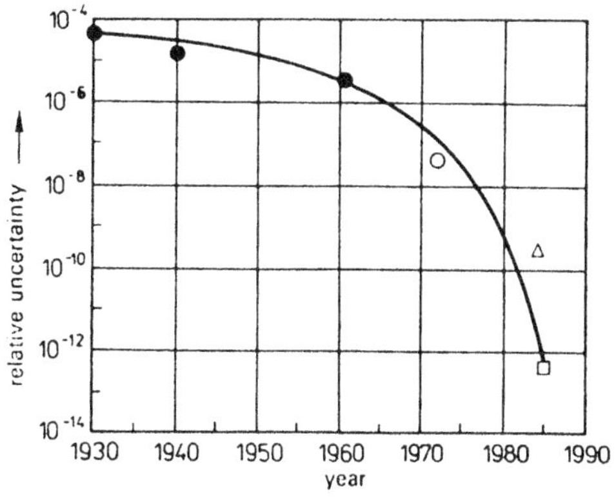

Fig. 2.1. The relative uncertainty of reproduction of 1 volt; black circle—Weston standard cell, white circle—Josephson junction, triangle—Josephson junction array at room temperature, square—junction array at helium temperature 4.2 K [162].

[148, 195]. The advantage of the quantum Hall effect is that the value of the measure of the standard is determined by fundamental physical constants. The value of voltage in the Josephson junction is frequency-dependent.

Progress in determination of fundamental physical constants is presented bv

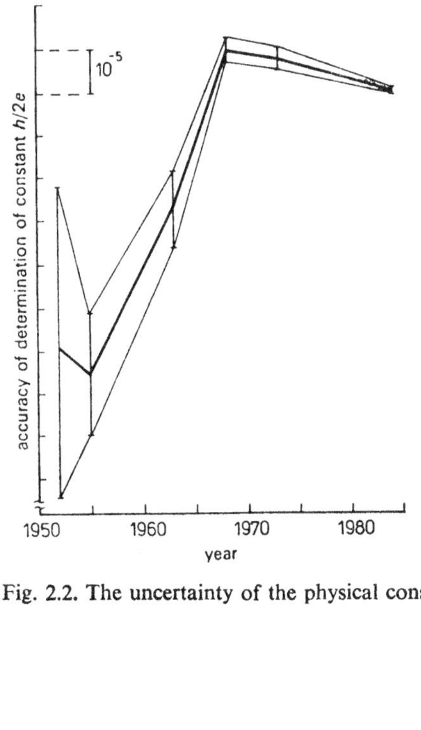

Fig. 2.2. The uncertainty of the physical constant $h/2e$ [162].

Kose and Wöger [162]. Figure 2.2 shows the improvement in accuracy of the constant $h/2e$.

A turning point can be shown to have occurred in the field of standards of measurement of electric quantities. Let us review its history.

The value of Φ_0 obtained from the EMF of a standard cell with an accuracy of 6×10^{-6} differed from the previous data by 3×10^{-5} [61]. The standard commonly applied for 90 years in the domain of electric quantities is the Weston cell. The life-time of the cell is estimated at about 20 years.[3]

The EMF of a cell may be derived from a model of the chemical reactions occurring in the cell. The physicochemical model of the cell is empirically proven, but knowledge of the physicochemical properties is not profound enough to determine the value of EMF theoretically. It is also impossible to determine the EMF through the standard current, because of the inferior accuracy of the latter.

In the field of electric quantities, a capacitance standard with an accuracy of about 1×10^{-7} may be applied. This is a so-called calculable standard, the measure of which is derived from a mathematical model by finding the dimensions of the standard. The construction of such a standard is shown in Fig. 2.3 [128].

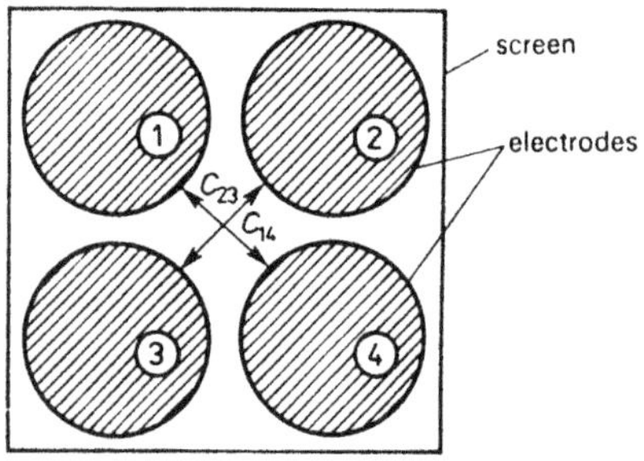

Fig. 2.3. Cross-section of the calculable Thompson–Lampard capacitance standard (cross-capacitor). Capacitance value per length unit is $C = (\varepsilon_0/\pi)\ln 2$ F/m [128].

Means are also available for the construction of a standard current by applying the effect of paramagnetic resonance. The accuracy of this standard would be 10^{-6} [49] to 4×10^{-7} [189].

Application of the Weston standard cell as the standard has been restricted by the difficulty of assuring the continuity of the reference state, because of the limited durability of the element. The method applied was later called the *group standard*. The group standard consists of a number of selected standards, and is

[3] As Hamer [31] states, in 1951 the group standard of NBS included the cells from 1906, 1933 and 1948. Thus the durability is even greater.

provided with new ones at the proper time. The standards whose measures depart from average are eliminated.

Selection of the elements to be included in the group standard is carried out on the basis of the metrological properties expressed as the class 1 characteristic (see Chapter 4), investigated for 1–2 years. At present, the EMF of a cell is measured with an uncertainty of single hundredths of a microvolt. The level of noise, in the form of oscillations and fluctuations, is thus very low for these elements. For example, the temperature is kept constant within 0.001°C. In Fig. 2.4, the

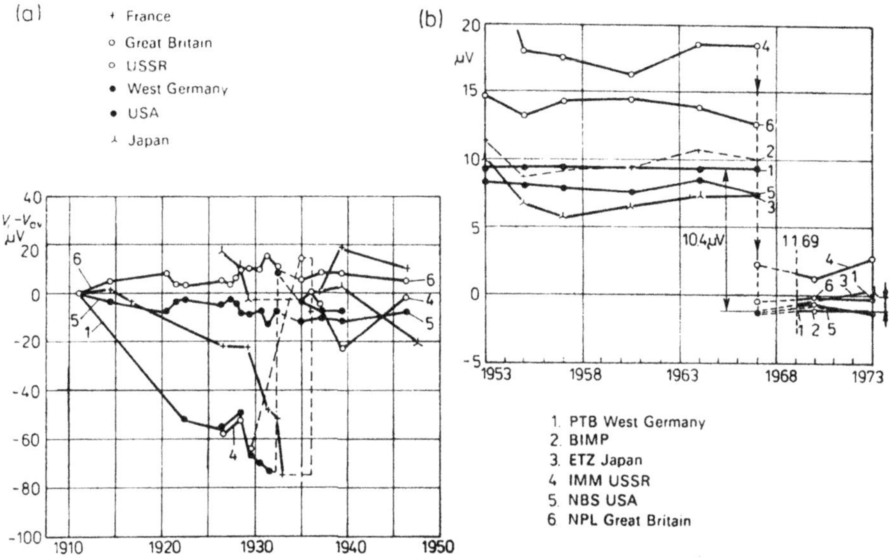

Fig. 2.4. The results of comparison of the national group voltage standards: (a) data according to Hamer [31], (b) data according to Froehlich [25] referred to the PTB standard.

variation of the nominal value of national group standards is shown. The corrections of the nominal values of the standards are indicated, as introduced in 1929–1936 and in 1969. According to M. Froehlich [25], the stability of the mean value of EMF of a group standard consisting of 38 cells was 1×10^{-8} V/year, as compared with the Josephson standard. Such good stability allows comparison over intervals of a few months [25]. The repeatability of the results is shown in Fig. 2.5.

The problem of maintaining the measure of a standard constant despite the limited durability of the standard has been solved by the application of a gradually restored group standard. The elements of the group standard are selected on the basis of their metrological properties. The criterion for the selection is the instability error not being too great.

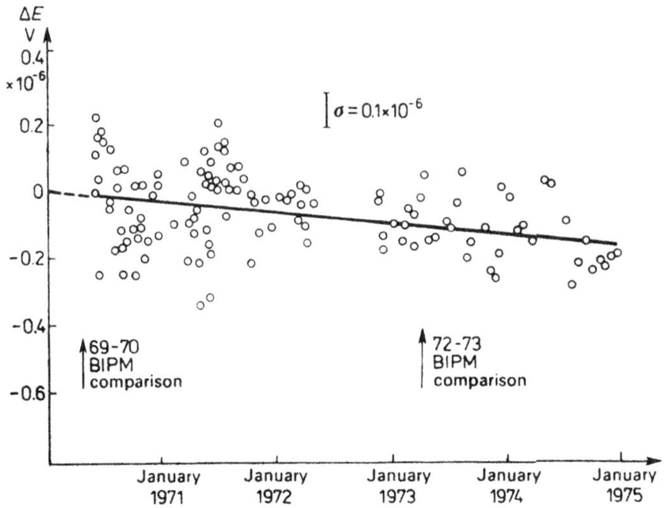

Fig. 2.5. The results of comparison of the NML group voltage standard (Australia) consisting of 7 cells, with Josephson effect standard [33].

The criterion is difficult to formulate for a small number of elements of the group standard, as well as for a small number of results of comparison with other standards. Thus, arbitrary decisions can be made here.

The state of the standards and the conditions of their reproduction are different for different physical quantities. For example, the mass standard was established a long time ago and, as may be supposed, the limits of its reproduction accuracy have already been reached. A fundamental change in accuracy and feasibility of reproduction occurred in length measurements when the prototype of metre was substituted by the atom effect standard. Changes occur as progress in science takes place. The prospects have been presented by Kose and Wöger [162].

2.2 Propagation of Standards

When two standards W_1 and W_2 are compared, it is stated a priori that one of them is the *primary standard* and the other is the *secondary standard*. The primary standard measure is transferred onto the secondary standard, and the unavoidable error is attributed to the secondary standard. For this reason, the standards of the same unit of measure, with the same value of measure, are classified at a proper level in the hierarchy of standards. The primary standard is placed at a higher level and the secondary standard at a lower one. The number of levels in the hierarchy is limited, for practical reasons. The lowest level includes utility tools, following inspection tools. The assignment of a measure to the standards to be accepted for use is called *calibration*, although in this case the measure of the standard is propagated and the term the *propagation of standards* is frequently used. The primary standard measure is transferred onto the secondary standard, and the uncertainty of the measure is determined from the course of the comparison process.

Let us consider the case of two compared standards having approximately the same measure. The primary standard W_1 has a *nominal measure* w_{10} established at the instant T_{01} of the *service time* of the primary standard. The *uncertainty of measure of the primary standard*

$$\Delta_1 = w_1 - w_{10} \tag{2.1}$$

is described by a *non-stationary random proces* $\{\Delta_1\}$:

$$\{\Delta_1(T)\} = g_1(T)\{\varepsilon_1\} + h_1(T), \tag{2.2}$$

where $g_1(T)$, $h_1(T)$ are time functions characterizing the *random* and *systematic errors*, respectively, $\{\varepsilon_1\}$ is a stationary random process with mean value equal to 0, unitary variance and the correlation function

$$K_\varepsilon(\tau) = E[\{\varepsilon_t\}\{\varepsilon_{t+\tau}\}] = \varrho(\tau), \tag{2.3a}$$

$$|\varrho(\tau)| \leqslant 1, \tag{2.3b}$$

T is the service time of the standard, t is the time, $t \in T$.

For obvious reasons, the nominal value of the standard was established at the instant T_{01} in order to avoid introducing a systematic error in the standard. The nominal value of the standard may vary during its life period by the value $h_{10}(T)$ (by a known rule) which may be predicted for a time $T > T_{01}$, based on checks done over the time interval $T < T_{01}$. Therefore,

$$w_{10}(T) = w_{10}(T_{01}) - \check{h}_{10}(T), \tag{2.4}$$

where $\check{h}_{10}(T)$ is the *prediction of the systematic error*. To simplify the notation, equations (2.1), (2.4) are further interpreted as[4]

$$\Delta_1(T) = w_1(T) - w_{10}(T). \tag{2.5}$$

2. The systematic error $h_1(T)$ in equation (2.2) does not include the component with the known value of $\check{h}_{10}(T)$. The error $h_1(T)$ is systematic in origin, but its value is unknown. The sources of this error are: the systematic error of the higher order standard (by which the measure w_{10} has been established), the comparison errors occurring in the process of establishing the measure of the standard, and

[4] This means that the measure corrected as in (2.4) is considered the nominal measure of the standard.

the estimation error of the component $\check{h}_{10}$ (prediction error). The process of establishing the measure of the secondary standard presented below will allow a glance at the process by which this error arises.

The properties of the *systematic error* $h_1(T)$ are as follows:

— its value

$$h_1(T) = h_1(T+\Delta T) \tag{2.6a}$$

is constant during the whole period of comparison with the secondary standard;

— the probability of realization of the value h is expressed by the probability density

$$p_h(h); \tag{2.6b}$$

— the value of h_1 is

$$h_1 = \int hp_h(h)\,\mathrm{d}h; \tag{2.6c}$$

— the value of this error depends on the time T because of the ageing process of the standard, also because of the dependence on T of the confidence interval of prediction, etc.;

— the value of h_1 is unknown and therefore may not be introduced as a correction to the measure of the standard;

— the error h_1 must be considered random, thus increasing the uncertainty of the measure of the standard.

The error of the standard which is *random* in origin, expressed by the component $g_1\{\varepsilon_1\}$, has its source in the intrinsic properties of the standard (e.g. oscillations and fluctuations). Its standard deviation is a function of time because of the changes of properties of the standard. The distribution of this error is

$$p_\varepsilon(\varepsilon). \tag{2.7}$$

The errors of a standard are conceptually difficult to describe. The opinion of the CIPM issued as Recommendation CI-1981 [178] is proof of this statement. The recommendation is not convincing, and polemical opinions have been published [164, 177]. Therefore the reader of the present interpretation should also familiarize himself with the physical causes of errors (Subsection 3.3.3) and the general model of error (3.136).

It should be stressed that the errors $h_1(T)$ and $g_1(T)\{\varepsilon_1\}$ are measurable using more accurate standards. Thus, the result presented in Fig. 2.6 may be interpreted as a systematic error distribution, namely the convolution of distributions (2.6b) and (2.7). In fact, it is the error distribution described by parameters (2.19) and (2.22), where part of some of the components may be negligibly small.

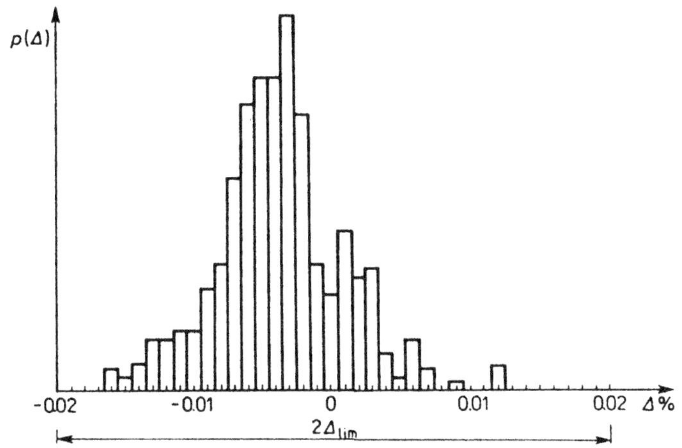

Fig. 2.6. The error distribution in a manufactured series of 410 standard normal resistors (1000 Ω, class 0.02) [100].

The *uncertainty of measure of the secondary standard* is similarly expressed as

$$\Delta_2 = w_2 - w_{20}, \tag{2.8a}$$

and

$$\{\Delta_2\} = g_2(T)\{\varepsilon_2\} + h_2(T). \tag{2.8b}$$

The location of the instant of comparison T in the scale of service time is different for each of the standards. For example, if the aim is to determine the measure of a secondary standard, then

$$T_1 > T_0, \quad T_2 < T_0, \tag{2.9a}$$

while if periodic checking is considered, then

$$T_1 > T_0, \quad T_2 > T_0, \tag{2.9b}$$

where T_0 is the instant when the nominal value of the standard is established, T_1 is the time of comparing against the standard W_1, and T_2 the time of comparing against the standard W_2.

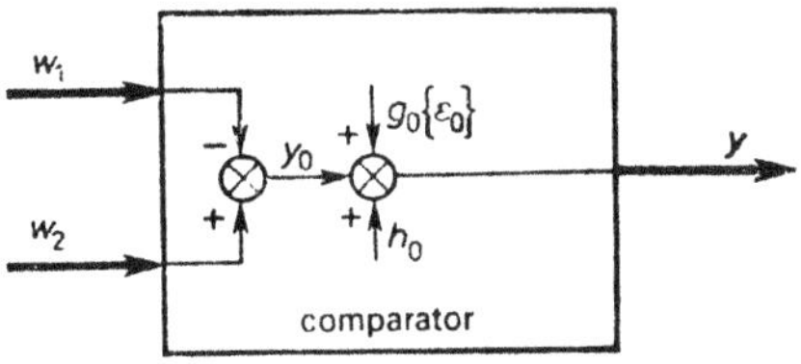

Fig. 2.7. A scheme for comparing two standards in the process of determining the measure of a secondary standard.

The standards are compared in the *comparator*, as shown in Fig. 2.7. The reading of the comparator is denoted as y, whereas the correct value y_0 of the reading is

$$y_0 = w_2 - w_1 \tag{2.10}$$

and is not known.

The comparator introduces an *error of comparison*[5]

$$\Delta = y - y_0, \tag{2.11}$$

which may be modelled by formula (2.2), and hence

$$\{\Delta\} = g_0(\varepsilon_0\} + h_0. \tag{2.12}$$

When establishing a secondary standard is being considered, the comparison result (measurement equation) is in the form[6]

$$w_2 = w_1 + y. \tag{2.13}$$

2.2.1 Single comparison

Random errors have one random realization each. Substituting expressions (2.1), (2.2), (2.8) and (2.12) in the measurement equation (2.13), we obtain

$$w_{20} + g_2\varepsilon_2 + h_2 = w_{10} + g_1\varepsilon_1 + h_1 + y_0 + g_0\varepsilon_0 + h_0. \tag{2.14}$$

The measure of the secondary standard may be derived from the known components of equation (2.14), including w_{10} and $y = y_0 + g_0\varepsilon_0 + h_0$. Hence

$$w_{20} = w_{10} + y = w_{10} + y_0 + g_0\varepsilon_0 + h_0. \tag{2.15}$$

The correct result should be

$$w_{20\,\mathrm{corr}} = w_{10} + y_0, \tag{2.16}$$

which means that the derived measure of the secondary standard is burdened with the comparator error.

The error equation is obtained by subtracting the correct value (i.e. equation (2.16)) from equation (2.14):

$$g_2\varepsilon_2 + h_2 = g_1\varepsilon_1 + h_1 + g_0\varepsilon_0 + h_0. \tag{2.17}$$

---

[5] Usually, when the uncertainty of the measure of each of the standards is δ, the difference between the measures of primary and secondary standards generally does not exceed a value of several δ. A comparator error of the order of 1% introduces a contribution of several hundredths of δ in the comparison error and is usually negligible. This feature results from the calculus of errors of indirect measurement, when the difference between the quantities being measured directly is very small [91].

[6] In the checking of two standards (2.9b),

$$y = w_2 - w_1.$$

After the averaging operation is performed, the equation for systematic errors is obtained according to the definition of systematic error (4.28). Since all the components in equation (2.17) are single realizations of random processes, their values are constant and $E(x) = x$. The value of the systematic error of the standard, calculated from equation (2.17), is

$$h_2 = h_1 + g_1 \varepsilon_1 - g_2 \varepsilon_2 + g_0 \varepsilon_0 + h_0. \tag{2.18}$$

Allowing for the properties of the sources of those errors which are the components in equation (2.18), the *expected value* and the *variance* of the systematic error of the secondary standard may be determined as

$$E(h_2) = E(h_1) + g_1 E(\varepsilon_1) - g_2 E(\varepsilon_2) + g_0 E(\varepsilon_0) + E(h_0) = \bar{h}_1 + \bar{h}_0, \tag{2.19}$$

because the mean values of the processes $\{\varepsilon\}$ were assumed to be zero, and

$$\operatorname{var}(h_2) = \operatorname{var}(h_1) + \operatorname{var}(h_0) + g_1^2 + g_2^2 + g_0^2. \tag{2.20}$$

These results need some comment. The value of the systematic error h_1 realized in the primary standard is an element of the set of systematic errors, characterizing the standards of the same type, under the same conditions and at the same time interval $T - T_0$ from the beginning of their use. Such characteristics are rarely available, partly because the power of the sets of identical standards is not large enough for the statistical characteristics to be reliable. An example may be the error distribution shown in Fig. 2.6. The expected value of the set of systematic errors is usually near zero, though in the example from Fig. 2.6 it differs markedly from this value. Also the variance of h_1 characterizes the set of errors of the standards of the same type, under the same conditions, at the same service time and for the same value of the reading y.

The result of variance estimation includes the variance of the random error of the secondary standard as one of the components. This is natural, since in the process of comparison this error was recognized and accounted for when the value of the error was determined. Expression (2.18), which is a sum of random processes, is the basis for determination of the probability distribution of the error, systematic in origin but with an unknown value, as the distributions of these errors are known (according to assumptions (2.6)). Therefore, for a given confidence level, the *interval of uncertainty of the measure* of a standard may be calculated.

Besides the uncertainty of measure mentioned, the secondary standard, when used, generates physically the random error

$$\{\Delta_{2r}\} = g_2 \{\varepsilon_2\}, \tag{2.21}$$

which causes the variance of the uncertainty of the measure of a standard to be

$$\begin{aligned} \operatorname{var}(w_2) &= \operatorname{var}\{\Delta_2\} = \operatorname{var}(h_2) + \operatorname{var}[g_2\{\varepsilon_2\}] = \operatorname{var}(h_2) + g_2^2 \\ &= \operatorname{var}(h_1) + \operatorname{var}(h_0) + g_1^2 + 2g_2^2 + g_0^2. \end{aligned} \tag{2.22}$$

Since the variance of the error of the primary standard is

$$\operatorname{var}(w_1) = \operatorname{var}\{\Delta_1\} = \operatorname{var}(h_1)+g_1^2, \tag{2.23}$$

a comparison of variances (2.22), (2.23) implies

$$\operatorname{var}(w_2)-\operatorname{var}(w_1) = \operatorname{var}(h_0)+g_0^2+2g_2^2 > 0, \tag{2.24}$$

which means that the uncertainty of measure of the secondary standard is greater than the uncertainty of the primary standard. This fact is illustrated in Fig. 2.8. Moreover, the measure of the secondary standard, when related to the measure of the primary standard, differs by the value of the comparator error.

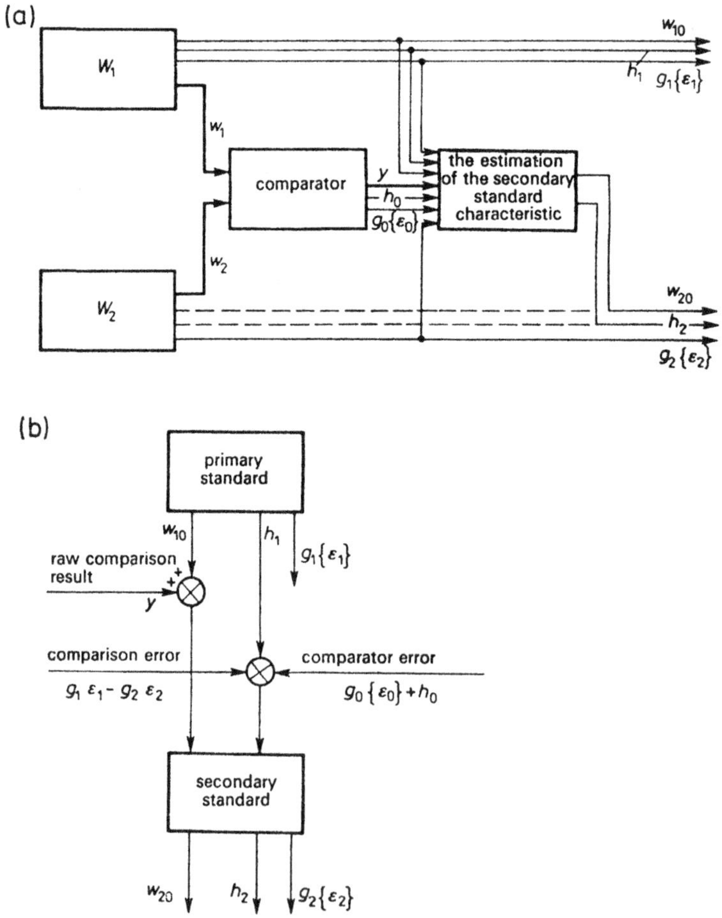

Fig. 2.8. A diagram of propagation of a measure and uncertainty of a primary standard onto a secondary standard: (a) the estimation course, (b) measure and error components.

2.2.2 *n-fold comparison with the same primary standard*

Let us consider a comparison process occurring over the time interval $\tau > \tau_{\text{lim}}$, for which $\varrho_1(\tau) = \varrho_2(\tau) = \varrho_0(\tau) = 0$, but $h_1(T+\tau) = h_1(T)$ and $g_1(T+\tau) = g_1(T)$ and similarly for h_2, g_2, h_0, g_0. If a mean value is accepted as a result of comparison, then instead of (2.14), n equations are obtained for the subsequent realizations of random processes $\{\varepsilon_1\}$, $\{\varepsilon_2\}$, $\{\varepsilon_0\}$:

$$
\begin{aligned}
w'_{20}+g_2\varepsilon_{21}+h'_2 &= w_{10}+g_1\varepsilon_{11}+h_1+y_{01}+g_0\varepsilon_{01}+h_0,\\
&\cdots\cdots\cdots\cdots\cdots\cdots\cdots\cdots \qquad (2.25)\\
w'_{20}+g_2\varepsilon_{2n}+h'_2 &= w_{10}+g_1\varepsilon_{1n}+h_1+y_{0n}+g_0\varepsilon_{0n}+h_0.
\end{aligned}
$$

By adding the equations and dividing by n, we obtain

$$
\begin{aligned}
&w'_{20}+g_2\frac{1}{n}\sum_{i=1}^{n}\varepsilon_{2i}+h'_2\\
&\quad = w_{10}+g_1\frac{1}{n}\sum_{i=1}^{n}\varepsilon_{1i}+h_1+g_0\frac{1}{n}\sum_{i=1}^{n}\varepsilon_{0i}+h_0+\frac{1}{n}\sum_{i=1}^{n}y_{0i}.
\end{aligned}
\tag{2.26}
$$

The nominal value of the secondary standard is not the same because

$$
w'_{20} = w_{10}+\frac{1}{n}\sum_{i=1}^{n}y_i = w_{10}+g_0\frac{1}{n}\sum_{i=1}^{n}\varepsilon_{0i}+h_0+\frac{1}{n}\sum_{i=1}^{n}y_{0i},
\tag{2.27}
$$

and for $n \to \infty$

$$
w'_{20} = w_{10}+\bar{y}_0+h_0.
\tag{2.28}
$$

After the correct value is subtracted from equation (2.26), the value of the systematic error is obtained

$$
h'_2 = g_1\frac{1}{n}\sum_{i=1}^{n}\varepsilon_{1i}+h_1-g_2\frac{1}{n}\sum_{i=1}^{n}\varepsilon_{2i}+\frac{1}{n}g_0\sum_{i=1}^{n}\varepsilon_{0i}+h_0.
\tag{2.29}
$$

The expected value of the error is

$$
E(h'_2) = \bar{h}_1+\bar{h}_0
\tag{2.30}
$$

and its variance is smaller than in the case of a single comparison, namely

$$
\operatorname{var}(h'_2) = \operatorname{var}(h_1)+\operatorname{var}(h_0)+\frac{1}{n}(g_1^2+g_2^2+g_0^2).
\tag{2.31}
$$

The error random in origin is $g_2\{\varepsilon_2\}$, so the variance of the uncertainty of measure of the secondary standard is

$$
\operatorname{var}(\Delta_2) = \operatorname{var}(h_1)+\operatorname{var}(h_0)+\frac{1}{n}(g_1^2+g_0^2)+\frac{n+1}{n}g_2^2.
\tag{2.32}
$$

Formula (2.8b), expressing the uncertainty of a secondary standard, is noteworthy after (2.29) is substituted in it. The error is the sum of $3n+1$ components which are independent realizations of random processes. The number n is considerable, and depends on the number of comparisons. At least $2n+1$ of them belong to similar populations.[7] This fact forms the basis for inferring that the distribution of error Δ_2 is close to normal. The inference is based on the central limit theorem of probability. If the value of $\bar{h}_1$ is significant, a normal distribution with a mean value different from zero is taken into consideration.

2.2.3 *m-fold comparison with different standards*

M-fold comparison with different standards having the nominal values w_{101}, $w_{102}, \ldots, w_{10m}$ gives interesting results. The subsequent standards are characterized by the error components

$$g_{11}, \ldots, g_{1m}; \quad \varepsilon_{11}, \ldots, \varepsilon_{1m}; \quad h_{11}, \ldots, h_{1m}. \tag{2.33a}$$

The components of the results of comparison $y_1, \ldots, y_m$ are

$$y_{01}, \ldots, y_{0m}; \quad g_0\varepsilon_{01}, \ldots, g_0\varepsilon_{0m}; \quad h_0. \tag{2.33b}$$

For subsequent comparisons, m equations are obtained

$$\begin{aligned} w''_{20}+g_2\varepsilon_{21}+h''_2 &= w_{101}+g_{11}\varepsilon_{11}+h_{11}+y_{01}+g_0\varepsilon_{01}+h_0, \\ &\cdots\cdots \\ w''_{20}+g_2\varepsilon_{2m}+h''_2 &= w_{10m}+g_{1m}\varepsilon_{1m}+h_{1m}+y_{0m}+g_0\varepsilon_{0m}+h_0. \end{aligned} \tag{2.34}$$

By adding the equations and then dividing by m, the result obtained is

$$\begin{aligned} w''_{20}+g_2\frac{1}{m}\sum_{j=1}^{m}\varepsilon_{2j}+h''_2 = \frac{1}{m}\sum_{j=1}^{m}w_{10j}+\frac{1}{m}\sum_{j=1}^{m}(g_{1j}\varepsilon_{1j})+\frac{1}{m}\sum_{j=1}^{m}h_{1j}+ \\ +\frac{1}{m}g_0\sum_{j=1}^{m}\varepsilon_{0j}+\frac{1}{m}\sum_{j=1}^{m}y_{0j}+h_0. \end{aligned} \tag{2.35}$$

We then have

$$\begin{aligned} w''_{20} &= \frac{1}{m}\sum_{j=1}^{m}w_{10j}+\frac{1}{m}\sum_{j=1}^{m}y_j \\ &= \frac{1}{m}\sum_{j=1}^{m}w_{10j}+\frac{1}{m}\sum_{j=1}^{m}y_{0j}+\frac{1}{m}g_0\sum_{j=1}^{m}\varepsilon_{0j}+h_0, \end{aligned} \tag{2.36}$$

[7] These are the components ε_{1i}, ε_{2i} and ε_2, generated by a secondary standard.

$$h_2'' = \frac{1}{m}\sum_{j=1}^{m} h_{1j} + \frac{1}{m}\sum_{j=1}^{m}(g_{1j}\varepsilon_{1j}) - g_2\frac{1}{m}\sum_{j=1}^{m}\varepsilon_{2j} + g_0\frac{1}{m}\sum_{j=1}^{m}\varepsilon_{0j} + h_0, \tag{2.37}$$

$$\operatorname{var}(h_2'') = \frac{1}{m^2}\sum_{j=1}^{m} g_{1j}^2 + \frac{1}{m}(g_2^2+g_0^2) + \frac{1}{m^2}\sum_{j=1}^{m}\operatorname{var}(h_{1j}) + \operatorname{var}(h_0). \tag{2.38}$$

A comparison with different standards causes the part h_1 of systematic error of the primary standard to be reduced $\sqrt{m}$ times. The m-fold realization of random errors also reduces their value $\sqrt{m}$ times.

2.2.4 *n-fold comparison with m different primary standards*

By repeating the calculations in Subsections 2.2.2 and 2.2.3, mn equations are obtained (n equations (2.34) for each $j = 1, \ldots, m$). By adding them and dividing by mn, the nominal measure of the secondary standard is derived:

$$w_{20}''' = \frac{1}{m}\sum_{j=1}^{m} w_{10j} + \frac{1}{mn}\sum_{j=1}^{m}\sum_{i=1}^{n} y_{ij}$$

$$= \frac{1}{m}\sum_{j=1}^{m} w_{10j} + \frac{1}{mn}\sum_{j=1}^{m}\sum_{i=1}^{n} y_{0ij} + \frac{1}{mn}\sum_{j=1}^{m} g_{0j}\sum_{i=1}^{n}\varepsilon_{0ij} + h_0. \tag{2.39}$$

The correct nominal value of the secondary standard is

$$w_{20\,\mathrm{corr}}''' = \frac{1}{m}\sum_{j=1}^{m} w_{10j} + \frac{1}{mn}\sum_{j=1}^{m}\sum_{i=1}^{n} y_{0ij}. \tag{2.40}$$

The value of the measure is burdened with the comparator error. The systematic error is calculated as before, giving the results

$$h_2''' = \frac{1}{mn}\sum_{j=1}^{m} g_{1j}\sum_{i=1}^{n}\varepsilon_{1ij} + \frac{1}{m}\sum_{j=1}^{m} h_{1j} - \frac{1}{mn}g_2\sum_{j=1}^{m}\sum_{i=1}^{n}\varepsilon_{2ij} +$$

$$+ \frac{1}{mn}g_0\sum_{j=1}^{m}\sum_{i=1}^{n}\varepsilon_{0ij} + h_0, \tag{2.41}$$

with expected value

$$E(h_2''') = \frac{1}{m}\sum_{j=1}^{m} h_{1j} + \bar{h}_0 \tag{2.42}$$

and variance

$$\operatorname{var}(h_2''') = \frac{1}{m^2 n}\sum_{j=1}^{m} g_{1j} + \frac{1}{mn}(g_2^2+g_0^2) + \frac{1}{m}\operatorname{var}(h_1) + \operatorname{var}(h_0). \tag{2.43}$$

Also,

$$\operatorname{var}(\Delta_2''') = \operatorname{var}(h_2''') + g_2^2. \tag{2.44}$$

The degree of reduction of the respective sources of errors may now be easily seen from the formulae.

2.2.5 *m times larger measure of a secondary standard*[8]

Let the measure of the primary standards be

$$w_{1j} = \frac{w_2}{m} \tag{2.45}$$

and let the metrological properties be described by the error

$$\{\Delta_{1j}\} = g_{1j}\{\varepsilon_{1j}\} + h_{1j} \tag{2.46}$$

whose components are identical with those in (2.2). The metrological properties of the standard w_2 are described by (2.8b), and the properties of the comparator by (2.12). The comparison

$$w_2 = \sum_{j=1}^{m} w_{1j} + y \tag{2.47}$$

implies

$$w_{20} + g_2\varepsilon_2 + h_2 = \sum_{j=1}^{m} w_{10j} + \sum_{j=1}^{m}(g_{1j}\varepsilon_{1j}) + \sum_{j=1}^{m} h_{1j} + y_0 + g_0\varepsilon_0 + h_0. \tag{2.48}$$

---

[8] Usually, the ratio of measures of two standards is approximately a natural number. There are two methods:

(i) the *comparison method* is applied, with the value of $k = m/n$ (*coincidence method*), i.e. $nw_{20} = mw_{10}$, where m, n are natural numbers [91];

(ii) the primary standard is compared with n identical secondary standards for which $w_{10} = nw_{20}$, or m identical primary standards are compared with one secondary standard, and then $mw_{10} \cong w_{20}$.

Physically realized operations of division or merging of the states of quantities are applied here. The operations are physically realized for a number of physical quantities, in accordance with the properties of metric scales. The properties of the first method are the consequence of the mode of measurement applied, as described in Chapter 5. For this reason, only the second method is discussed here.

The value of the nominal measure of the secondary standard is

$$w_{20} = \sum_{j=1}^{n} w_{10j} + y = \sum_{j=1}^{m} w_{10j} + y_0 + g_0\varepsilon_0 + h_0. \tag{2.49}$$

The systematic error, with an unknown value, is then

$$h_2 = \sum_{j=1}^{m} (g_{1j}\varepsilon_{1j}) + \sum_{j=1}^{m} h_{1j} - g_2\varepsilon_2 + g_0\varepsilon_0 + h_0, \tag{2.50}$$

and the random error is $g_2\{\varepsilon_2\}$. If the primary standards are of the same type and the conditions

$$\bar{g}_{1j} = \bar{g}_1, \quad \bar{h}_{1j} = \bar{h}_1 \quad \text{for} \quad j = 1, \ldots, m, \tag{2.51}$$

are met, i.e., the errors come from the same population and

$$\frac{\bar{g}_2}{\bar{g}_1} \approx \frac{\bar{h}_2}{\bar{h}_1} \approx \frac{w_2}{w_1} \approx m, \tag{2.52}$$

then

$$E(h_2) = \sum_{j=1}^{m} \bar{h}_{1j} + h_0 = m\bar{h}_1 + h_0 = \bar{h}_1' + h_0, \tag{2.53}$$

and, writing $\bar{h}_1' = m\bar{h}_1$, $g_1' = mg_1$ and $\text{var}\,h_1' = m^2 \text{var}\,h_1$,

$$\begin{aligned} \text{var}(h_2) &= mg_1^2 + m\,\text{var}(h_1) + g_2^2 + g_0^2 + \text{var}(h_0) \\ &= \frac{1}{m}[m^2 g_1^2 + m^2 \text{var}(h_1)] + g_2^2 + g_0^2 + \text{var}(h_0) \\ &= \frac{1}{m}[g_1'^2 + \text{var}(h_1')] + g_2^2 + g_0^2 + \text{var}(h_0). \end{aligned} \tag{2.54}$$

The results given above are interpreted as follows. If the relative error of the primary standards and the secondary standard are close, i.e., the absolute errors differ m times (2.52), the expected value of the systematic error of the secondary standard will be equal to the sum of the expected value of the errors of the primary standards (2.53). The case $\bar{h}_1 = 0$ is very advantageous. In the case when a small value of the measure of primary standards is transferred onto a secondary standard, the variance of the error, systematic in nature, decreases m times (the first component in formula (2.54)). For example, if the limiting error of $m = 10$ primary standards is 0.01%, the part of this error in the error of the secondary standard is $0.01/\sqrt{10} = 0.003\%$. The effect is better accuracy of the secondary standard as compared with the accuracy of the primary standards. This effect arises when $\bar{h}_{1j} = 0$, but this condition is hardly verifiable. In other cases the effect is not so strong (see Fig. 2.9).

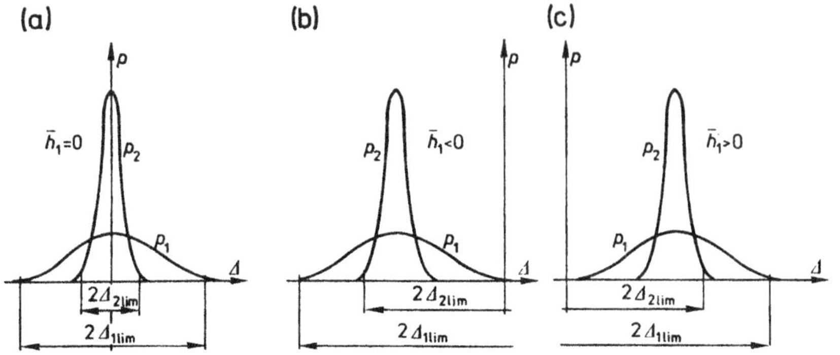

Fig. 2.9. The influence of the systematic error of a primary standard on the error of the secondary standard. The mean value of the primary standard is (a) zero, (b), (c) different from zero.

Multiple comparisons with the same set of equipment leads to n equations (2.48) if the subsequent comparisons are performed at time intervals $\tau > \tau_{lim}$ assuring independent realizations of random errors. By making calculations as in Subsections 2.2.2 and 2.2.5, the nominal measure of the standard may be calculated:

$$w'_{20} = \sum_{j=1}^{m} w_{10j} + \frac{1}{n}\sum_{i=1}^{n} y_i. \tag{2.55}$$

Also the systematic error

$$h'_2 = \sum_{j=1}^{m} h_{1j} + h_0 + \frac{1}{n}\sum_{j=1}^{m} g_{1j}\sum_{i=1}^{n} \varepsilon_{1i} - \frac{1}{n} g_2 \sum_{i=1}^{n} \varepsilon_{2i} + \frac{1}{n} g_0 \sum_{i=1}^{n} \varepsilon_{0i}, \tag{2.56}$$

as well as its expected value and variance may be calculated.

2.2.6 *m times smaller measure of a secondary standard*

The measure of a secondary standard is

$$w_{2j} m \cong w_1. \tag{2.57}$$

The comparison procedure is more complex here. M values of nominal measures of secondary standards should be determined, together with m estimates of the uncertainty of the measures. A minimum of m independent comparisons must be made in order to obtain m independent equations for calculating the results. Because of condition (2.57), it is necessary to perform the comparison described by the measurement equation

$$\sum_{j=1}^{m} w_{2j} = w_1 + y_1, \tag{2.58}$$

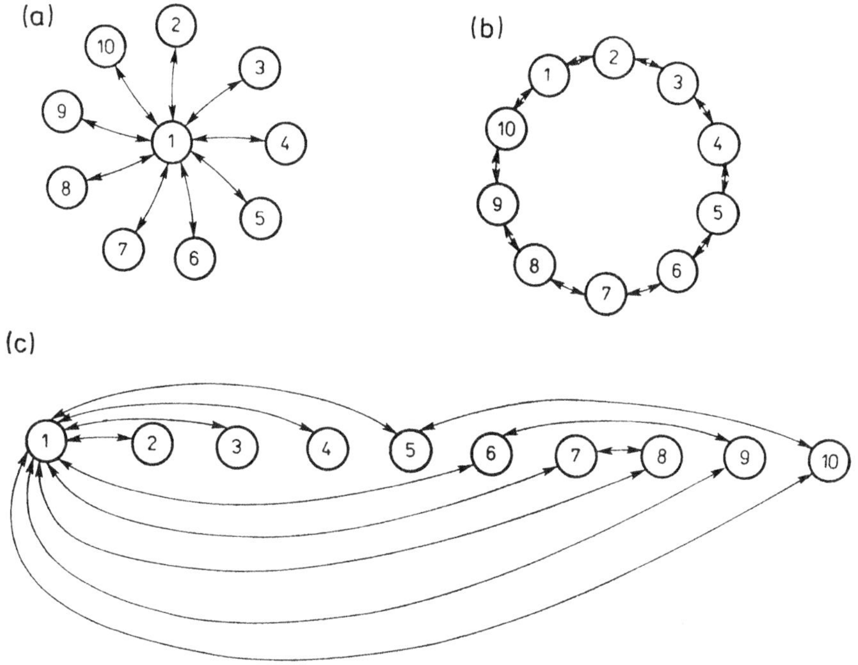

Fig. 2.10. A scheme of comparisons made to obtain the measure of m secondary standards: (a) as related to the first standard, (b) comparisons of subsequent standards, (c) another combination of comparisons.

together with $m-1$ other comparisons. Examples of comparison schemes are shown in Fig. 2.10.

For the scheme shown in Fig. 2.10a, where each secondary standard is compared with the secondary standard denoted by $j = 1$, the system of equations is

$$w_{21} = w_{2j}+y_j, \quad j = 2, 3, \ldots, m. \tag{2.59}$$

To avoid any improper mathematical transformations, the index i ($i = 1, \ldots$ $\ldots, m$) is introduced, denoting the number of comparison. This allows distinct individual realizations of random errors of the standard W_{21}. For $i = 1$, the comparison (according to equation 2.58) may be described as follows:

$$\sum_{j=1}^{m} w_{20j}+\sum_{j=1}^{m} (g_{2j}\varepsilon_{2j})+\sum_{j=1}^{m} h_{2j} = w_{10}+g_1\varepsilon_1+h_1+y_{01}+g_{01}\varepsilon_{01}+h_{01}. \tag{2.60}$$

From (2.59) we derive $m-1$ equations describing the comparisons $i = 2, 3, \ldots$ $\ldots, m$:

$$w_{20j}+g_{2j}\varepsilon_{2ji}+h_{2ji} = w_{201}+g_{21}\varepsilon_{21i}+h_{21}-y_{0i}-g_{0i}\varepsilon_{0i}-h_{0i}. \tag{2.61}$$

After substituting $m-1$ equations (2.61) solved for w_{20j} into (2.60), we obtain

$$mw_{201}+g_{21}\sum_{i=1}^{m}\varepsilon_{21i}+mh_{21}+\sum_{j=2}^{m}(g_{2j}\varepsilon_{2j1})-\sum_{i=j=2}^{m}(g_{2i}\varepsilon_{2ji})+\sum_{j=2}^{m}h_{2j1}-$$
$$-\sum_{i=j=2}^{m}h_{2ji}=w_{10}+g_1\varepsilon_1+h_1+\sum_{i=1}^{m}(y_{0i}+g_{0i}\varepsilon_{0i}+h_{0i}). \tag{2.62}$$

The assumed properties of the systematic error imply $h_{2ij}=h_{2j}$, $h_{21i}=h_{21}$, and for $i=2,3,\ldots,m$ we have $h_{0i}=h_0$ and $g_{0i}=g_0$. The last two components on the left-hand side are reduced. The nominal measure of the standard W_{21} is

$$w_{201}=\frac{1}{m}w_{10}+\frac{1}{m}\sum_{i=1}^{m}y_i$$
$$=\frac{1}{m}\left[w_{10}+\sum_{i=1}^{m}y_{0i}+g_{01}\varepsilon_{01}+h_{01}+g_0\sum_{i=2}^{m}\varepsilon_{0i}+(m-1)h_0\right] \tag{2.63}$$

and the correct value is

$$w_{201\text{corr}}=\frac{1}{m}w_{10}+\frac{1}{m}\sum_{i=1}^{m}y_{i0}. \tag{2.64}$$

In formula (2.63) it was accepted that the values w_1 compared in the first measurement are m times greater than the values compared in the rest of the measurements, for which the properties of the comparator are different. The values of h_{01} and h_0 are different as well as the values of g_{01} and g_0. The value of the systematic error h_{21} is obtained by subtracting correct value (2.64) from equation (2.62):

$$h_{21}=\frac{1}{m}\left[h_1+g_1\varepsilon_1+g_{01}\varepsilon_{01}+h_{01}+\sum_{i=2}^{m}(g_0\varepsilon_{0i})+(m-1)h_0+\right.$$
$$\left.+\sum_{j=i=2}^{m}+(g_{2j}\varepsilon_{2ji})-\sum_{j=1}^{m}(g_{2j}\varepsilon_{2j1})-g_{21}\sum_{i=2}^{m}\varepsilon_{21i}\right]. \tag{2.65}$$

The expected value of the systematic error is

$$E(h_{21})=\frac{1}{m}\bar{h}_1+\frac{1}{m}\bar{h}_{01}+\frac{m-1}{m}\bar{h}_0=\bar{h}_{21}, \tag{2.66}$$

and the variance is

$$\mathrm{var}(h_{21})=\frac{1}{m^2}[\mathrm{var}(h_1)+\mathrm{var}(h_{01})+(m-1)^2\mathrm{var}(h_0)+$$
$$+(m-1)g_0^2+g_{01}^2+g_1^2+(3m-2)g_2^2]. \tag{2.67}$$

The parameters of the standards $W_{22}, W_{23}, \ldots, W_{2m}$ are calculated from equation (2.61), as in Subsection 2.2.1, because of the properties already determined for the standard W_{21}.

For $j = 2, 3, \ldots, m$,

$$w_{20j} = w_{201} - y_j, \tag{2.68}$$

$$h_{2j} = h_{21} - h_0 + g_{21}\varepsilon_{21i} - g_{2j}\varepsilon_{2ji} - g_0\varepsilon_{0i}. \tag{2.69}$$

Thence

$$E(h_{2j}) = \bar{h}_{21} - \bar{h}_0 \tag{2.69a}$$

and

$$\mathrm{var}(h_{2j}) = \mathrm{var}(h_{21}) + \mathrm{var}(h_0) + 2g_2^2 + g_0^2, \tag{2.69b}$$

where h_{21} and $\mathrm{var}(h_{21})$ are expressed by formulae (2.66) and (2.67), respectively. Also

$$\mathrm{var}(\Delta_{2j}) = \mathrm{var}(h_{21}) + \mathrm{var}(h_0) + 3g_2^2 + g_0^2. \tag{2.70}$$

The scheme of comparisons applied prefers to some extent the secondary standard W_{21}, since the uncertainty of this standard is smaller by the random error of the compared secondary standards and the random error of the comparator in a single comparison.

On interpretation of (2.67), it is seen that the random errors occurring at every comparison are a significant part of the uncertainty of the standard. Therefore, the secondary standard No. 1 should be the standard with the smallest random error, i.e., with the smallest value of g_2. Multiple comparisons by the same scheme will cause part of the random errors to diminish, by a ratio which can be determined from the derived formulae, knowing the values of the components of error.

To determine the measure of the secondary standard, the other scheme of comparisons may be applied, as in Figs. 2.10b, c. The scheme may include $n > m$ comparisons by any method. In this way, more results than unknowns are obtained, and the accuracy of the measures of secondary standards determined and of their errors is improved. In the case discussed, groups of standards containing 2, 3 or more elements may be compared. Thus, one comparison causes several sources of random errors to be realized at the same time. The effect of diminishing the variance of random error will depend on the properties of the error of the sum, as shown in the example (see Tables 4.6 and 4.7).

The problem discussed is connected with the mutual comparison of the elements of a group standard, e.g. the comparison of the elements within one standard. It is known that such intercomparison does not allow estimation of changes of measure of a group standard. However, it allows estimation of the change of measure of elements, referred to the group measure [190]. Such information is sufficient for checking the elements of the group standard and for elimination of the elements with measures departing from the mean value. The practical procedure [168, 177] prefers comparison schemes of the type 'each element with each other'. Such a scheme produces symmetrical results, easy to process.

As an effect of applying the above ways of transferring the measure of a standard, a *system of standards* with different values of measure may be created based on one standard only. The uncertainty of measure in this case is determined with respect to the way in which the measure is established. Figure 2.11 shows an

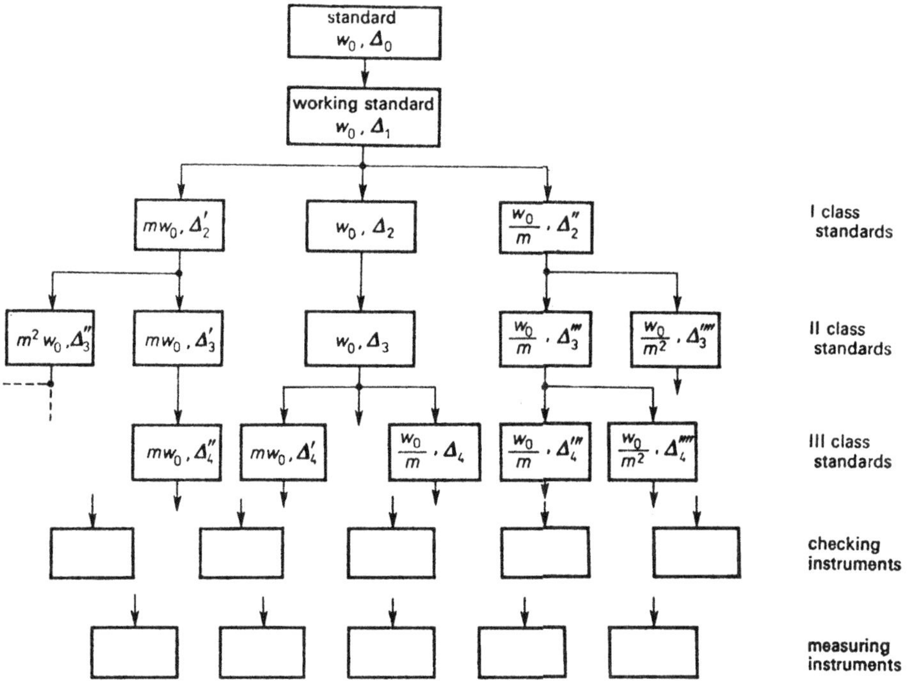

Fig. 2.11. An example of a hierarchy of standards.

example of the structure of a system of standards of one quantity, for which the manner of division and merging described in Subsections 2.2.5 and 2.2.6 was applied. The diagram is simplified. One standard is shown at each level, while the *national standard* (*etalon*) consists of several standards:

— the fundamental standard,
— the comparison standard,
— the witnessing standard,
— the reference (working) standard.

No physical limitations, other comparison methods, etc. have been taken into account. Such a structure of the system of standards is characteristic of additive quantities.

Another group consists of quantities with a complex structure of the system of standards. For example, temperature reproduced in the International Practical Temperature Scale (IPTS) has several reference states distinguished as the defining fixed points. The continuous standards used for reproduction are the platinum resistance thermometer, PtRh–Pt thermocouple and the radiation effect. Other examples are the system of time standards and the system of inductivity standards, with different frequencies and comparison methods.

The third group is made up of derived quantities. The system of standards is based here on a standard of a given quantity and on the additive structure of lower-level standards, as well as on standards whose measure is determined by an indirect method, as in a definition of the quantity. There are at least two different ways for determining the measure of a standard at a given level in the hierarchy. An example may be the volume standard: the measure may be determined by comparison with a higher-level standard or by an indirect method, using length standards. A similar situation occurs for the pressure standard, for the calculable capacitance standard, etc. A number of physical and technical limitations to the realization of selected reference states exist. The problem of working out the real structure of the hierarchy of standards is complex, because of the difficulty of formulation of the optimality criterion and because of changes of the data on the properties of standards, the maintenance costs, verification costs, etc. The problem of the multiplication factor of the uncertainty of measure of the standards at neighbouring levels in the hierarchy and the problem of the respective number of levels in the hierarchy are the most frequently discussed questions [13, 112]. The first question may be answered on the basis of formulae derived above if the ratio of the systematic and random errors is known. For the ratio 1:1 and for negligible comparator errors, it may be seen (for example, from formula (2.32) that the ratio of variances

$$\frac{\operatorname{var}(\Delta_2)}{\operatorname{var}(\Delta_1)} = \frac{\operatorname{var}(h_1)+\operatorname{var}(h_0)+(1/n)(g_1^2+g_0^2)+[(n+1)/n]g_2^2}{\operatorname{var}(h_1)+g_1^2} \tag{2.71}$$

varies from 2 (for $n = 1$) to 1 (for $n \to \infty$) as $g_1 = g_2$. This problem has been discussed in detail by Crown [14], but for another aspect of errors. For a known quotient of errors and for a required accuracy of instruments, a number of levels in the hierarchy may be derived [107].

Other authors have analysed the problem of propagation of errors in the aspect of erroneous estimation of an allowable error of the secondary standard [53, 134]. The number of levels in the hierarchy of standards is calculated from the assumed level of hazard. The problem is more definite when the costs of calibration are taken into account [107]. Examples of the hierarchies established for specific applications of instruments may be found in the bibliography [34, 65].

Keeping the measures unified and conforming to international agreements is a task for the state, which is performed by the state administration and authorized institutions. The state administrations of measures maintain the national standards, perform comparisons and attestations of standards, establish systems of standards, etc. In the regulations published by the state administration of measures, diagrams showing the hierarchy of standards are periodically included. According to needs, the measurement staff also perform the tests and check that measurement apparatus is consistent with the national system of standards. In this way, a reference to international standards is established.

2.3 The Certified Reference Materials

The formulation of images of reality must include qualitative and quantitative phenomena. The *qualitative image of matter* is expressed in the following categories:

— the elements forming matter on a macroscopic scale;
— chemical compounds;
— mixtures of different elements and compounds;
— organic and biological bodies;
— particles and quanta building atoms and atomic nuclei;
— the phenomena of interaction of matter at rest or in motion;
— forms of energy and the transformations of energy;
— the phenomena connected with the influence of energy on matter;
— different qualities of matter and energy described by the definitions of physical quantities.

Some of the qualitative categories are easy to identify, while others are not. The measure of distinction of the qualitative categories is the number of physical qualities and the states which these qualities may assume. The qualities must be observable and measurable. The existing qualitative categories have their images in the domain of abstraction. The images are expressed by names and different symbols, and they are one-to-one mappings of reality. A qualitative category is described by a denumerable number N of the qualities $Q_1, \dots, Q_N$, and by the sets of states $\{q_1\}, \dots, \{q_N\}$ assumed by these qualities. The number N increases as the reality is recognized.

Two categories A and B:

$$A = A(Q_1, \dots, Q_i, \dots, Q_N, q_1, \dots, q_N), \tag{2.72a}$$

$$B = B(Q_1, \dots, Q_j, \dots, Q_M, q_1, \dots, q_M). \tag{2.72b}$$

are distinct when for at least one and the same quality $Q_{iA} = Q_{jB}$, $i = 1, \ldots, N$, $j = 1, \ldots, M$, the sets of the states are disjoint, which may be written as

$$\{q_{iA}\} \cap \{q_{jB}\} = \emptyset; \quad Q_{iA} \equiv Q_{jB}. \tag{2.73}$$

Condition (2.73) may be fulfilled by more than one quality. The categories are then easier to distinguish. For readily understandable practical purposes, the qualities which are more easily observed and measured and which satisfy condition (2.73) are used to distinguish the qualitative categories.

In intentional human activity, it is a necessary to apply standards for a considerable number of qualitative categories in the places where different materials are manufactured and used, for quality control, routine checks and scientific research. The standards are samples of materials properly prepared for use, called *certified reference materials* (CRM). Because of the activity of the physical environment and because of the ageing process, the life of these standards is limited. Moreover, in most cases the measurement of qualitative features causes some consumption of the standard material. Thus, it is necessary that the CRM can be restored to their former state and consumption can be compensated.

According to the classification by Plebański [100], there is a demand for the following kinds of standards:

— analytical standards,

— standards of ultra-pure substances,

— standards of physical properties,

— standards of technical properties,

— standards for health protection and natural environment protection, as well as for agriculture (e.g. standards for clinical laboratories, standards of pesticide mixtures, biological standards, etc.).

The certified reference materials are the physical bodies which reproduce, with a determined accuracy, those of their properties for which the materials were certified by a suitable institution [100]. Although the definition is similar to the one used for the standards of measurement units, the rank of the CRM is not as high. The reasons for this fact are as follows.

1. The qualitative category reproduced by a certified reference material is defined through measurable qualities and through the states of those of the qualities for which the standards of units of measure exist.

2. No primary certified reference material is necessary to determine a qualitative category or a measure of a certified reference material which expresses this category, because the measure may be determined by measurement of the physical properties of the material.

3. A certified reference material reproduces the measure of the standard over a limited range of physical and chemical properties. These properties are certified.

This fact limits the applications of the CRM to an instrument in which the certified property is the basis for identification.

The CRM and the standards of units of measure are alike in:

(1) the way in which the measure of standard and the accuracy of measure are expressed;

(2) propagation of some of the CRM standards by comparison of primary and secondary standards;

(3) the purpose and the mode of applying the standards and the procedure of attesting and certification.

The qualitative category of a certified reference material is expressed verbally by the name of the material, or by generally accepted symbols denoting the chemical constitution of the material, and by the state of a non-dimensional quantitative feature (one, or a few) from the set $\langle 0,1 \rangle$.

The state of a quantitative feature may be expressed as a percentage or a quotient of quantitative states, for example, as a quotient of mass and volume. The quantitative feature is defined and indirectly measurable by an absolute method. The *measure of a CRM standard* is a quantitative feature called the concentration or mole (or molar) fraction (of weight or of volume). The measure includes every component of the certified reference material existing within the uncertainty of measurement. The properties of this measure are identical with the properties of the measure of physical quantities, and may be treated as a vector measure. The restrictions imposed on a measure of CRM standard have the form of component balance:

$$\sum_{i=1}^{n} q_i = 1, \tag{2.74}$$

where q_i is the fraction of the ith component of the material. The inaccuracy of measure of a CRM standard concerns the components $i \in n$ and the components $i > n$.[9]

The inaccuracy of measure Δ_{qi} may be modelled by a multidimensional non-stationary random process (3.136)

$$\{\Delta\} = g\{\varepsilon\}+h, \tag{2.75}$$

in which g, h are the functions of such quantities as spatial dimensions of the sample, time, temperature, etc., i.e, describe the changes of the properties of the sample material as an effect of diffusion, dissolving of the materials, oxidation, etc., and $\{\varepsilon\}$ is a stationary random process describing the non-homogeneity of the sample material along the chosen coordinates of the sample, e.g. the surface

[9] This way of presenting the measure of a CRM standard is valid for some standards only. In other ways, for example when the components are added to stabilize the properties, the full constitution is not given.

or volume coordinates. Thus the uncertainty of measure of a CRM standard changes after a time interval T.

The inaccuracy of the measure of a CRM standard is expressed by metrological characteristics, of the classes 4 or 5 (see Section 4.4). A small (usually negligible) fraction of the random error expressed by $g\{\varepsilon\}$ is characteristic for the CRM. The value of this error may not always be determined unambiguously. The reason for this is that in the process of reproduction of the state of the CRM standard, only a part of the surface or volume of the standard is active and the measure is the mean value for the active area. The reproduction of a measure in an instrument with a different active area of the standard makes the result of averaging different.

The systematic error h has two components:

— the error of nominal measure of the CRM standard (the error is systematic in origin, but its value remains unknown);

— the instability error caused by ageing of the standard.

The ways of determining the measure of the CRM standard and its inaccuracy may differ, depending on the kind of CRM standard considered and its manufacturing technology. The most typical cases will be examined here.

Standard gases. The standard gases, which are the standards of physical properties with various applications, are prepared by measuring [110]:

(a) volume at constant pressure;

(b) pressure at constant volume;

(c) flow rate

of components A, B, C of a mixture, and by mixing of the components.

The components of the mixture should have a known chemical constitution. The measure of this standard is the volumetric fraction of the components of the mixture, expressed as percentages. The measure is derived from the quantities measured directly

$$X_A = \frac{V_A}{V} \times 100,\ \%A, \quad \text{or} \quad X_A = \frac{P_A}{P} \times 100, \quad \text{or} \tag{2.76a}$$

$$X_A = \frac{\dot{V}_A}{\dot{V}} \times 100, \quad \text{or} \quad X_B = \frac{V_B}{V} \times 100,\ \%B, \quad \text{etc.,} \tag{2.76b}$$

where

$$V_A + V_B + V_C + \ldots = V, \tag{2.77a}$$

$$P_A + P_B + P_C + \ldots = P. \tag{2.77b}$$

The uncertainty of measure of a CRM standard is determined on the basis of errors of measurement of the quantities defining the measure of the standard.

This is done as in the method for indirect measurement, because the uncertainties of the quantities measured directly are known. The sources of uncertainty of measure are:

(1) the uncertainties of the quantities measured directly;

(2) contamination of the component gases;

(3) changes of temperature during the process of measurement;

(4) departures from Dalton's law at higher pressures [122];

(5) badly mixed components, lamination of the mixture (when the densities of the components differ considerably), condensation of the components, etc.

In most of the cases, the direct finding of the measure of the standard gas ensures better accuracy than in case of direct measurement of its constitution by analytical methods.

The electrolyte standard. The most common low-concentration electrolyte standard is an aqueous solution of potassium chloride [100]. A weighed amount of KCl is added to a measured volume of distilled water. The measure of the standard is calculated from the formula

$$x = c = \frac{m}{V}, \quad \frac{\text{mg KCl}}{1\ \text{H}_2\text{O}}. \tag{2.78}$$

The uncertainty of the measure is the effect of:

(1) uncertainty of the direct measurements of mass and volume. As the defining equation (2.78) is in the form of a quotient,

$$\delta_c = \delta_m + \delta_v, \tag{2.79}$$

where δ is the relative error of the quantities denoted by the indices;

(2) uncertainty of the purity of the components;

(3) instability of the chemical constitution, caused by dissolving the atmospheric components (mainly CO_2) and the vessel material.

Standards obtained by preparative chromatography. The effect of the eluate concentration as a function of time is applied here, as shown in Fig. 2.12. Over the

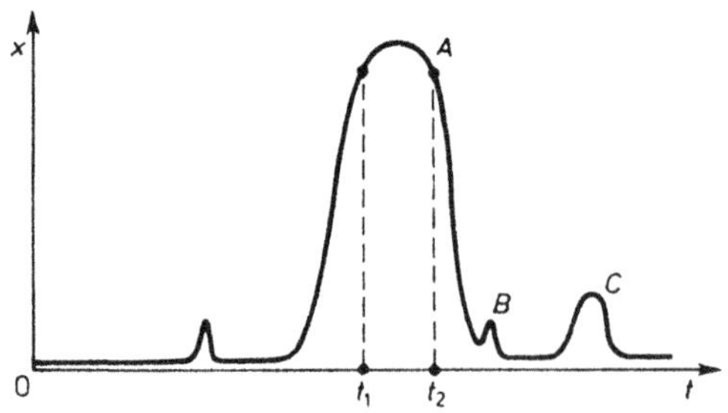

Fig. 2.12. The principle of extracting pure components from a chromatographic eluate. *A*, *B*, *C* are the components of the output sample.

time interval $t_1 - t_2$, the eluate is the component A of the mixture which is being separated in the chromatographic column. The purity of the component depends on the time interval $\langle t_1, t_2 \rangle$, since for $t < t_1$ and $t > t_2$ the eluate contains the carrier gas or another component, due to the dissipation of the interface in the equipment beyond the column and because the interface in the column is not sufficiently sharp. A pure component of the mixture being separated is obtained as a CRM standard. The kind of component obtained is determined by the value of its retention time. The purity of the standard may be verified by chromatographic methods.

Determination of the measure of a CRM standard based on a sample. Some CRM standards are obtained by metal or metal alloy melts. The number of samples taken from one melt is N. The measure of a standard is determined from investigation of $n \ll N$ samples [119]. The output material is characterized by the spatial heterogeneity of its constitution and properties (the coordinates of this space are usually geometric ones). The heterogenetity may be described by a model (2.75), i.e., a multidimensional, non-stationary random process (more precisely, by as great a number of processes as the number of qualities characterizing the standard). For example, in the case of a metal alloy, the measure of a standard is determined by m components and by their concentrations. The sample is described by m 3-dimensional processes

$$\{q_j\} = g_j(x, y, z)\{\varepsilon_j\} + h_j(x, y, z), \quad j = 1, \ldots, m. \tag{2.80}$$

The processes are usually correlated. The stationary process $\{\varepsilon_j\}$ describes the fluctuations of concentration due, for example, to wrong mixing of the components. The function h describes various stratifications due to sedimentation, degasing, etc., occurring before freezing of the alloy. The realization of the process q_j is one sample, and the geometric dimensions of the sample are the limitation for the horizon of observation of the processes $\{\varepsilon_j\}$. The number of realizations depends on the expected character of the function h_j.

To identify the heterogeneity, the methodology described in Subsection 4.5.1 may be applied. The measure of a standard is the mean value of measures of the samples, corrected by the interpolated value of the systematic heterogeneity $\hat{h}$. The measure of uncertainty is the sum of the component $g_j\{\varepsilon_j\}$, of the uncertainty of determination of the function h_j and of the measurement error of the states of quality q_j, i.e.,

$$\{\Delta_r\} = g_j\{\varepsilon_j\} + \{\Delta h_j\} + \{\Delta q_j\}. \tag{2.81}$$

After adding these processes, the uncertainty limit is calculated as a confidence interval at a given confidence level. This uncertainty is characteristic of a certified reference material, which is a single realization of the process $\{\Delta_j\}$. The error of the CRM standard is systematic in origin, but has an unknown value. The way of calculating the expected value of uncertainty and its variance is similar to that described by formulae (2.19), (2.20).

Application of the CRM standards created with this method allows reduction of the uncertainty of measurement. This may be done by:

1. Application of n CRM standards from the same population, and taking the mean value as a result. The uncertainty of the measure introduced by components 1, 2 and the part $g_q\{\varepsilon_q\}$ of component 3 in formula (2.81) decreases n times. The component h_q remains unchanged. The mean values of elements 1 and 2 are zero and $E\{\Delta q_j\} = h_q$.

2. The component h_q may be reduced by applying m standards whose measures are determined with m sets of instruments for which $\bar{h}_q = 0$. There is an analogy here to the examples described in Subsections 2.2.1–2.2.6.

The international ISO[10] requirements [73] were issued for certified reference materials, like the IUPAC[11] requirements [11] which apply to pure substances. Institutions certifying the CRM (in order to eliminate the systematic errors of instruments) conduct comparisons of CRM standards, like those for the standards of measures.

Certified reference materials play an important role in metrology. Not only do they make the measurements simpler, but also some of them are the sources of constant and strictly determined physical quantities. Quite frequently these states are defined on the basis of a type of pure material (e.g. water, pure metals for reproduction of temperature at fixed points, etc.). The certified reference materials make studies of the instability of some measurement systems possible [96], even when a nominal measure is not known precisely. It is sufficient to ensure that the given quantity is constant. Therefore, the certified reference materials may be regarded as pseudo-standards of some of the physical quantities, i.e., as pseudo-standards of measures.

---

[10] The International Organization for Standardization.

[11] The International Union of Pure and Applied Chemistry.

Chapter 3

Modelling of Measurement Systems

3.1 The General Principles of Modelling

Modelling is the process of formulating a mathematical model which is an approximate description of the phenomena occurring in a measurement system (MS). The above statement is accepted as an axiom.

The *mathematical model* is a concept which expands the notion of a physical law. The discovery of phenomena of a random nature led to relationships between physical quantities that appeared to be ambiguous. The meaning of these phenomena is stochastic, and the equations describing these relationships are not laws but mathematical models.

Two types of modelling are used, of signals and of systems. The modelling of signals has an ample bibliography [23, 60, 3] and will not be discussed here. Metrology is concerned with determinate and stochastic signals, presented either in time or frequency domains.

The concept of a *signal* is usually associated with a physical quantity which is an energy carrier. However, a quantity which is not an energy carrier, but characterizes a state of matter, can also be a signal. There is a certain relationship between a model of a signal and a model of an MS: a model of an MS is the relationship between signals.

Measurement systems consist of two types of elements, active and passive. The *active* are those generating physical quantities which are energy carriers; these elements are characterized by one, two or more physical quantities. The *passive* have properties which depend on the state of matter, and are expressed by quantities which are not energy carriers.

Examples of active elements include cells and batteries, power supplies, sources of light and electromagnetic and corpuscular radiation, masses in gravitational fields, compressed air, electrical signal amplifiers, pneumatic amplifiers, etc.

Among passive elements are the elements of mechanical construction characterized by their dimensions, mass and properties, optical elements such as filters, prisms, lenses and mirrors, electrical elements such as resistors, capacitors and inductors, molecule-permeable and ion-permeable membranes, etc.

The properties of both single construction elements and the assemblies are described by the appropriate equations known from physics or some of the technical sciences. The description used in physics is the most detailed and coherent. Various approaches are used in the different technical sciences, and since they are sometimes not coherent, the modelling is more difficult when an MS is built of different types of elements.

The ways of formulating descriptions are known from physics, and need not be repeated here. Apart from those applied in physics, no other methods need to be used in metrology for describing the properties of the elements of an MS.

Modelling in metrology differs from the modelling conducted in other sciences in three ways:

(1) the purpose of modelling is specific, which means that different importance is assigned to various phenomena and different properties of elements must be disclosed;

(2) much greater accuracy of modelling is required; this may be stated in another way, in that only those elements or phenomena whose models are known sufficiently well are used in constructing the MSs;

(3) the mathematical model of an MS is always determined, i.e., an instrument with an unknown model cannot be used as a measuring instrument. Besides the model itself, the accuracy of the model is also calculated.

The discussion of the above points is best begun from the last one.

An MS as a system is characterized by its inputs and outputs. The input is the state of a measurand fed into the MS. The input is also the set of influencing quantities describing the operating conditions of the MS. The output is the indication (reading) of the instrument, or the state of output quantities recognized by human senses. Three goals are formulated within the system of interpretation of different phenomena:

(1) the input and system model are known, and the output is to be found;

(2) the input and output are known, the model is to be found;

(3) the model and output are known, the input is required.

Every measurement is a case of the third goal. In order to be able to reproduce the state of the input quantity (*measurand*) on the basis of the output state, it is necessary to know the model. The above statement points to the reason for creating models. In this study we are interested in input-output types of models.

One of the essential properties in metrology is unambiguous mapping, and knowing the limits of its ambiguity. This property is strictly connected with the aim of modelling and the reproducibility of the input state.

The reasons why MS models are required to be the most accurate arise from the fact that the MS is used to identify the models of other devices, and for recognizing different phenomena. The error of an MS model which involves uncertainty burdens the model of the phenomenon investigated. The phenomena applied in metrology are described with an accuracy of 10^{-2} to 10^{-6}, and in the case of some quantities and standards even 10^{-8} to 10^{-12}. No other devices are modelled with an accuracy like that.

The purpose of modelling was stated above. A prerequisite for the use of an MS for measurements is that the model must be known. All the characteristic sources of ambiguity of processing, all the properties of elements, etc. are modelled. Both physical and statistical models are in use.

Only a limited number of existing phenomena are registered by human senses. Measuring instruments (measurement systems) are used for this reason for measurements. The instruments responding to a given phenomenon transform a measured quality x into a quality recognizable by human senses. Therefore, the measurand x is the input quantity of the MS, while the output quantity is received by the observer (man).

The principles of mapping presented in Chapter 1 concern the measured quantity x and the output quantity y individually; in an MS the principles are more complex. Since the task of an MS is not to measure the output quantity y, but to measure the quantity x, the mapping of the states y into the values of the measured quantity x must be derived. The mapped values are denoted as x^*.

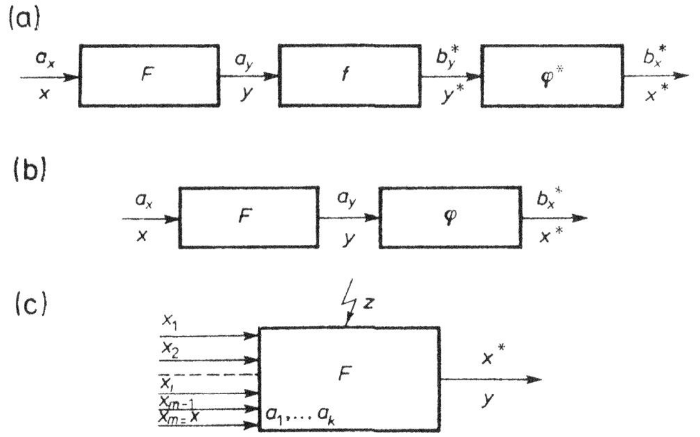

Fig. 3.1. The structure of a mathematical model of an MS: (a) formally applicable mappings, (b) practical mappings, (c) general model.

The mappings realized in MS are shown in Fig. 3.1. The double notation introduced in Chapter 1 and used in the following chapters is also applied here. The mappings are as follows.

1. State a_x of a measurand x into a state a_y of an output quantity of measurement system

$$a_x \to a_y \quad \text{or} \quad x \to y.$$

This mapping is physically realized by the physical conversion described by the *conversion equation*

$$a_y = F(a_x) \quad \text{or} \quad y = F(x) \tag{3.1}$$

2. The state a_y of the output quantity y into the value b_y^* of this quantity

$$a_y \to b_y^* \quad \text{or} \quad y \to y^*.$$

This mapping is discussed in Chapter 1 and the mapping function is described by equation (1.35).

3. The value b_y^* of the output quantity y of the MS into the value b_x^* of the measurand x

$$b_y^* \to b_x^* \quad \text{or} \quad y^* \to x^*.$$

This mapping, realized in the domain of abstraction, is described by the procedure of reproducing the values of the measurand based on the value of the output quantity

$$b_x^* = \varphi^*(b_y^*) \quad \text{or} \quad x^* = \varphi^*(y^*). \tag{3.2}$$

The above subsequent mappings may be written conjointly as

$$b_x^* = \varphi^* \{f_y[F(a_x)]\} \tag{3.3a}$$

or

$$x^* = \varphi^* \{f_y[F(x)\}. \tag{3.3b}$$

Since the measurement scale of the measurand is defined by the function f_x as in (1.35)

$$b_x = f_x(a_x) \quad \text{or} \quad x^* = f_x(x),$$

the equality

$$\varphi^*[f_y(F)] = f_x \tag{3.4a}$$

must be satisfied, or

$$\varphi^* f_y F = f_x. \tag{3.4b}$$

where f, φ^*, F are functions or operators. The above relationship may be interpreted in two ways.

(i) The scale f_x of the quantity x measured with a given measurement system depends on the scale f_y of the output quantity y and on the mathematical model of the MS. The form of the scale may be affected by selection of the properties of the elements of the MS described by the model F and by selection of the procedure φ^* of reproduction of the measurand.

(ii) Since the scales f_x, f_y of the measurable quantities x and y are generally established (the metric scales), the condition of correct measurement (for $f_x = f_y = f$) is a procedure such that

$$\varphi^* f F = f, \tag{3.4c}$$

and for linear operators

$$\varphi^* = F^{-1}. \tag{3.4d}$$

In practice, the notion of *calibration procedure* is used. In the calibration procedure, the measure b_x^* of the measurand x is attributed to the states a_y of the output quantity. Thus, the procedure describes conjointly the last two mappings from Fig. 3.1a:

$$a_y \to b_x^* \quad \text{or} \quad y \to x^*.$$

As the procedure is a function or an operator φ, thus

$$b_x^* = \varphi(a_y) \quad \text{or} \quad x^* = \varphi(y). \tag{3.5}$$

Taking into consideration the physical conversion $x \to y$ occurring in the MS and described by the operator F (3.1), one obtains

$$b_x^* = \varphi[F(a_x)] \tag{3.6a}$$

or

$$x^* = \varphi[F(x)]. \tag{3.6b}$$

The domain of mapping (3.6) is the reality in which the states a_x of a measured quality x exist. The region is the real axis in which the value b_x^* of the measurand x is indicated by a pointer or another element of the measurement system construction. Mapping (3.6) is equivalent to (1.35) and determines the scale of a quantity measured in a given MS. In the last notation (3.6), the condition of isomorphism between the relations of the states of the measured quantity and the values of this quantity indicated by the MS is

$$\varphi F = f_x, \tag{3.6c}$$

where f_x is mapping (1.35) generating the scale of the quantity x. The condition is fulfilled if

$$\varphi = f_x F^{-1}. \tag{3.6d}$$

Generally, mappings (3.3) and (3.6b) are called the *mathematical model of MS*. The model consists of two or three parts, as shown in Figs. 3.1a, b. The first part describes the performance of a technical device to which the measured quantity is fed. The characteristic state y of the device may be received by a human, either directly or indirectly. The second part of the model of an MS as in Fig. 3.1b is the calibration procedure. When the calibration procedure is fixed, a device becomes a measurement system.

For reasons discussed later, conditions (3.4d) and (3.6d) cannot be fulfilled exactly or completely. The effect of this is an error of the MS:

$$\Delta = x^* - x \neq 0. \tag{3.7}$$

One of the causes of errors is the influence of ambient conditions on the properties of the MS, described by conversion equation (3.1). The ambient conditions are characterized by such *influencing quantities* as temperature, supply voltage, humidity, position, etc., generally denoted as $x_1, x_2, \ldots, x_{m-1}$. Also by denoting $x = x_m$, *m input quantities* are obtained, as shown in Fig. 3.1c. The *output quantity* is the reading of the MS, expressed in the values x^* of the measured quantity (the reading may also be denoted as y).

Variability of the influencing quantities causes the relationship $y = F(x)$ to be ambiguous and thus causes errors (3.7) to appear.

3.2 Modelling of the Characteristics of Measurement Systems

The following types of models are distinguished:

— linear and nonlinear,
— static and dynamic,
— deterministic and probabilistic.

Depending on the properties of the domain of the model, the models are

— continuous or with discrete time intervals,
— analog or digital, i.e., the domain of values may be a dense or a discrete set.

Depending on the way in which they are determined, the models are

— physical (physicochemical),
— statistical.

As shown in Fig. 3.1c, the mathematical model of an MS has the general form

$$y = F(x_1, \ldots, x_{m-1}, x_m, T, a_1, \ldots, a_k, z). \tag{3.8}$$

The quantities $x_1, \ldots, x_m$ are random processes with either known or unknown characteristics. Depending on the purpose and application conditions of an MS, they are measurable or unmeasurable, controllable or uncontrollable. T is the service time of the MS. If it is not necessary, time T is neglected in the model. The quantities $a_1, \ldots, a_k$ are the construction parameters of the MS. The quantity z represents an inadequacy of the model, together with ambient influences (neglected or unrecognizable) on the MS. By its nature z is unmeasurable and uncontrollable, and it is often assumed that $E(z) = 0$. The quantity y is the output quantity of the MS. The model may often be limited to a case of one output quantity.

The operator F and the construction parameters a form a mathematical model of the MS, describing the relationship between the input and output quantities. The parameters a_j ($j = 1, \ldots, k$) characterize the properties of the elements building the MS.

3.2.1 The static model

The model is *static* if the input quantities are realized as single points. The operator F is then a function f, namely

$$y = f(x_1, \ldots, x_m). \tag{3.9}$$

Such commonly applied static models are deterministic, i.e., the construction parameters $a_1, \ldots, a_k$ are assumed to be constant and it is assumed that $z \equiv 0$. Each point of the input quantities corresponds to one point of an output quantity. The points form a set in $(m+1)$-dimensional metric space, whose cross-sections are the *characteristics* of the MS. The most important relationship is

$$y_0 = f_0(x) = f(x_{10}, \ldots, x_{(m-1)0}, x) \tag{3.10}$$

determined in reference conditions x_{i0} for $i = 1, 2, \ldots, m-1$. The relationship is called a characteristic of an MS.

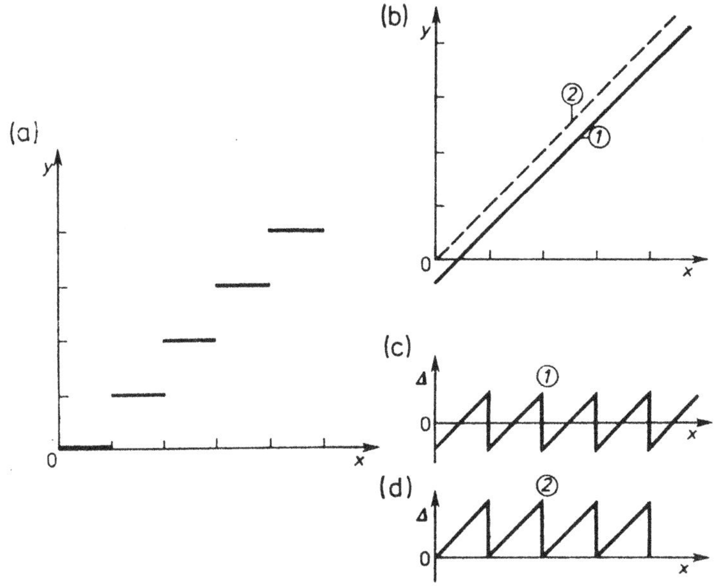

Fig. 3.2. A linear approximation of the characteristic of an A/D converter: (a) true characteristic, (b) approximation, (c), (d) approximation error.

The function f_0 may be linear or nonlinear, but should be either non-decreasing or non-increasing. For some of the nonlinear characteristics, a functional transformation is applied to the variables in order to obtain a linear characteristic. The transformation may take the forms

$$u = +\sqrt{y}; \quad \log y; \quad \arcsin y; \quad \text{etc.}$$

Nonlinear models are often replaced by linear or piecewise-linear models, as shown in Figs. 3.2 and 3.3. Figure 3.3 shows an opposite case: the real piecewise-linear characteristic is approximated by a square characteristic and the sign of the approximation error is then opposite.

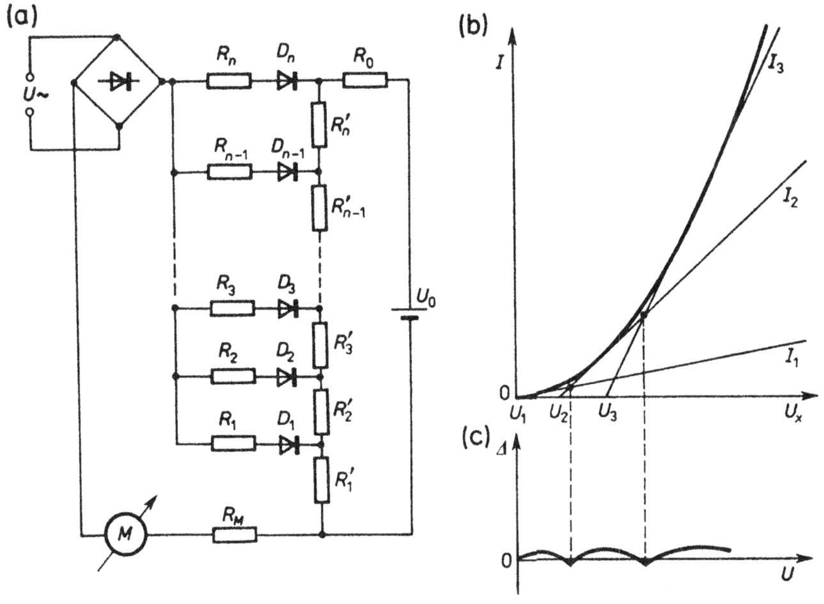

Fig. 3.3. A piecewise-linear approx imation of a square characteristic: (a) diode squaring circuit (b) true (piecewise-linear) and ideal (continuous) characteristics, (c) approximation error.

An approximation of a model necessitates estimation of the approximation error (which depends on the type of approximation). For example, a piecewise-linear approximation may be performed as in Fig. 3.4, and the approximation errors are different for different cases. An adequate value of the approximation error is obtained by selection of the approximation type, as in Fig. 3.4, and depends on the number of pieces.

The piecewise-linear approximation is applied increasingly often, because its simple numerical notation allows convenient digital modelling with computers.

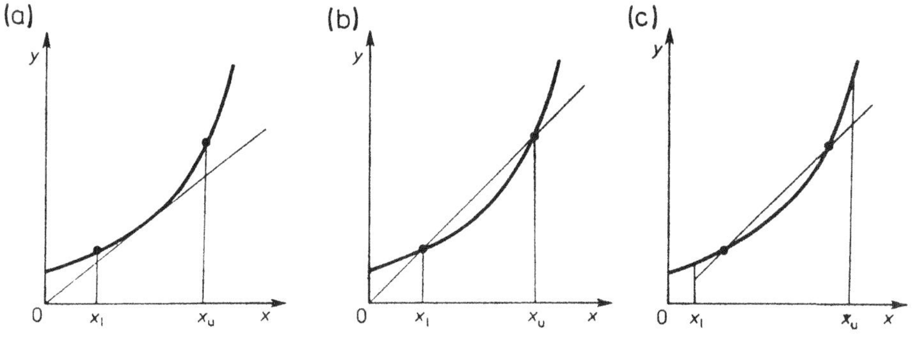

Fig. 3.4. Types of nonlinear continuous approximation: (a) tangent, (b) chord, (c) secant.

The characteristic is recorded in the form of the coordinates of the nodes, and the values between the nodes are calculated each time they are needed. In this way, the required memory space is smaller than in the case storing in array. The deterministic models are applied for describing the properties of an ideal MS or as a first approximation of the description of a real MS.

The fact that an MS is chosen to measure the quantity x does not always weaken the dependence of the output quantity y on the influencing quantities $x_1, \ldots, x_{m-1}$. For example, sensitivity to disturbing ions in ion-selective electrodes is often greater than the sensitivity to the basic ion. In thermoconductometric gas analysers the resistance of a single heater is influenced more strongly by the supply current than by the thermal conductivity, but this influence is reduced by the use of a differential circuit and by stabilization of the current. The models of an MS may therefore be the functions, usually nonlinear, of many arguments. For this reason, various approximations are applied, such as

$$f(x_1, \ldots, x_m) = f_1(x_1)f_2(x_2)\ldots f_m(x). \tag{3.11}$$

A special case is

$$f(x_1, \ldots, x_m) = \sum_{i=1}^{m-1} A_i(x_i - x_{i0}) + k_0 x \prod_{i=1}^{m-1} \left(\frac{x_i}{x_{i0}}\right)^{\delta_i} + b\xi(x_1, \ldots, x_{m-1})x^2 \ldots, \tag{3.12}$$

where the function $f_m(x)$ is developed into a power series. The coefficients of the series have special properties: the first coefficient is the sum of linear functions of the arguments x_i, the second is the product of the power functions of x_i. Usually the values of A_i, k_0, δ_i are constant and depend on the construction parameters a_j, i.e.,

$$A_i, k_0, \delta_i = \Psi(a_1, \ldots, a_k). \tag{3.13}$$

The latter form of the model is very convenient for approximation of a statistical model derived empirically.

In the models presented above, the properties of the set of input and output quantities are not bound by any limitations. These sets are usually continuous (dense), i.e., they are analog quantities. As shown in Fig. 3.2, the output set (also the input sets) may be discrete. The occurrence of discrete sets does not complicate the notation of the characteristic of an MS. The characteristic is still a non-decreasing or non-increasing function of x. It is only the domain of a given quantity that must be determined precisely. When an output quantity is discrete, the model of an MS is called *digital*. The output set is then a set of integer numbers (or its subset), multiplied by a quantum q of the quantity, namely

$$y = q\mathrm{Ent}[f_0(x)] = Nq, \quad N = 0, \pm 1, \pm 2, \ldots, q > 0. \tag{3.14}$$

The measured quantity x may be discrete, e.g. $x = Nq_x$ and $y = Nq_y$. The relation

$$q_y = f_0' q_x \tag{3.15}$$

holds if the derivative $f_0' = \partial y/\partial x$ is constant within the quantum.

The quanta of x and y need not be constant. In a digital MS, the quantity y is realized in its natural or coded forms. The physical state and the notation depend on the type of code. Generally

$$N = \boldsymbol{y}^T \boldsymbol{w}, \tag{3.16}$$

where, for the binary code,

$$\boldsymbol{y} = [y_n \quad y_{n-1} \quad \ldots \quad y_1 \quad y_0], \tag{3.17}$$

$$\boldsymbol{w} = [2^n \quad 2^{n-1} \quad \ldots \quad 2^1 \quad 2^0], \tag{3.18}$$

and the values of the components of the vector $\boldsymbol{y}$ are 0 or 1. For the BCD (Binary Coded Decimal) code with weights 1, 2, 2, 4, we have

$$\boldsymbol{y} = [y_{mn} \quad y_{m(n-1)} \quad \ldots \quad y_{m0} \quad y_{(m-1)n} \quad \ldots \quad y_{(m-1)0} \quad \ldots], \tag{3.19}$$

$$\boldsymbol{w} = [1\times 10^m \quad 2\times 10^m \quad 2\times 10^m \quad 4\times 10^m \quad 1\times 10^{m-1} \quad \ldots \quad 4\times 10^{m-1} \quad 1\times 10^0 \quad 4\times 10^0], \tag{3.20}$$

where $m = 3, 2, 1, 0$ (a value of a decade), $n = 1, 2, 2, 4$ (weights of the code). The components y_{mn} are 0 or 1.

Classification of the MS with modulated output signals is somewhat difficult. The models of these systems are considered dynamic, according to the rule that the properties of an output quantity decide the class of the model.

3.2.2 Dynamic models

The model is dynamic if the input quantities are realized as sets of elements in a strictly determined order. The input quantities are then functions of continuous time or of discrete time intervals. An output quantity is also a set. One element of a set is realized at a given moment of time.

The variety of dynamic models is a consequence of the different ways in which the models are formulated. Since the model is derived from physical equations, the equations of the model are differential. The starting points are usually the Lagrange, Maxwell or Kirchhoff equations, the laws of thermodynamics, energy conservation equations, etc., or some more specific physical equations. When the construction parameters $a_1, \ldots, a_n$ are constant and $z \equiv 0$, the model is deterministic.

For a single construction element of an MS, the output quantity is a function of an input signal x and of the derivatives of x, y, namely

$$y = f(y, \dot{y}, \ddot{y}, x, \dot{x}, \ddot{x}). \tag{3.21}$$

The interaction of many elements causes the order of the differential equation to increase. The equation becomes implicit and one or several equations of the type

$$F(y, \dot{y}, \ddot{y}, \ldots, y^{(n)}, x, \dot{x}, \ddot{x}, \ldots, x^{(p)}) = 0 \tag{3.22}$$

are obtained. Generally, the model is nonlinear.

A particular class is formed by linear models in which the differential equation is linear with constant coefficients. To solve this equation, Laplace or Fourier transformations may be applied and the properties of the MS may be described by a transfer function for a frequency response.

The problems of linear models are widely presented in the bibliography [116, 18, 129].

A convenient form of presenting a dynamic model is a state equation of an MS

$$\frac{\mathrm{d}y}{\mathrm{d}t} = f(y, x), \tag{3.23}$$

which may be written in a matrix notation with continuous time as

$$\dot{\boldsymbol{y}} = \boldsymbol{A}\boldsymbol{y} + \boldsymbol{B}\boldsymbol{x}, \tag{3.24}$$

or, in a discrete-time case,

$$\boldsymbol{y}(n+1) = \boldsymbol{A}\boldsymbol{y}(n) + \boldsymbol{B}\boldsymbol{x}(n), \tag{3.25}$$

where n denotes the subsequent time instants. (The exact notation is $y(n\Delta t)$, $n = 0, 1, 2, \ldots$)

The advantages of the state equation are the convenience of reproduction of the measured quantity x (with known y), and the rich bibliography concerning automatic control [18, 116].

Another way of formulating a model comes from the mathematical approach.

The mapping of the set X into the set Y is realized using an operator F. A linear *Volterra operator* is applied for linear systems [59, 18]:

$$y(t) = F_{V}[X(t)] = \int_0^t k(t, \tau)x(\tau)\mathrm{d}\tau, \tag{3.26}$$

where $k(t, \tau)$ is the *kernel of the operator*, which is the response of a system to a pulse excitation.

The nonlinear systems are described by nonlinear operators. For example, the Hammerstein operator [59, 18]

$$y(t) = F_{H}[X(t)] = \int_0^t k(t, \tau)f[x(\tau), \tau]\mathrm{d}\tau \tag{3.27}$$

describes systems which may be presented as the serial connection of a nonlinear static model and a linear dynamic element whose response to the pulse excitation is $k(t, \tau)$. The model is shown in Fig. 3.5.

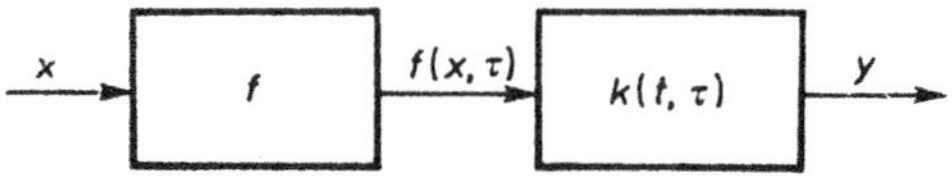

Fig. 3.5. Diagram of a model described by the Hammerstein operator.

A broader class of nonlinear systems may be described by the Lichtenstein–Lyapunov operator [59][1]

$$y(t) = F_{LL}[x(t)] = \sum_{i=1}^{\infty}\int_0^t \dots \int_0^t k_i(t, \tau_1, \tau_2, \dots, \tau_i)x(\tau_1)x(\tau_2)\dots x(\tau_i)\mathrm{d}\tau_1 \dots \mathrm{d}\tau_i. \tag{3.28}$$

The description of an MS applying these operators makes the analysis and interpretation of the results of an experimental identification of the system more convenient. The response to a pulse excitation (or unit step function) is easy to observe and record.

---

[1] The Volterra series

$$y(t) = \int_0^t k_1(\tau)x(t-\tau)\mathrm{d}\tau + \int_0^t\int_0^t k_2(\tau_1, \tau_2)x(t-\tau_1)x(t-\tau_2)\mathrm{d}\tau_1\mathrm{d}\tau_2 +$$

$$+\int_0^t\int_0^t\int_0^t k_3(\tau_1, \tau_2, \tau_3)x(t-\tau_1)x(t-\tau_2)x(t-\tau_3)\mathrm{d}\tau_1\mathrm{d}\tau_2\mathrm{d}\tau_3 + \dots$$

has a similar structure but a simpler physical interpretation [18].

An important group is formed by the nonlinear models in which the signals are modelled and coded. The models of such elements of an MS may be considered in two categories:

(1) a physical model described by input and output quantities whose form and structure are identical with that in the described element;

(2) a physical model of a mapping of a measured quantity into the set of input and output values of a given element (i.e., the model is presented in the domain of the values of the measured quantity).[2]

Models of the latter category are especially important for the determination of metrological properties, because the errors are expressed in the domain of the measured quantity. The physical phenomena occurring in an MS may be described only by a model from the first category, thus both kinds of models are considered in common, as shown in Fig. 3.6. The symbol u denotes a natural input

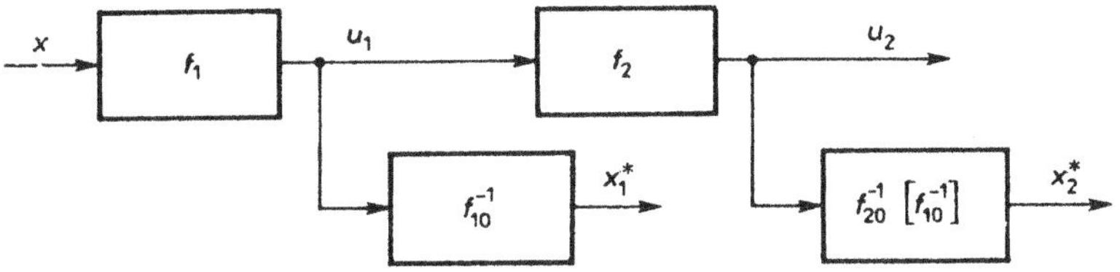

Fig. 3.6. Signal mapping of a measured quantity.

or output signal of the element, while x and $y = x^*$ are the mappings in the domain of a measured quantity. The physical conversion is described by a conversion equation (3.1) and the mapping into the domain of a measured quantity is described by procedure (3.5). Some practice in dealing with these models makes it possible to simplify the notation: physical signals are expressed in the domain of a measured quantity. Such a method of modelling concerns analog nonlinear systems with modulation as well as digital systems operating either continuously or periodically. The examples below illustrate this mode of modelling.

Amplitude modulator. A block diagram and a structural diagram of a modulator are shown in Fig. 3.7. In Fig. 3.7a the symbol f denotes the conversion of a quantity x into a quantity y by the equation

$$y = xu, \tag{3.29}$$

where[3]

$$u = A\sin\omega_0 t. \tag{3.30}$$

[2] The first category are called the signal models and the second the value (quantity) models [40].

[3] In the case of pulse amplitude modulation, the modulated signal is $u = A\delta(t - i\Delta t)$.

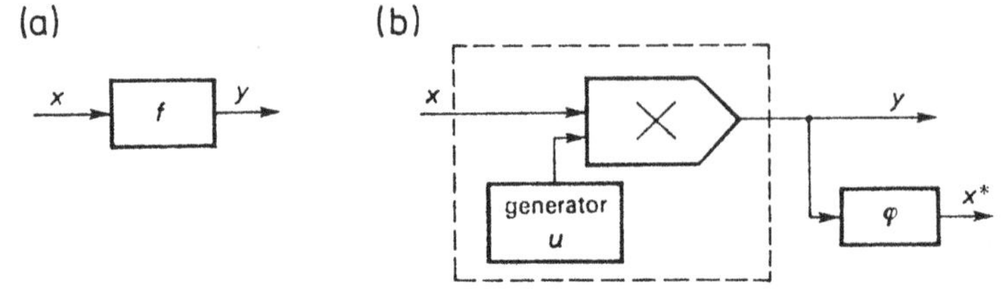

Fig. 3.7. Amplitude modulator: (a) block diagram, (b) structural diagram.

Thus,

$$y(t) = x(t)A\sin\omega_0 t = f_0(x) \tag{3.31}$$

is the ideal model of modulation. The real model is slightly different, namely $f(x)$. The result of restoring an input state is

$$x^* = \varphi(y) = f_0^{-1}[f(x)]. \tag{3.32}$$

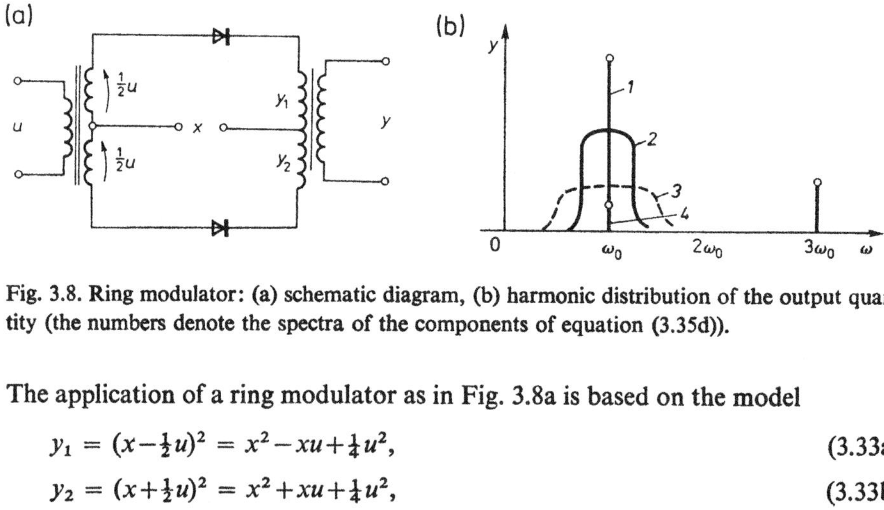

Fig. 3.8. Ring modulator: (a) schematic diagram, (b) harmonic distribution of the output quantity (the numbers denote the spectra of the components of equation (3.35d)).

The application of a ring modulator as in Fig. 3.8a is based on the model

$$y_1 = (x-\tfrac{1}{2}u)^2 = x^2-xu+\tfrac{1}{4}u^2, \tag{3.33a}$$

$$y_2 = (x+\tfrac{1}{2}u)^2 = x^2+xu+\tfrac{1}{4}u^2, \tag{3.33b}$$

$$y_0 = y_2-y_1 = 2xu. \tag{3.33c}$$

The real characteristics of the rectifiers are not quadratic and thus the form of the characteristic is

$$y_i = v^m = b_1 v+b_2 v^2+b_3 v^3 \tag{3.34}$$

(instead of $m = 2$). Thus, after substituting $v = x \pm u/2$, we obtain

$$y_1 = b_1 x-\tfrac{1}{2}b_1 u+b_2 x^2-b_2 xu+\tfrac{1}{4}b_2 u^2+b_3 x^3-\tfrac{3}{2}b_3 x^2 u+\tfrac{3}{4}b_3 xu^2-\tfrac{1}{8}b_3 u^3, \tag{3.35a}$$

$$y_2 = b_1 x + \tfrac{1}{2} b_1 u + b_2 x^2 + b_2 xu + \tfrac{1}{4} b_2 u^2 + b_3 x^3 + \tfrac{3}{2} b_3 x^2 u + \tfrac{3}{4} b_3 xu^2 + \tfrac{1}{8} b_3 u^3, \tag{3.35b}$$

$$y = y_2 - y_1 = b_1 u + 2b_2 xu + 3b_3 x^2 u + \tfrac{1}{4} b_3 u^3, \tag{3.35c}$$

$$y = b_1 A \sin\omega_0 t + 2b_2 Ax(t) \sin\omega_0 t + 3b_3 Ax^2(t) \sin\omega_0 t + \tfrac{1}{4} b_3 A^3 \sin^3\omega_0 t. \tag{3.35d}$$

The output quantity y may also be expressed in a frequency domain, as shown in Fig. 3.8b, where *1*, *2*, *3*, *4* denote subsequent components in equation (3.35d). Thus[4]

$$x^* = \frac{y}{2u} = (\tfrac{1}{2} b_1 + \tfrac{1}{8} b_3 u^2) + b_2 x + \tfrac{3}{2} b_3 x^2$$
$$= (\tfrac{1}{2} b_1 + \tfrac{1}{8} b_3 A^2 \sin^2\omega_0 t) + b_2 x + \tfrac{3}{2} b_3 x^2. \tag{3.36}$$

In the next stages of processing the signal y may be filtered (the $3\omega_0$ component introduced by the u^3 term is usually subjected to filtration), the component independent of x may be cut off, etc. This treatment improves the properties of a modulator.

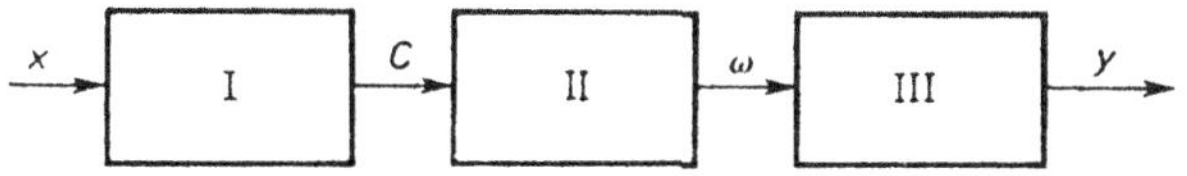

Fig. 3.9. A frequency modulator block diagram.

Frequency modulator. The block diagram of a frequency modulator is shown in Fig. 3.9. The form of conversion equation of an ideal modulator is

$$y_0 = A \sin[(\omega_0 + x)t], \quad |x| \ll \omega_0, \tag{3.37}$$

where $y = A \sin\omega t$.

The procedure for restoring x from a real output signal y comes from (3.37) namely

$$x^* = a\left[\frac{1}{t} \arcsin\frac{y}{A} - \omega_0\right], \tag{3.38}$$

where a is the degree of freedom of the calibration procedure. If a varactor diode with the characteristic (first approximation)

$$C = C_0 + b_1 x, \tag{3.39}$$

[4] The procedure (3.32) $x^* = \varphi(y) = y/2u$ does not describe a physical demodulation. It is a mathematical operation.

is applied in the modulator and the second stage is described by the equation

$$\omega = \frac{1}{2\pi\sqrt{LC}} = \frac{(1+b_1x/C_0)^{-1/2}}{2\pi\sqrt{LC_0}} = \omega_0\left(1+\frac{b_1x}{C_0}\right)^{-1/2}, \tag{3.40}$$

then, by developing the formula into a power series, the conversion equation is obtained:

$$y = A\sin\left[\omega_0 t\left(1-\frac{1}{2}\frac{b_1}{C_0}x+\frac{1\cdot 3}{2\cdot 4}\left(\frac{b_1}{C_0}\right)^2 x^2-\frac{1\cdot 3\cdot 5}{2\cdot 4\cdot 6}\left(\frac{b_1}{C_0}\right)^3 x^3+\right.\right.$$
$$\left.\left.+\frac{1\cdot 3\cdot 5\cdot 7}{2\cdot 4\cdot 6\cdot 8}\left(\frac{b_1}{C_0}\right)^4 x^4+\ldots\right)\right]. \tag{3.41}$$

Application of procedure (3.38) gives

$$x^* = x-2\frac{1\cdot 3}{2\cdot 4}\frac{b_1}{C_0}x^2+\frac{1\cdot 3\cdot 5}{2\cdot 4\cdot 6}\left(\frac{b_1}{C_0}\right)^2 x^3-\ \ldots, \tag{3.42}$$

when $-\frac{1}{2}a(b_1/C_0)\,\omega_0 = 1$, i.e., $a = -2C_0/b_1\omega_0$.

3.2.3 Discrete-time Models of Measurement Systems

Many measurement systems work cyclically and the output measurement result takes the form

$$y(t) = y(t_j)[\mathbf{1}(t-jT)-\mathbf{1}(t-(j+1)T)] = y(t_j)H_j, \tag{3.43}$$

where $j = 0, 1, 2, \ldots;\ t_j = jT$. The symbol H_j denotes the function

$$H_j = \mathbf{1}(t-jT)-\mathbf{1}[t-(j+1)T] \tag{3.44}$$

which describes the performance of the output processing stage of a digital MS, called a *register*. The register, controlled by the MS controller, holds the memory of a measurement result for one measurement cycle. During the first measurement, the MS outputs the result y_0, i.e., the state present before the measurement. During the second measurement, the first result y_1 is displayed, etc. The procedure

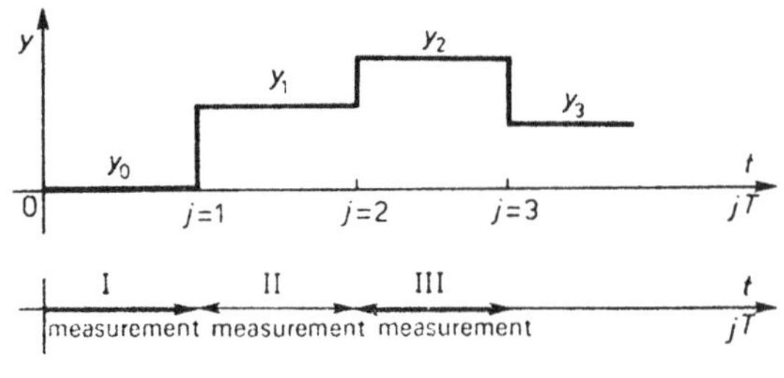

Fig. 3.10. Measurement result produced by a digital MS.

is shown in Fig. 3.10. The index j is a measurement number and T is the measurement period (time). The measurement period is divided into elementary quanta of time Δt, numbered by the index i. The measurements start at quantum $i = 0$. The following changes may occur at the instants $i\Delta t$:

— change of state of the elements of an MS or change of the quantities describing this state;

— change of structure of the MS.

The state of an element is described by the quantity u_i or by the quantities u_{ki} at the instants $i\Delta t$. The function H_i (3.44) describes the fact that the state is held unchanged for the time quantum, as in (3.43). Examples of modelling of an MS with discrete time will be presented below.

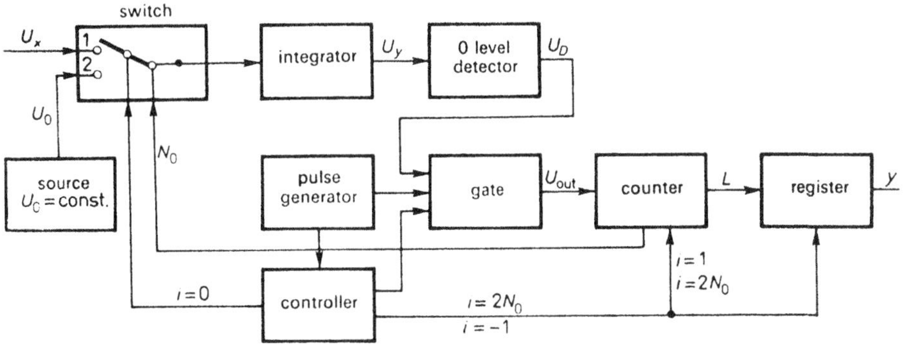

Fig. 3.11. The structural diagram of a dual-slope digital voltmeter.

The dual-slope digital voltmeter. The structural diagram of a digital voltmeter is shown in Fig. 3.11 and the signals at different stages of processing in Fig. 3.12. In the first phase of processing, the voltage U_x is integrated and the output voltage of the integrator is

$$U_{y1} = k_1 \int_0^t U_x \mathrm{d}t. \tag{3.45}$$

At the instant $i = 0$, the counter starts counting the pulses from the generator

$$U_G = \delta(t - i\Delta t), \quad i = \ldots, -2, -1, 0, 1, 2, 3, \ldots \tag{3.46}$$

and the current state of the counter is

$$L = \int_0^t U_G \mathrm{d}t = \int_0^t \delta(t - i\Delta t)\mathrm{d}t = \sum_{i=1}^{N} \mathbf{1}(t - i\Delta t), \tag{3.47}$$

where

$$N\Delta t < t < (N+1)\Delta t. \tag{3.48}$$

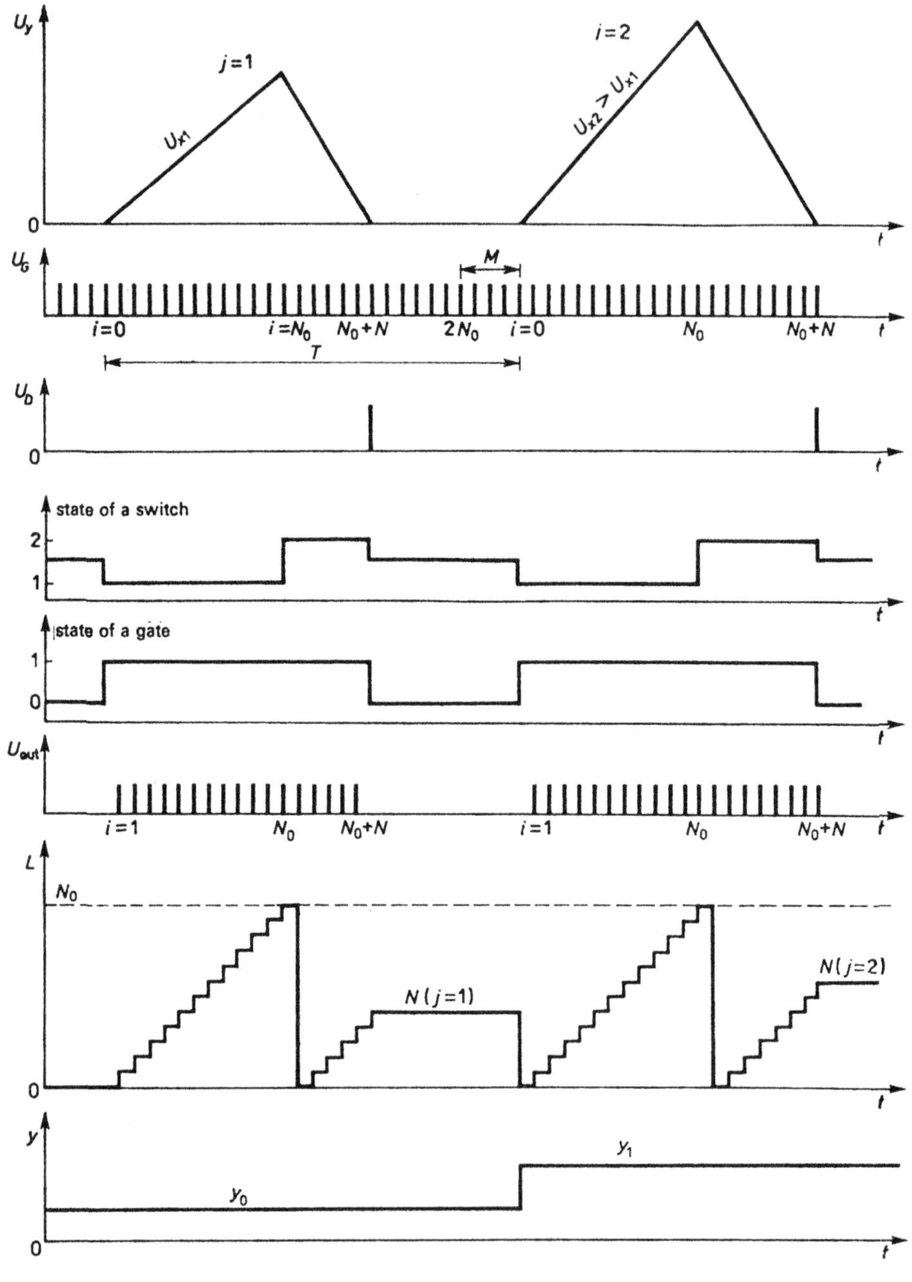

Fig. 3.12. Timing diagram of a dual-slope digital voltmeter.

When, after a period t_1, counter overflow occurs, the input voltage U_x is disconnected from the integrator and the reference voltage U_0 is supplied to the integrator input. At the instant

$$t_1 = \sum_{i=1}^{N_0} \Delta t_i = N_0 \Delta t_{av} = \frac{N_0}{f_{1av}}, \tag{3.49}$$

the first phase of the measurement is over. At that time, we have

$$U_1 = U_{y1} = k_1 \int_0^{t_0} U_x \mathrm{d}t = t_1 k_1 \frac{1}{t_1} \int_0^{t_1} U_x \mathrm{d}t = k_1 t_1 U_{x\,av} = \frac{k_1 N_0}{f_{1av}} U_{x\,av}. \tag{3.50}$$

In the second phase $j = 1$ and the output voltage of the integrator decreases

$$U_{y2} = U_1 - k_2 \int_{t_1}^{t} U_0 \mathrm{d}t, \tag{3.51}$$

and reaches zero at the instant t_2. At this moment, the detector generates the pulse

$$U_D = \delta(U_y) = \delta(t-t_2), \tag{3.52}$$

which closes the gate; from this moment no more pulses are counted by the counter. The integration ends when the output voltage of the integrator reaches the threshold level U_D of the detector

$$U_1 - k_2 \int_{t_1}^{t_2} U_0 \mathrm{d}t + U_D = U_1 - k_2 U_0 (t_2 - t_1) + U_D = 0. \tag{3.53}$$

Then with (3.50) we get

$$\begin{aligned} t_2 - t_1 &= \frac{U_1 + U_D}{k_2 U_0} = \frac{k_1}{k_2} \frac{N_0}{f_{1av}} \frac{U_{x\,av}}{U_0} + \frac{U_D}{k_2 U_0} \\ &= \frac{k_1}{k_2} \frac{N_0}{f_{1av}} \frac{U_{x\,av} + U_{Dx}}{U_0}, \end{aligned} \tag{3.54}$$

where U_{Dx} is the value of the dead band of the balance detector brought to the input of the digital voltmeter before the integration

$$k_1 \int_{t_0}^{t_1} U_{Dx} \mathrm{d}t = k_1 t_1 U_{Dx} = U_D. \tag{3.54a}$$

At the instant t_2, the state of the counter is

$$L = \sum_{i=1}^{N_0+N} \mathbf{1}(t - i\Delta t) = N_0 + N, \tag{3.55}$$

and the value of N meets the condition

$$(N_0+N)\Delta t < t_2 < (N_0+N+1)\Delta t, \tag{3.56}$$

at the instant of closing the gate. Measurement of the time t_2 is performed by counting the quanta Δt_i, so that

$$t_2 = \sum_{i=0}^{N_0} \Delta t_i + \sum_{i=N_0}^{N_0+N} \Delta t_i + \Delta_2 = t_1 + \frac{N}{f_{2\,\mathrm{av}}} + \Delta_2, \tag{3.57}$$

where

$$0 < \Delta_2 \leqslant \Delta t_{N+1} \tag{3.57a}$$

or

$$0 < \Delta_2 \cdot f_{2N} \leqslant 1. \tag{3.57b}$$

After substituting (3.54) in (3.57), we obtain

$$N = (t_2 - t_1 + \Delta_2) f_{2\,\mathrm{av}} = \frac{k_1}{k_2} \frac{f_{2\,\mathrm{av}}}{f_{1\,\mathrm{av}}} N_0 \frac{U_{x\,\mathrm{av}} + U_{Dx}}{U_0} + \Delta_2 f_{2N}. \tag{3.58}$$

The above equation is the conversion equation of an integrating voltmeter. The measured quantity is an average voltage $x = U_{x\,\mathrm{av}}$ and the output quantity is $y = N$. The quantity Δ_2 is the counting error whose value is random and falls into an interval (0–1) quantum. The model presented includes some real properties of a digital voltmeter (DVM), such as the dead band of the detector and the facts that both the integration constants and the pulse generator frequency values are not the same in the first and the second phase of measurement. The calibration procedure follows the ideal model, i.e., equation (3.58) for $k_1 = k_2$, $f_{1\,\mathrm{av}} = f_{2\,\mathrm{av}}$, $U_D = 0$:

$$\frac{N}{N_0} U_0 < U^*_{x\,\mathrm{av}} < \frac{N+1}{N_0} U_0. \tag{3.59}$$

By substituting the real model (3.58) in this procedure we obtain

$$\frac{k_1}{k_2} \frac{f_{2\,\mathrm{av}}}{f_{1\,\mathrm{av}}} (U_{x\,\mathrm{av}} + U_D) < U^*_{x\,\mathrm{av}} < \frac{k_1}{k_2} \frac{f_{2\,\mathrm{av}}}{f_{1\,\mathrm{av}}} (U_{x\,\mathrm{av}} + U_D) + \frac{U_0}{N_0}. \tag{3.60}$$

The duration of the measurement cycle is $T = 2N_0\Delta t + M\Delta t$, where M is the number of quanta required for sending the counter contents (state) into the register and for clearing the integrator and the counter.

The conversion equations (3.45), (3.50), (3.47), (3.48) make it possible to determine the state of the elements of DVM at any time t; the model presented is dynamic.

The ultrasonic flowmeter [28]. The diagram of a typical single-channel ultrasonic flowmeter based on the principle of frequency difference measurement is

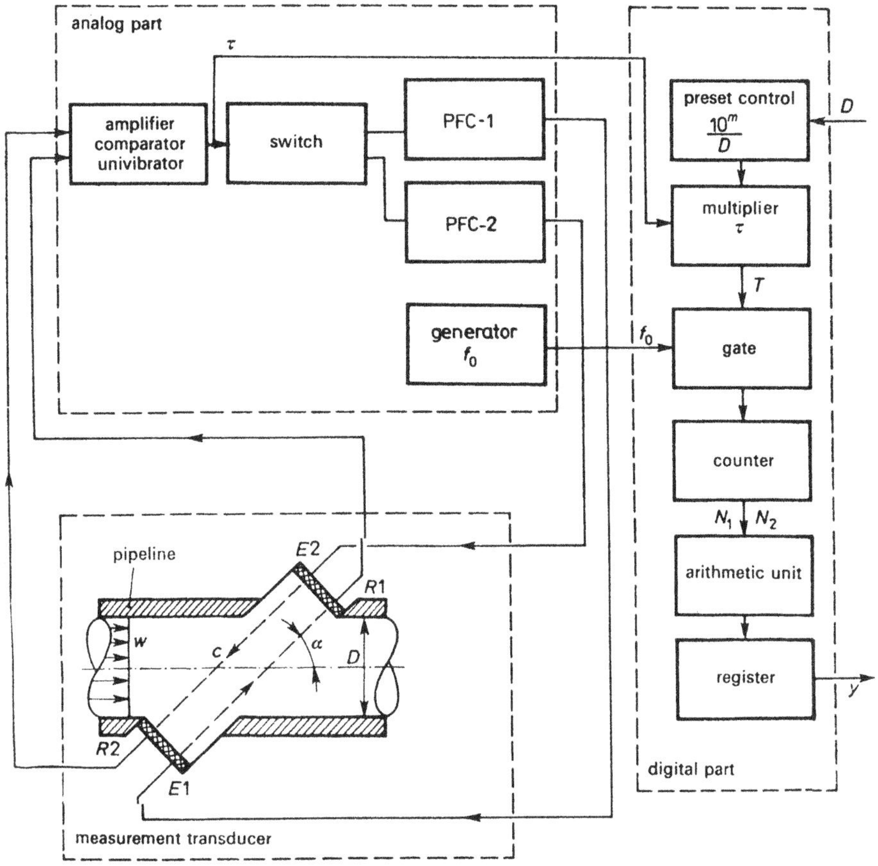

Fig. 3.13. The structural diagram of an ultrasonic flowmeter with frequency difference measurement.

shown in Fig. 3.13. The instrument works in three cycles. In the first cycle, the pulse-forming circuit PFC-1 activates the emitter $E1$. The ultrasonic pulse goes through the liquid and is received in the receiver $R1$. The received signal is amplified and formed, and then it reactivates the forming circuit PFC-1 so that pulses are now generated with the frequency f_1. In this way, the time interval T_1 = constant/f_1 is generated and during this time the generator pulses are counted. Next, the second cycle starts. The circuit PFC-2, the emitter $E2$ and the receiver $R2$ are active. The pulses are generated at a frequency f_2 and the counter counts the pulses for the time T_2 = constant/f_2. The result of this counting is N_2. In the third cycle, the arithmetic operations are performed and the result is sent to the register. The next measurement is performed after clearing the counters.

The frequencies f_1, f_2 are, respectively,

$$f_1 = \frac{\cos\alpha(c+w\sin\alpha)}{D}, \tag{3.61a}$$

$$f_2 = \frac{\cos\alpha(c-w\sin\alpha)}{D}, \tag{3.61b}$$

$$f = f_1 - f_2 = \frac{\sin 2\alpha}{D}\, w, \tag{3.61c}$$

where D is the pipe diameter, α the angle between the direction of the ultrasonic beam and the pipe axis, c the ultrasonic beam velocity and w the velocity of the liquid. The velocities are added in the circuit E–$R1$ and substracted in the circuit E–$R2$.

To increase the accuracy of measurement, the frequencies f_1, f_2 are measured by an indirect method. The values of the periods $\tau_1 = 1/f_1$, $\tau_2 = 1/f_2$ are multiplied by the factor $d = 10^m/D$ (thus, the pipe diameter data are included). The counting results are, respectively,

$$N_1 = [T_1 f_0] = \left[\frac{10^m}{D}\frac{f_0}{f_1}\right], \tag{3.62a}$$

$$N_2 = [T_2 f_0] = \left[\frac{10^m}{D}\frac{f_0}{f_2}\right]. \tag{3.62b}$$

In the arithmetic unit, the frequencies f_1, f_2 and their difference are calculated. The measurement result is

$$y = \left[\frac{10^p}{N_1}\right] - \left[\frac{10^p}{N_2}\right] = M, \tag{3.63}$$

where p is a natural number. In formulae (3.62) and (3.63) the brackets denote the integer part of a number (entier function); this results from the properties of the arithmetic unit. From the above equations we obtain

$$y = \frac{10^p}{10^m}\frac{\sin 2\alpha}{f_0}\, w. \tag{3.64}$$

The form of the calibration procedure is

$$w^* = ky, \tag{3.65}$$

where the value of k should be matched:

$$k = \frac{10^m}{10^p}\frac{f_0}{\sin 2\alpha}.$$

The errors of the flowmeter result mainly from rounding off in the arithmetic operations and in the counting process.

The digitally processed results make the final result rounded as in (3.62). The rounding error is the quantization error. Its distribution is uniform over a single-quantum interval. This error is modelled by adding a random component Δ_k at the input of the quantization stage. Therefore, the value $x+\Delta_k$ is measured and the result is $N+\Delta_k$. As shown in Fig. 3.2c, d, the quantization error assumes values within the interval $\langle -q, 0\rangle$ or $\langle -0.5q; 0.5q\rangle$. The variance of the quantization error, called Shepard's correcting factor [161], is $q^2/12$.

3.3 The Modelling of Error Sources. Probabilistic Models

When errors are modelled, the real properties of the construction elements of an MS should be taken into account. A number of phenomena have a molecular nature and are particularly difficult to describe. An MS cannot be abstracted from its environment and some of the environmental factors exert a barely perceptible influence on the phenomena occurring within an MS. The construction elements of an MS age and wear out, and their properties suffer random changes. Effects such as these are not accounted for in a model, and the coefficients in the conversion equation are assumed to be constant. Some examples follow.

1. A beam of particles emitted by an isotope source consists of individual particles emitted at time intervals subject to the Poisson distribution. The velocities of the particles are not equal and are described approximately by the Maxwell distribution. In deterministic models, the concept of mean emission rate $\bar{N}$ is applied. Figure 3.14 illustrates how the random character of the effect vanishes

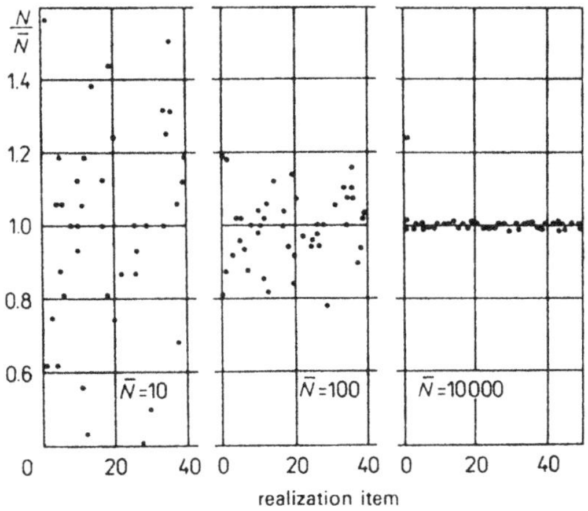

Fig. 3.14. Distribution of values of a number N of particles for different mean values $\bar{N}$.

as the averaging time increases (or as the activity of the radiation sources increases).

2. Gas particles striking the thread of a torsion balance or of a galvanometer cause fluctuations of the moving organ, resulting in the deflection $\Delta\varphi$. The root mean square value of deflection is [39]

$$\Delta\varphi_{rms} = \sqrt{\frac{kT}{a}}, \tag{3.66}$$

where $a = \pi^2 r^4 G/2l$, and G is the rigidity modulus of the material, r the thread cross-section radius, l its length, k is the Boltzmann constant (universal gas constant) and T is the temperature in the Kelvin scale.

3. The thermal noise of resistors is described by the Nyquist formula

$$E_{rms} = \sqrt{4kTR\Delta f}, \tag{3.67}$$

where E_{rms} is the root mean square value of the EMF over the frequency interval Δf and R is the resistance value of the resistor.

4. The shot noise of photomultipliers and of electron tubes is an effect of the cathode emission; it is characterized by the root mean square emission current

$$i_{rms} = \sqrt{2ei_0\Delta f} \quad \text{or} \quad \Delta i = \sqrt{\frac{ei_0}{t}}, \tag{3.68}$$

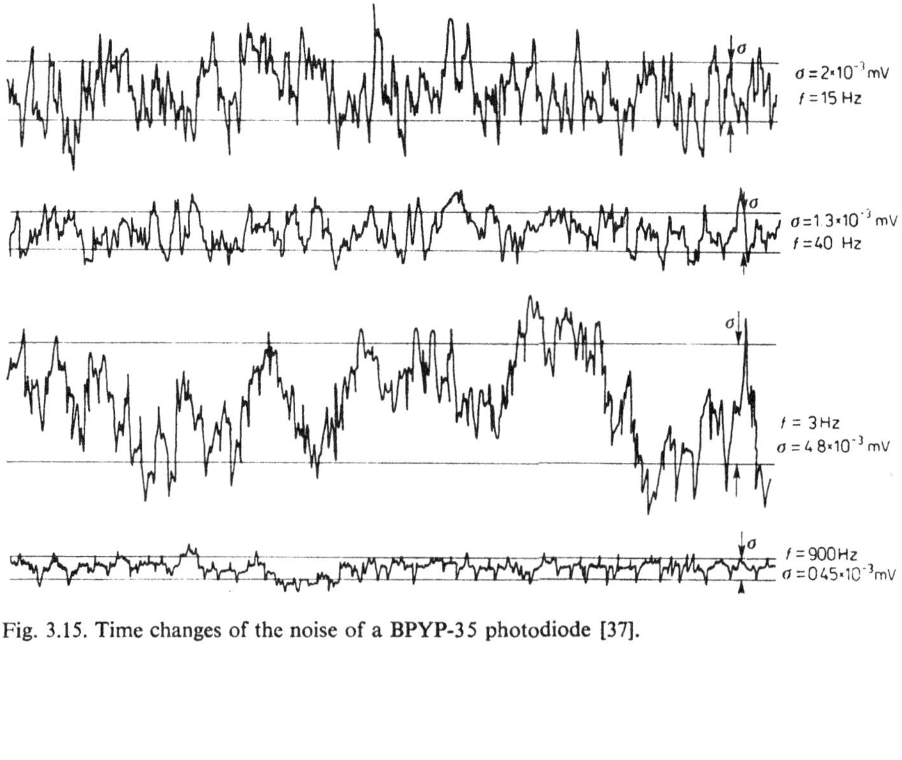

Fig. 3.15. Time changes of the noise of a BPYP-35 photodiode [37].

where i_0 is the mean value of the current in the period t and e is the electron charge.

5. Current noise, resulting from the molecular nature of electricity, shows up in low-current measurement sensors such as radiation detectors, thin-layer resistors, semiconductors, etc. The root mean square value of this current is described by the empirical formula

$$\Delta V_{\mathrm{rms}} = cRi\sqrt{\frac{\Delta f}{f}}, \tag{3.69}$$

where c is a constant value, R is resistance, i current and f frequency. An example of the noise voltage of a BPYP-35 photodiode with a BPW-10 FET amplifier (for frequency band $\Delta f = 1$ Hz) for different frequencies is shown in Fig. 3.15 [37]. The current noise $1/f$ and frequency-independent thermal and shot noises, together with high-frequency parasitic noises, appear in electronic circuits. The power distribution $S_u(\omega)$ of the noise in an electronic circuit with a BPYP-35 photodiode is shown in Fig. 3.16 [37].

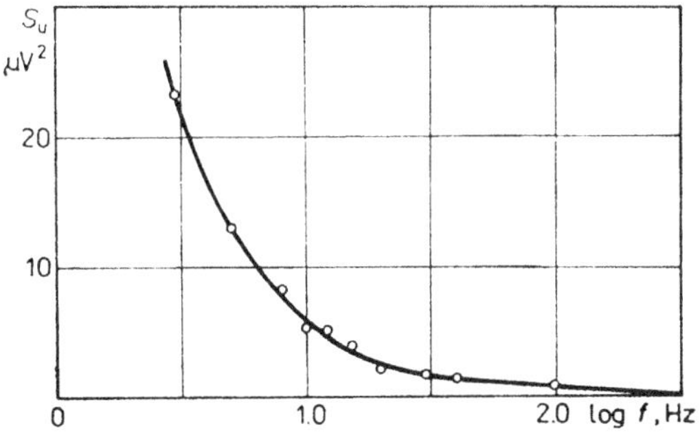

Fig. 3.16. The power distribution of the noise of a BPYP-35 photodiode over a 1 Hz frequency band [37].

Heat exchange by conduction, convection and radiation, fluid pressure, and many other phenomena have a molecular nature.

The environmental influence on an MS is the result of the properties of construction elements being dependent on the environmental conditions. Almost all of the properties of materials depend on the temperature, devices are energized from power supplies, etc. Relationships which have been revealed and recognized are accounted for in mathematical models: the dependences of an output quantity on the influencing quantities $x_1, \ldots, x_{m-1}$ are determined for different values of x. The dynamic effects may be considered as well. It is impossible, however, to include all the phenomena in a model. For example, the residual influence of changes in the supply voltage will not be eliminated; the influence of solar radiation may be reduced, but it is still very hard to specify and eliminate it. Under average conditions such influences as pressure changes in a room resulting from wind force are also hard to account for. The wind velocity and direction change, as is shown in Fig. 3.17. These effects of residual environmental influence are

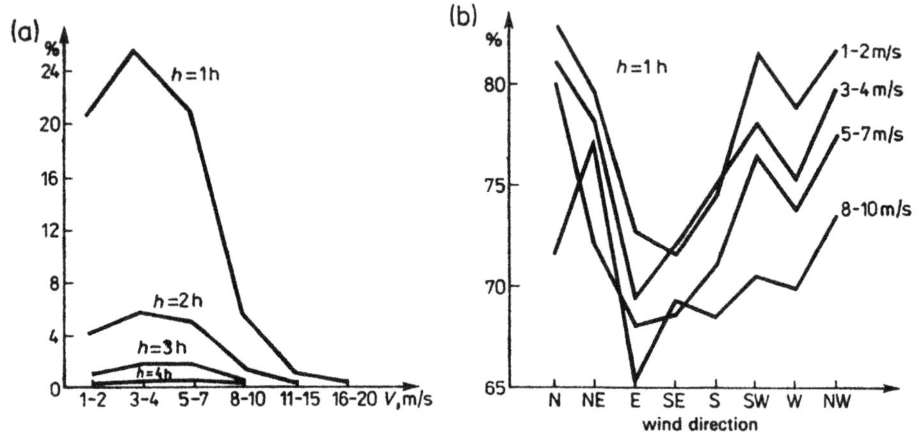

Fig. 3.17. The time variability of wind velocity (a) and direction (b) [121].

included in the model by introducing a variable z. Let us look at an example. The readings of a thermomagnetic oxygen analyser KTM-7 for a constant value of the measured quantity, constant temperature, supply current and gas flow rate, vary with time as shown in Fig. 3.18 [71] (the gas pressure is not stabilized). The day-by-day variation is clearly seen in this case.

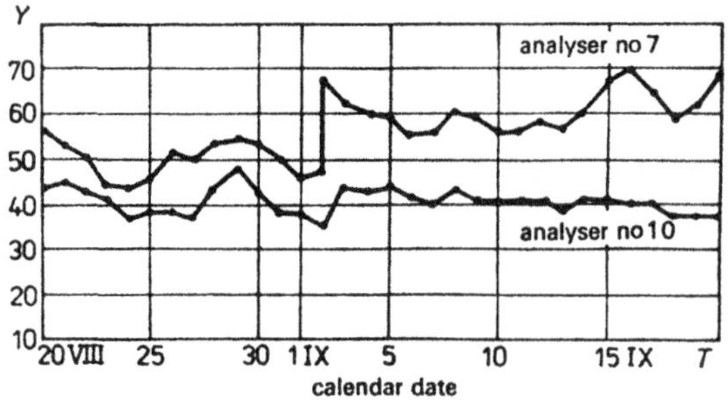

Fig. 3.18. The time variability of the indications of a thermomagnetic analyser KTM-7 [71].

On the other hand, the ageing and wearing out of materials are slowly varying processes, usually monotonic. For example, insulation ageing is described by the Arrhenius formula [116]:

$$v = v_{\infty} e^{-\Delta V/RT} = v_{\infty} e^{-B/T}, \tag{3.70}$$

where v is the chemical reaction rate, R the universal gas constant, T the temperature, ΔV the activation energy, v_{∞} a reaction constant and B = constant.

The evaporation rate of a material of a hot filament is described by the formula [118]

$$G = \frac{1}{\sqrt{2\pi R}} p \sqrt{\frac{M}{T}}, \tag{3.71}$$

where p is the pressure, M the molecular weight of the material, R the gas constant and T the temperature.

The wearing and grinding times of moving parts may be estimated similarly [118]. Contamination of the surface of optical, current-conducting and other elements, active in various senses, is progressive, but its analytical description is difficult. In measurement systems, a number of elements are subjected to wearing and ageing. The observed effect, not always monotonic, is a result of all of these processes. The results of measurements of water-meter errors caused by wear of the elements are presented as an example in Fig. 3.19 [29].

The starting point of modelling of the sources of errors is the model of an MS already determined. The coefficients $b_1, \ldots, b_l$ of the model are unambiguously related to the *construction parameters*[5] $a_1, \ldots, a_k$

$$b_l = \psi_l(a_1, \ldots, a_k). \tag{3.72}$$

Each construction parameter may express the molecular effects or wearing of the materials. For this reason, each parameter a_j includes a constant component a_{j0} and a random component, namely

$$a_j = a_{j0} + \Delta a_j. \tag{3.73}$$

Some of the phenomena must be included by introducing a source of error as a new quantity in the model, since they are not implied by the model. The thermal voltage of a resistor, the resistance of contacts, etc. can be named as examples. Each source of error, being a random variable, is characterized by its probability distribution and spectral density. If we know the position of an error source in a structure [91], together with the source characteristics and properties of the

[5] For the model in its reference conditions, the following designations have been applied:

$$y = F_0(x) = b_1 f_1(x) + b_2 f_2(x) + \ldots + b_l f_l(x),$$

as in the model (3.13).

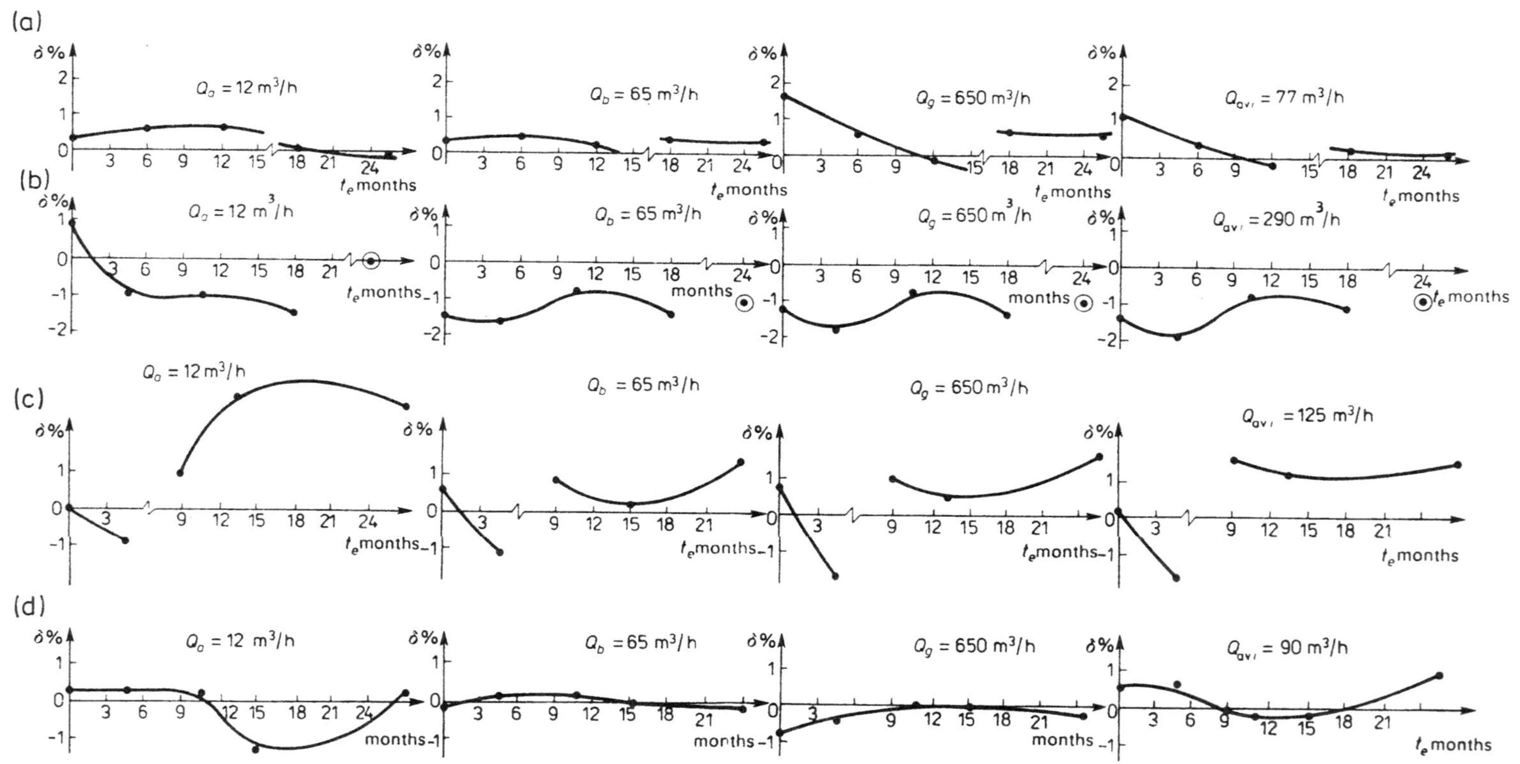

Fig. 3.19. MZ-200 water-meter error measurement results: (a) water meter 4, (b) water meter 6, (c) water meter 14, (d) water meter 18 [29]. A scale break denotes adjustment of the water meter. Q_a, Q_b, Q_g—nominal flow rates, δ—relative error.

model, we may calculated the component of an output quantity arising from this error source. The problem is complex, and is presented in more detail below than for the static and dynamic models.

First of all, despite the constant values of the input quantities $x_1, \ldots, x_m$, which permit a static model to be taken into consideration, sources of random variables exist within the MS. Hence, a dynamic model should be used to synthesize these sources.

Secondly, the frequency band of the internal error sources is rather wide. The high frequencies appear as the result of molecular effects and the low frequencies from the ageing processes. The internal error sources are described by various different probability distributions. The ageing processes are the reason why the resultant effect may be a non-stationary process.

Thirdly, nonlinear elements may cause the noise spectrum to expand greatly [90, 52] and the averaging elements may limit this effect. Therefore, approximate estimations may be erroneous.

Before the mode of modelling of the error sources is presented, some properties of the processing of the sources of errors will be shown.

3.3.1 Nonlinear elements

Nonlinear inertialess (non-dynamic) elements are taken into account. At the element input, an error source is described by a probability distribution $p(x)$ and by a spectral density $S_x(\omega)$. The random variable X is therefore stationary. A stationary random variable Y is obtained at the output of an element with the characteristic $y = f(x)$.

Determination of the probability distribution. The initial parameters used for calculations are the construction parameters $a_1, \ldots, a_k$, a multidimensional probability distribution $p(a_1, \ldots, a_k)$ and the limiting probability distributions $p_1(a_1), \ldots, p_k(a_k)$. The coefficients of the model are the quantities $b_1, \ldots, b_l$, expressed by formula (3.72) with $l = 1, 2, \ldots, l$, namely

$$b_l = \psi_l(a_1, \ldots, a_k).$$

The function ψ_l may be rewritten (in the manner of subsequent substitutions) in the form of elementary functions: a sum, a difference, a quotient and a product of two primary variables a_j or a single-argument function. The computational formulae for the elementary functions are given in Table 3.1 [22].

Determination of the spectral density. The initial parameters used in the calculations are the spectral density $S_x(\omega)$ or correlation function $K_x(\tau)$ of a stationary error source and the characteristic of an element.[6]

[6] As indicated above, the input quantities are the construction parameters, denoted as x.

TABLE 3.1

Computational formulae of the probability distributions of continuous random variables [22]

Item	Function	Variables X, Y, dependent data $p(x, y)$	Variables X, Y, independent data $p(x, y) = p_1(x)p_2(y)$
1	single-argument $Z = f(X)$ if $X = \varphi(Z)$ and φ' exist	$p_z(z) = p_x[\varphi(z)]\,|\varphi'(z)|$	
2	$Z = X+Y$	$p_z(z) = \int_{-\infty}^{\infty} p(x, z-x)\,\mathrm{d}x$	$p_z(z) = \int_{-\infty}^{\infty} p_1(x)p_2(z-x)\,\mathrm{d}x$
3	$Z = X-Y$	$p_z(z) = \int_{-\infty}^{\infty} p(z+y, y)\,\mathrm{d}y$	$p_z(z) = \int_{-\infty}^{\infty} p_1(z+y)p_2(y)\,\mathrm{d}y$
4	$Z = XY$	$p_z(z) = \int_{-\infty}^{\infty} p\left(z, \frac{z}{x}\right)\frac{1}{\lvert x\rvert}\,\mathrm{d}x$	$p_z(z) = \int_{-\infty}^{\infty} \frac{1}{\lvert x\rvert} p_1(x)p_2\left(\frac{z}{x}\right)\mathrm{d}x$
5	$Z = X/Y$	$p_z(z) = \int_{-\infty}^{\infty} p(yz, y)\lvert y\rvert\,\mathrm{d}y$	$p_z(z) = \int_{-\infty}^{\infty} p_1(yz)p_2(y)\lvert y\rvert\,\mathrm{d}y$
6	Generally $U_1 = g_1(X, Y)$ $U_2 = g_2(X, Y)$ $X = h_1(U_1, U_2)$ $Y = h_2(U_1, U_2)$ $p(x, y)$—given g_1, g_2—isomorphic functions	$p_u(u_1, u_2) = p[h_1(u_1, u_2), h_2(u_1, u_2)]\lvert J\rvert$	$J = \begin{vmatrix} \frac{\partial X}{\partial U_1} & \frac{\partial X}{\partial U_2} \\ \frac{\partial Y}{\partial U_1} & \frac{\partial Y}{\partial U_2} \end{vmatrix}$

For a linear element, a relationship occurs such that

$$S_y(\omega) = k^2 S_x(\omega), \tag{3.74}$$

$$K_y(\tau) = k^2 K_x(\tau). \tag{3.75}$$

The correlation function at the outputs of elements with quadratic or polynomial characteristics may be calculated relatively simply from its definition. For this reason it is convenient to approximate different characteristics by polynomials.

The formulae given below relate to a random variable x with a mean value zero ($E[x] = 0$) and a normal probability distribution. When

$$Y = X^2, \tag{3.76}$$

and after denoting

$$X_1 = X(t), \quad X_2 = X(t+\tau), \tag{3.77}$$

the correlation function may be calculated from the definition as

$$K_y(\tau) = E\{[X_1^2 - E(X_1^2)][X_2^2 - E(X_2^2)]\}. \tag{3.78}$$

The covariance matrix of a bidimensional random variable is of the form

$$\operatorname{cov}(X_1, X_2) = \begin{bmatrix} K_x(0) & K_x(\tau) \\ K_x(\tau) & K_x(0) \end{bmatrix}. \tag{3.79}$$

After the multiplication in (3.78) is performed, the mean values of separate components may be calculated using the characteristic function. Thus [122]

$$E(X_1^2 X_2^2) = \left. \frac{\partial^4 \varphi(z_1, z_2)}{\partial z_1^2 \partial z_2^2} \right|_{z_1 = z_2 = 0} \tag{3.80}$$

while

$$\varphi(z_1, z_2) = \exp[-\tfrac{1}{2}(k_{11} z_1^2 + k_{22} z_2^2 + 2k_{12} z_1 z_2)], \tag{3.81}$$

where φ is the characteristic function of a bidimensional normal distribution and k_{ij} are the elements of the covariance matrix (3.79). The result of the calculations is

$$K_y(\tau) = 2[K_x(\tau)]^2. \tag{3.82}$$

The spectral density is calculated from the Wiener–Khinchin formula

$$S_y(\omega) = \frac{1}{2\pi} \int_{-\infty}^{\infty} 2[K_x(\tau)]^2 e^{-j\omega\tau} d\tau. \tag{3.83}$$

Sveshnikov [122] proposes calculating integral (3.83) from the formula

$$\frac{1}{2\pi} \int_{-\infty}^{\infty} 2[K_x(\tau)]^2 e^{-j\omega\tau} d\tau = \int_{-\infty}^{\infty} S_x(\omega_1) S_x(\omega - \omega_1) d\omega_1 . \tag{3.84}$$

which makes the calculations much simpler.

If the form of a characteristic of an element is

$$Y = b_0 + b_1 X + b_2 X^2 \tag{3.85}$$

and X is a random variable with normal distribution and zero mean value, then

$$K_y(\tau) = 2b_2^2 [K_x(\tau)]^2 + b_1^2 K_x(\tau). \tag{3.86}$$

The result of using formula (3.84) in (3.86) is

$$S_y(\omega) = (24 b_1 b_2 K_x(0) + 2b_2^2) \int_{-\infty}^{\infty} S_x(\omega_1) S_x(\omega - \omega_1) d\omega_1 + b_1^2 S_x(\omega). \tag{3.87}$$

Given a nonlinear element with the characteristic

$$Y = XU, \tag{3.88}$$

when $E(X) = E(U) = 0$, the distributions of both variables X, U are normal and their characteristics are known, then

$$K_y(\tau) = K_x(\tau)K_u(\tau) + [K_{xu}(\tau)]^2. \tag{3.89}$$

Considering that a product of time functions corresponds to a convolution of transforms, we obtain [122]

$$S_y(\omega) = \int_{-\infty}^{\infty} [S_x(\omega_1)S_u(\omega-\omega_1) + S_{xu}(\omega_1)S_{xu}(\omega_1-\omega)]\,\mathrm{d}\omega_1. \tag{3.90}$$

Examples (3.85), (3.88) show how to calculate, step by step, some more complex characteristics of particular elements. However, the calculations become more complicated when the probability distribution of an output quantity is no longer normal after the signal has passed a nonlinear element.

3.3.2. Dynamic elements

The model of elements described using the Hammerstein operator (3.27) consists of two stages, as shown in Fig. 3.5. The dynamic part of the element has linear properties described by the Volterra operator (3.26). The solutions are simple for a normally distributed input quantity. The linear properties of the Volterra operator (3.26) make the distribution of the output quantity normal as well. Therefore, it is sufficient to derive the first two moments of the distribution

$$\bar{y}(t) = E\{F_V[X(t)]\} = E\left\{\int_0^t k(t, \tau)x(\tau)\,\mathrm{d}\tau\right\}, \tag{3.91}$$

$$K_y(t, \tau) = F_V^* F_V[K_x(t, \tau)] = \int_0^t\int_0^t k(t, \tau_1)k(t, \tau_2)\,\mathrm{d}\tau_1\,\mathrm{d}\tau_2. \tag{3.92}$$

This leads to the formulae

$$S_y(\omega) = |K(\mathrm{j}\omega)|^2 S_x(\omega), \tag{3.93}$$

or, for p independent inputs,

$$S_y(\omega) = \sum_{i=1}^{p} |K_i(\mathrm{j}\omega)|^2 S_i(\omega) \tag{3.94}$$

and

$$\sigma_y^2 = \int_0^{\infty} S_y(\omega)\,\mathrm{d}\omega. \tag{3.95}$$

The solution may be extended for variables with different distributions. The higher-order central moments may be calculated [122] as well

$$\mu_3 = \int_0^t\int_0^t\int_0^t k(t, \tau_1)k(t, \tau_2)\, k(t, \tau_3)K_x(\tau_1, \tau_2, \tau_3)\mathrm{d}\tau_1\mathrm{d}\tau_2\mathrm{d}\tau_3, \tag{3.96}$$

where

$$K_x(\tau_1, \tau_2, \tau_3) = E\{[X(\tau_1)-\bar{x}(\tau_1)][X(\tau_2)-\bar{x}(\tau_2)][X(\tau_3)-\bar{x}(\tau_3)]\}. \tag{3.97}$$

Then μ_4 and the higher-order moments are calculated similarly.

Thus the multidimensional spectral density may be calculated from the Wiener–Khinchin formula [122] and the probability density function is

$$p(y|t) \approx \varphi\left(\frac{y-\bar{y}}{\sigma}\right) - \frac{a}{3!}\varphi'''\left(\frac{y-\bar{y}}{\sigma}\right) + \frac{\varepsilon}{4!}\varphi^{(4)}\left(\frac{y-\bar{y}}{\sigma}\right), \tag{3.98}$$

where

$$a = \frac{\mu_3}{\mu_2^{3/2}}, \qquad \varepsilon = \frac{\mu_4}{\mu_2^2} - 3, \qquad \mu_2 = \sigma_y^2, \tag{3.99}$$

while $\varphi = \varphi(u) = \left(1/\sqrt{2\pi}\right)\exp(-u^2/2)$ is the Gauss function and φ''', $\varphi^{(4)}$ are its derivatives.

Let us look at an example. The quantity $X = U^2$ is applied at the input of an element, the distribution $p(u)$ is normal,

$$E(U) = 0$$

and

$$K_u(\tau) = A\mathrm{e}^{-a|\tau|}. \tag{3.100}$$

The correlation function of the quantity X is described by (3.82) and the probability distribution may be calculated according to the formula from Table 3.1 as

$$p(x) = \frac{1}{2\sqrt{2\pi A}}\,\frac{1}{\sqrt{x}}\,\mathrm{e}^{-x/2A}. \tag{3.101}$$

The element is dynamic and its characteristic is

$$\frac{\mathrm{d}Y(t)}{\mathrm{d}t} + \frac{1}{T}Y(t) = X(t). \tag{3.102}$$

The kernel of the Volterra operator is

$$k(t, \tau) = \mathrm{e}^{-(t-\tau)/T}, \tag{3.103}$$

therefore

$$Y(t) = \int_0^t \mathrm{e}^{-(t-\tau)/T}X(\tau)\mathrm{d}\tau. \tag{3.104}$$

The respective correlation functions, as in (3.97), are [122]

$$K_x(\tau_1) = 2K_u^2(\tau_1), \tag{3.105a}$$

$$K_x(\tau_1, \tau_2) = 8K_u(\tau_1)K_u(\tau_2), \tag{3.105b}$$

$$K_x(\tau_1, \tau_2, \tau_3) = 16K_u(\tau_1)K_u(\tau_2)K_u(\tau_3). \tag{3.105c}$$

For $t \gg \tau$, after using (3.105a) and the value of $K_u(\tau)$ as in (3.100), the moments of the output quantity are

$$\mu_1 = \bar{y} = \int_0^t e^{-(t-\tau)/T}[E(u^2)]\,d\tau = \int_0^t e^{-(t-\tau)/T}K_u(0)\,d\tau = AT(1-e^{-t/T}) \approx AT, \tag{3.106}$$

$$\mu_2 = \sigma_y^2 = Y(t) = \int_0^t\int_0^t e^{-(t-\tau_1)/T}e^{-(t-\tau_2)/T}K_x(\tau_2-\tau_1)\,d\tau_1\,d\tau_2$$

$$= \frac{4A^2T^2}{1-4a^2T^2}[1-e^{-t(1+2aT)/T}] - \frac{2A^2T^2}{1-2aT}[1-e^{-2t/T}] \approx \frac{2A^2T^2}{1+2aT}. \tag{3.107}$$

Similarly, the calculated higher-order moments are

$$\mu_3 \approx \frac{8A^3T^3}{(1+aT)(1+2aT)}, \tag{3.108}$$

$$\mu_4 \approx \frac{12A^4T^4(15+25aT+2a^2T^2)}{(1+aT)(1+2aT)^2(3+2aT)}. \tag{3.109}$$

After coefficients (3.99) were calculated from (3.106)–(3.109), the distribution $p(y)$ may be calculated from formula (3.98). The distributions of u, x, y are shown in Fig. 3.20.

3.3.3 Non-stationary error sources

Non-stationary effects arise from the ageing and wearing of materials. They manifest themselves in the variability and time-dependence of the construction parameters of an MS and, hence, in the variability of the coefficients of a static or dynamic mathematical model.

Non-stationary element. An element is non-stationary when its parameters vary in a random way and are non-stationary quantities. As mentioned above, in such a case a given parameter is treated as an independent source of instability and as an independent source of an input quantity for a given stage of processing. Since every construction parameter is a potential source of instability, the number of independent input random variables is $m+k-1$ ($m-1$ influencing quantities and k construction parameters).

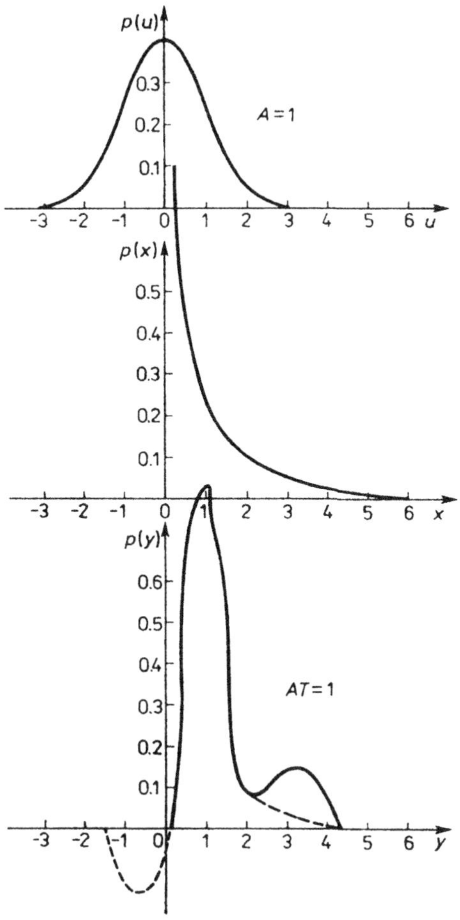

Fig. 3.20. Distribution of the quantities u, x, y.

In analysis of the physical causes of instability of construction parameters, the following components should be distinguished.

1. $\Delta\tilde{a}_{j0}$—uncertainty of the nominal values of construction parameters. We introduce the nominal value a_{j0} of a given parameter into a mathematical model based on the measurement results of this parameter, with associated uncertainty Δa_{j0}. In the MS analysed, one constant value of this uncertainty is realized (one element of the set) and thus the value of the error is constant. Nevertheless, this uncertainty should be considered in the model.

2. $\Delta\tilde{a}_{jd}$—expresses the dependence of a given parameter on measurable ambient conditions, e.g. on temperature. The cause shows up in the model by considering the ith quantity as an influencing quantity. The dependence of the output quantity

on the influencing quantity x_i, as met in model (3.8), is the combined effect of all construction parameters, i.e.,

$$dy_i = \left[\frac{\partial F(x_1, \ldots, x_{m-1}, x)}{\partial x_i} + \sum_{j=1}^{k} \frac{\partial F}{\partial a_j} \frac{\partial a_j}{\partial x_i}\right] dx_i, \tag{3.110}$$

or

$$\Delta y_i = \int_{x_{i0}}^{x_i} \frac{\partial F}{\partial x_i} dx_i + \int_{x_{i0}}^{x_i} \sum_{j=1}^{k} \frac{\partial F}{\partial a_j} \frac{\partial a_j}{\partial x_i} dx_i. \tag{3.111}$$

The factor $\partial a_j/\partial x_i$ arises from the properties of materials, while $\partial F/\partial a_j$ is derived from model (3.8) of an MS.

3. $\Delta\tilde{a}_{jin}$—time instability of the parameter a_j. The instability may be caused by:

— Molecular effects, otherwise called *oscillations*. By convention, the frequency range $f > 1$ Hz is regarded as characteristic for these effects, i.e., the oscillations are imperceptible by the observer and efficiently filtered by analog meters.

— Unnoticeable changes of influencing quantities, e.g. the change of atmospheric pressure due to wind pressure. The changes are not detectable because they have been neglected in the model or because they occur below the detection level of a given influencing quantity (e.g. temperature changes in a thermostat); the characteristic frequency range is in this case $f = 10^{-5}$ to 1 Hz. The infimum is the effect of a day cycle of a number of phenomena. These effects are called *fluctuations*.

— Ageing and wearing of materials. This is the so-called *drift*. The frequency of changes is small, $f < 10^{-5}$ Hz. Drift is the main cause of non-stationary effects [27].

Munroe [157] presents the noise characteristic in the frequency range 10^{-8}–10^{8} Hz. The noise distribution confirms the adequacy of the classification into oscillations, fluctuations and drift. In the case of oscillations in an industrial environment (in the chemical industry), noise with a frequency up to 700 Hz is characterized by different probability distributions [173], viz. normal, uniform or sinusoidal distributions.

According to (3.73), the jth construction parameter is described by the variable

$$a = a_0 + \widetilde{\Delta a} = a_0 + \widetilde{\Delta a_0} + \widetilde{\Delta a_d} + \widetilde{\Delta a_{in}}. \tag{3.112}$$

The sources of errors are shown in Fig. 3.21. The symbol a_{corr} denotes the correct value of the parameter. The index 0 denotes the reference conditions, T is the service time. The number of parameters affecting the jth parameter may be greater.

Non-stationary source. The component $\widetilde{\Delta a_{in}}$ in formula (3.112) is an internal,

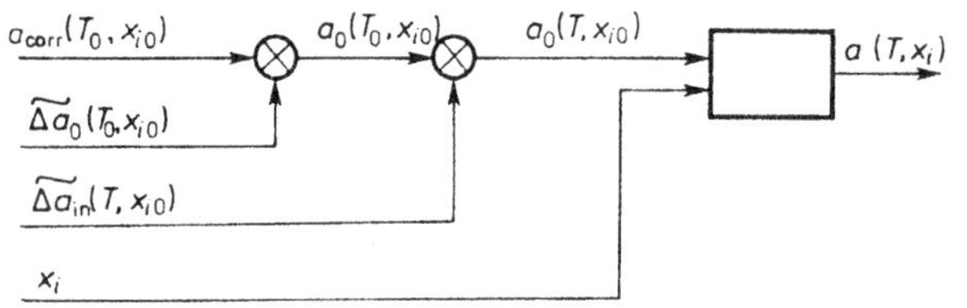

Fig. 3.21. Components of the error of a construction parameter.

non-stationary error source of an MS. Let us denote it as $u(t)$ and let its characteristic be known, for example

$$u(t) = g(t)\,\varepsilon(t), \tag{3.113}$$

where $g(t)$ or $g(T)$ is a known function of time and $\varepsilon(t)$ is a stationary random process. In this case, the output quantity $y(t)$ of an element incorporating the source $u(t)$ is a non-stationary random process.

There is a relatively wide class with error sources for which

$$g(t) = t^m, \qquad m = 1, 2, \ldots, m, \tag{3.114}$$

or

$$g(t) = e^{pt}, \tag{3.115}$$

or there is a linear combination of these sources: $c_r t^{m_r} e^{p_r t}$. The solution is relatively simple in these cases [122].

A stationary quantity may be expressed by the Fourier integral

$$\varepsilon(t) = \int_{-\infty}^{\infty} e^{j\omega t} d\Phi(\omega) + \bar{\varepsilon}(t). \tag{3.116}$$

Hence, (3.113) may be rewritten as

$$u(t) = \int_{-\infty}^{\infty} g(t) e^{j\omega t} d\Phi(\omega) + g(t)\bar{\varepsilon}(t). \tag{3.117}$$

According to (3.26), we obtain as an output of an element

$$y_1(t) = \int_{-\infty}^{\infty} P(\omega, t) e^{j\omega t} d\Phi(\omega) + \bar{y}_1(t), \tag{3.118}$$

where $P(\omega, t)$ is the Fourier transform of the kernel $k(t, \tau)$ of the operator. The convolution operation (3.26) was replaced by multiplication of the Fourier transforms. When $g(t) = t^m$, then, in a particular solution of (3.118), $P(\omega, t)$ is an mth order polynomial of the variable t. The coefficients $l_0, \ldots, l_m$ of the polynomial $P_m(\omega, t)$ are derived by comparing the expression

$$y(\omega, t) = (l_0 + l_1 t + l_2 t^2 + \ldots + l_m t^m) e^{j\omega t} \tag{3.119}$$

with the solution of (3.118). From the particular solution of (3.118) the correlation functions are derived by definition

$$K_y(t_1, t_2) = \int_{-\infty}^{\infty} P_m^*(\omega, t_1) P_m(\omega, t_2) e^{j\omega(t_2 - t_1)} S_\varepsilon(\omega) d\omega . \tag{3.120}$$

In case when $g(t) = e^{pt}$ we have

$$y(\omega, t) = A(\omega) e^{(p+j\omega)t}, \tag{3.121}$$

$$A(j\omega) = \frac{1}{a_0 + a_1(p+j\omega) + \ldots + a_n(p+j\omega)^n} \tag{3.122}$$

and

$$y_1(t) = \int_{-\infty}^{\infty} A(\omega) e^{(p+j\omega)t} d\Phi(\omega) + \bar{y}_1(t). \tag{3.123}$$

Thus

$$K_y(t_1, t_2) = e^{pt_2 + p^* t_1} \int_{-\infty}^{\infty} |A(j\omega)|^2 e^{j\omega(t_2 - t_1)} S_\varepsilon(\omega) d\omega , \tag{3.124}$$

and the spectral density is

$$S_y(\omega) = |A(j\omega)|^2 S_u(\omega - p). \tag{3.125}$$

A particular solution is stationary when p is an imaginary number.

3.4 The General Model of Error

3.4.1 The continuous-time model

The mathematical model of an MS is shown in Fig. 3.1b. The form of the first part of the model, the so-called conversion equation [91] is (3.8):

$$y = F(x_1, \ldots, x_{m-1}, x_m, T, a_1, \ldots, a_k, z). \tag{3.126}$$

The arguments of the operator F are the influencing quantities, the random variables or non-stationary random processes. The conversion equation is a stochastic relationship.

The second part of the model, called the calibration procedure (3.5), is

$$x^* = \varphi(y) = F_0^{-1}(y), \tag{3.127}$$

where

$$F_0 = F_0(x_1, \ldots, x_{m-1}, x_m, a_{10}, \ldots, a_{k0}) \tag{3.128}$$

is a function of nominal values of the construction parameters (with constant or determined values) and of measurable input quantities.

The error of an MS is defined (see (3.7)) as the difference between the value of the measured quantity reproduced from the readings of the MS and the value of the measurand

$$\Delta = x^* - x, \tag{3.129}$$

i.e.,

$$\begin{aligned}\Delta &= \varphi[F(x_1, \ldots, x_{m-1}, x, T, a_1, \ldots, a_k, z)] - x \\ &= F_0^{-1}[F(x_1, \ldots, x_{m-1}, x, T, a_1, \ldots, a_k, z)] - x.\end{aligned} \tag{3.130}$$

In the best case, when $F = F_0$, the minuend in the error equation is not equal to x, as the arguments of the operators F and F_0 are different. The arguments of F_0 are the values of measurable determined quantities, while the arguments of F are the random variables.

Consider the following example. The conversion equation of a magnetoelectric meter [91] takes the form

$$\alpha = \frac{\Phi z l}{D} J = \frac{\tilde{a}_1 \tilde{a}_2 \tilde{a}_3}{\tilde{a}_4} x = \tilde{k} x, \tag{3.131}$$

where $x = J$, $y = \alpha$, $a_1 = \Phi$, $a_2 = z$, $a_3 = l$, $a_4 = D$. As the construction parameters may be random variables, they have been marked with the tilde. The calibration procedure is

$$x^* = \varphi(\alpha) = c\alpha = \frac{1}{k_0} \alpha, \tag{3.132}$$

where c is a constant, usually $c = 1/k_0$. Hence

$$\Delta = x^* - x = \frac{1}{k_0} \tilde{k} x - x = \left(\frac{\tilde{k}}{k_0} - 1\right) x = \frac{\tilde{k} - k_0}{k_0} x. \tag{3.133}$$

The quotient $\tilde{k}/k_0 \neq 1$, for $\tilde{k}$ is a random variable. Because Δ, like y, is a function of the error sources a_j ($j = 1, \ldots, k$) which are non-stationary random processes, Δ as a function of a_j is also a non-stationary random process. The interpretation of the properties of a construction parameter as shown in Fig. 3.21 is based on the internal variability of a_j, expressed by the time-dependent component $\Delta \tilde{a}_j$. The external influence of the MS operating conditions, expressed as influencing quantities, is included as well. Thus, the error Δ is a multidimensional random process and its arguments are the input quantities $x_i \in \langle x_{il}, x_{iu} \rangle$ and the service time T of an MS, namely

$$\{\Delta\} = \{\Delta_k(x_1, \ldots, x_m, T)\}, \tag{3.134}$$

where the index k is a realization number for the process. The process $\{\Delta\}$ is realized in the real time $t \in T_k \in T$. The physical interpretation of the error $\{\Delta\}$ as a random process is that the MS is located in an $(m+1)$-dimensional random field with coordinates $x_1, \ldots, x_{m+1}$ (where $x_{m+1} = T$). Depending on the position of the MS in this field, different realizations of errors are obtained [97]. For example, if $x_i = x_{i0} =$ constant for $i = 1, \ldots, m-1$, $T =$ constant, the error process is unidimensional $\{\Delta\} = \{\Delta(x)\}$.

Figure 3.22 shows the error distributions of electric analog meters of magnetoelectric and electromagnetic types. The characteristic of an electromagnetic meter is quadratic: $y = ax^2$. The error distribution of a magnetoelectric meter is normal, with the variance being constant for a given x ($\sigma^2 =$ constant), while the mean value $\underset{k}{E}(\Delta_k\} = m(x)$ depends on the value of x (in Fig. 3.22 this has

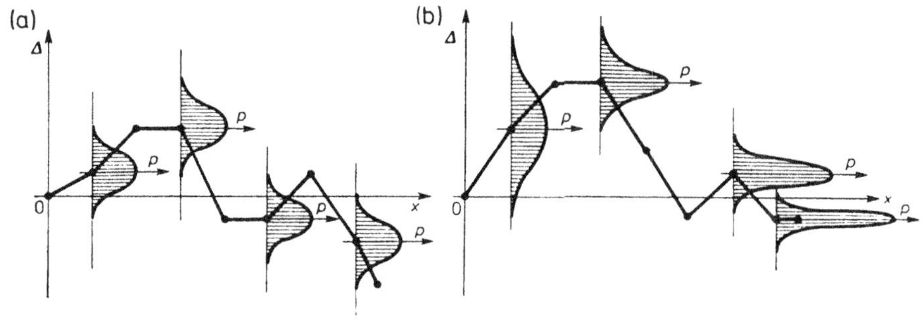

Fig. 3.22. Error distributions of electromechanical meters: (a) magnetoelectric, (b) electromagnetic.

been marked with a piecewise linear line). The condition that the distribution function of $P(\Delta)$ depends on the difference of arguments $\xi = x_1 - x_2$ and not on the argument itself (the condition of being stationary in a narrow sense) [22] is not fulfilled here, as $P[\Delta(x)] = P[\Delta(x_1, x_2)] \neq P[\Delta(\xi)]$. The property of the error distribution of an electromagnetic meter is that both the mean value $m(x)$ and the variance $\sigma^2(x)$ depend on x. The error is non-stationary, both in a broad and a narrow sense. For example, if all the arguments of the error process (3.134) are constant, except for the ambient temperature ϑ (which is an influencing quantity), then $\{\Delta\} = \{\Delta(\vartheta)\}$. The distribution of the systematic error and of the variance of random error of a telemetric transducer type PT is shown in Fig. 3.23 [106]. The moments of the distribution of error are dependent on the argument ϑ, which means that the process is non-stationary.

For example, with constant values of the input quantities $x_1, \ldots, x_m$, the error

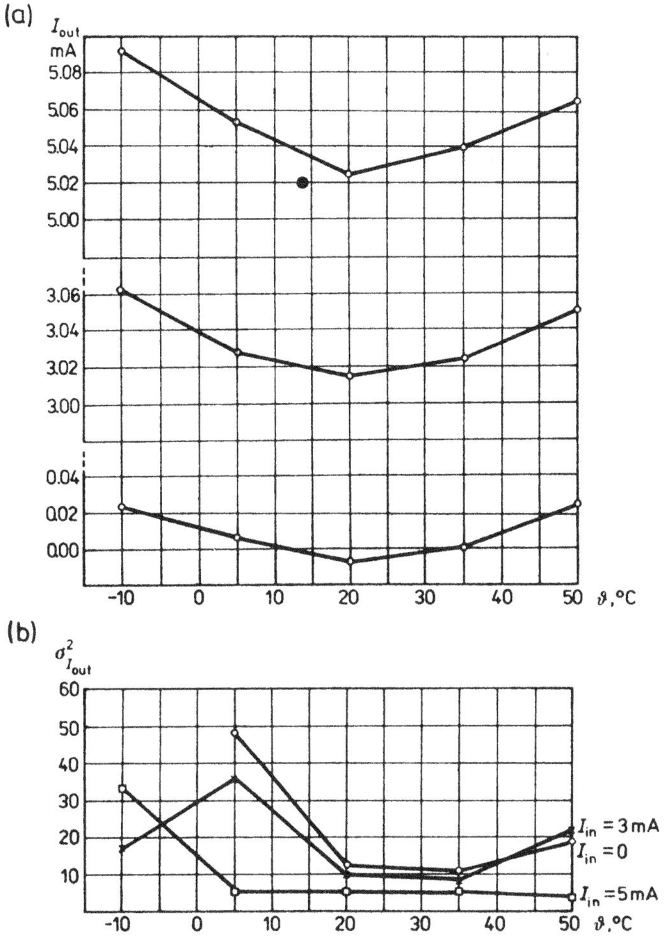

Fig. 3.23. Temperature dependence of the error of a telemetric transducer PT: (a) mean value of error, (b) variance of error.

is time-dependent, as shown in Fig. 3.24. By averaging the error realizations as in Fig. 3.24, we obtain

$$E[\Delta_k] = m(T_i, T_j) \neq m(T_j - T_i), \tag{3.135a}$$

$$\mathrm{var}[\Delta_k] = \sigma^2(T_i, T_j) \neq \sigma^2(T_j - T_i), \tag{3.135b}$$

which means that the process is non-stationary.

The general model of error is then a multidimensional, non-stationary random process (3.134). In some specific cases, however, the errors of some MS are described by a stationary process. For example, the systematic error of a magne-

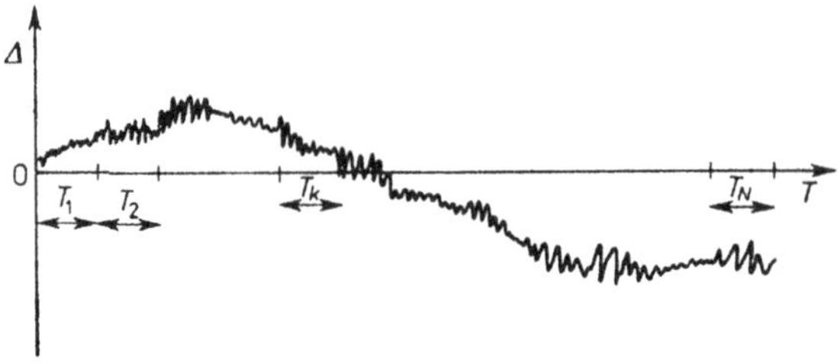

Fig. 3.24. Time variability of error.

toelectric meter may be negligibly small and the variance of the random error may be constant, so in this case the properties of this meter in reference conditions may be described by a stationary process.

In many cases, a non-stationary error process $\{\Delta\}$ may be expressed as a function of a stationary process $\{\varepsilon\}$ and the functions $h(x_1, \ldots, x_m, T)$, $g(x_1, \ldots, x_m, T)$ of measurable parameters, so that [175, 99, 180]

$$\{\Delta\} = g(x_1, \ldots, x_m, T)\{\varepsilon\}+h(x_1, \ldots, x_m, T). \tag{3.136}$$

The properties of the random process $\{\varepsilon\}$ are

$$E\{\varepsilon\} = 0, \quad E\{\varepsilon_t^2\} = 1, \tag{3.137}$$

$$E[\{\varepsilon_t\}\{\varepsilon_{t+\tau}\}] = E[\{\varepsilon_t\}\{\varepsilon_{t-\tau}\}] = \varrho(\tau). \tag{3.138}$$

When the subsequent realizations occur at time intervals greater than τ_{lim} (after that time $\varrho(\tau) = 0$), the above assumptions imply

$$\underset{k}{E}\{\Delta_k\} = \underset{k}{E}[g(\cdot)\{\varepsilon\}+h(\cdot)] = g(\cdot)\underset{k}{E}\{\varepsilon\}+h(\cdot) = h(x_1, \ldots, x_m, T), \tag{3.139}$$

and

$$\text{var}\{\Delta_k\} = \underset{k}{E}[(\{\Delta_k\}-E\{\Delta_k\})^2] = g^2(\cdot)\underset{k}{E}[\{\varepsilon\}^2] = g^2(x_1, \ldots, x_m, T). \tag{3.140}$$

The expected values are calculated in the sets of realizations.

The general model of errors (3.136) may be given theoretical grounds to some extent. Assuming that the error process is a Wiener process with independent increments, the probability density may be described by the prospective Kolmogorov equation [22, 199]

$$\frac{\partial p}{\partial t} = -m(t)\frac{\partial p}{\partial \Delta}+\frac{1}{2}\sigma^2(t)\frac{\partial^2 p}{\partial \Delta^2}. \tag{3.141}$$

The coefficients m and σ^2 correspond to the expected value and variance around the point (Δ, t), respectively. The general solution of this equation is unknown. In a particular case, the solution is a density function

$$p(t_0, t, \Delta) = \frac{1}{B(t_0, t)\sqrt{2\pi}}\exp\left\{-\frac{[\Delta-M(t_0, t)]^2}{2B^2(t_0, t)}\right\}, \tag{3.142}$$

where

$$M(t_0, t) = \int_{t_0}^{t} m(\alpha)\mathrm{d}\alpha, \tag{3.142a}$$

$$B^2(t_0, t) = \int_{t_0}^{t} \sigma^2(\alpha)\mathrm{d}\alpha, \quad \sigma^2 > 0, \tag{3.142b}$$

$$0 \leqslant t_0 < t < \infty. \tag{3.142c}$$

The increments of the process Δ are independent, because

$$B^2(t_1, t_n) = \sum_{k=1}^{n-1} B^2(t_k, t_{k+1}). \tag{3.143}$$

In relation to the model (3.136), we have $m(\alpha) = h(t)$ and $\sigma(\alpha) = g(T)$. Assuming that the probability distribution in the initial conditions is $p(\Delta, t_0)$, the distribution at point t may be calculated. Such calculations were performed by Isayev and Mechannikov [198] for normal, δ, uniform, binomial and sinusoidal distributions. The time after which the error process leaves a predetermined interval has been calculated as well. However, according to this author, the Wiener model does not give a physical interpretation of error and the condition $\sigma^2 > 0$ causes the variance of the process to increase during operation. Some cases of decreasing variance of error are known, for example in electric analog meters. These facts prove that the Wiener process may be a model only of some of the error-causing phenomena in measuring instruments.

There are no serious obstacles to either theoretical or experimental determination of the functions g and h. For example, in a magnetoelectric meter, improper balancing of a measuring element makes an additional moment arise. The moment is proportional to the mass m of the measuring element, the gravitional acceleration g, the distance R of the mass centre from the axis of the measurement element and to the sines of the respective angles φ_0, β, shown in Fig. 3.25. Hence the moment is [95, 58]

$$M_G = mgR\sin\beta\sin(\varphi+\varphi_0). \tag{3.144}$$

The moment equation $\Sigma M_i = 0$ leads to the meter model

$$\alpha = \tilde{k}J - b\sin(\varphi-\varphi_0), \tag{3.145a}$$

where J is the current, $\tilde{k}$ the sensitivity, $b = mgR\sin\beta/D =$ constant and φ stands for the angular position of the measuring element, corresponding to the deflection α. The values R, φ_0 are random realizations in a given meter and $R \in \langle 0, R_{\mathrm{all}}\rangle$, $\varphi_0 \in \langle 0, 2\pi\rangle$. If the form of the calibration procedure is (3.132)

$$J^* = c\alpha \quad \text{and} \quad c = \frac{1}{k_0},$$

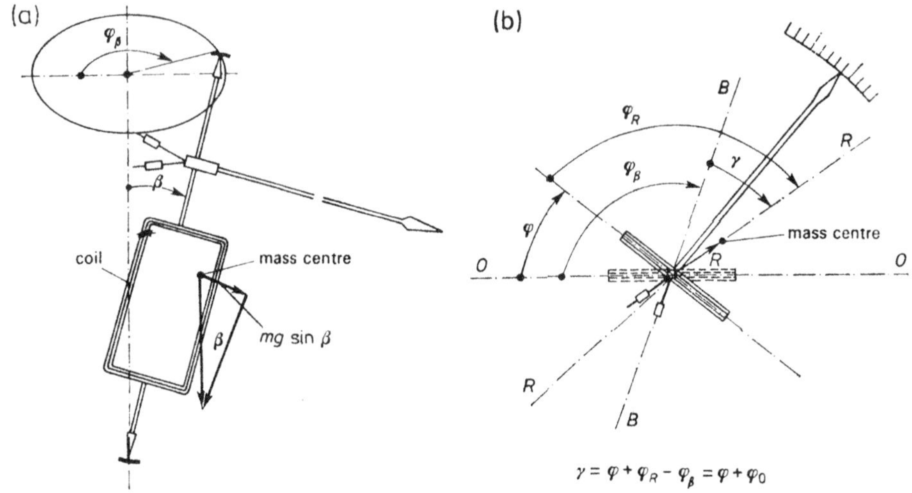

Fig. 3.25. The quantities describing the moment due to improper balance of the measuring element of an analog meter. (a) Schematic view of the measuring element, (b) view along the axis direction. *O–O*—null position of the measuring element, *R–R*—the plane including the axis and the mass centre, φ_R—the angle of the plane *R–R*, *B–B*—the plane of the deflected axis of the measuring element, φ_β—the angle of the plane *B–B*, β—deflection angle of the measuring element, $R\sin\gamma$—the arm of the turning force, $mg\sin\beta$—gravity force component, turning the measuring element.

then

$$\Delta = J^* - J = c[\tilde{k}J - b\sin(\varphi+\varphi_0)] - J = \frac{\tilde{k}-k_0}{k_0}J - \frac{b}{k_0}\sin(\varphi+\varphi_0) \quad (3.145b)$$

over the interval $\varphi = \varphi_z = \varphi_u - \varphi_1$, determined by the measurement range. Applying definitions (3.139), (3.140) to formula (3.145b), we obtain

$$h(x) = -\frac{b}{k_0}\sin(\varphi+\varphi_0) = -\frac{b}{k_0}\sin(k_1 J + \varphi_0), \quad (3.145c)$$

as $\varphi = k_1 J$ in the ideal model and

$$g^2(x) = \operatorname{var}\left[\frac{\tilde{k}-k_0}{k_0}J\right] = \left(\frac{J}{k_0}\right)^2 \sigma_{\tilde{k}}^2, \quad (3.145d)$$

when k_0, J are constants and $\tilde{k}$ is a random variable with mean value k_0 and variance $\sigma_{\tilde{k}}^2$. The distributions of $h(x)$, $x = J$ for different values of φ_0 are given in Fig. 4.16. The calibration procedure (3.132) may be selected so that the systematic error is reduced (e.g. $c = c_1 \neq 1/k$).

In some cases, an error model may be derived theoretically. For example, a source of error may be the thermal noise of a resistor, described by the Nyquist formula (3.67). In this case, $g(\vartheta)$ directly equals $2\sqrt{k\vartheta R}$ and depends on the temperature ϑ of the resistor. The variance of the noise of a diode, whose properties are shown in Fig. 3.16, is $\sigma_u^2 = 3000\ \mu V^2$.

The experimental determination of functions g and h is a complex problem. First of all, to determine g at one point, the number of realizations considered should be great enough for the variance estimator to be negligible. Let this number be n. If an MS has a property g = constant, the number of measurements may be reduced significantly. Secondly, the realization of each argument in q points gives the necessary number of points

$$N = (m+1)^q. \tag{3.146}$$

Proper selection of the points affects the cost research. Some of the measurements, e.g. physicochemical ones, require use and consumption of the standards, and the measurements at varying temperature are time-consuming. The need for simplification is then obvious.

3.4.2 The Discrete-time model

Let us first assume that the quantities $x_1, \ldots, x_m$ are constant. This makes the error process unidimensional with time as an argument. When the functions $g(T)$ and $h(T)$ are determined experimentally, the form of the general model may be [7, 2]

$$Z_t = \sum_{i=1}^{p} \varphi_i Z_{t-i} + \sum_{j=0}^{q} \gamma_j t^j + e_t, \tag{3.147}$$

where φ_i, γ_j are constant coefficients of the model, t is time, index t denotes discrete time instants, which are multiples of a constant period ΔT, $t \in \langle 1, N \rangle$, i, j are current indices, Z_{t-i} is a short notation for the value of a variable (error Δ) at the instant $t - i\Delta T$, i.e., $Z_{t-i} = Z(t - i\Delta T)$, and e_t is a stationary random process described by the equation $e_t = a_t - \theta_1 a_{t-1} - \ldots - \theta_r a_{t-r}$, where a_t is a white noise with mean value equal to zero and constant variance σ_a^2.

The variable Z_t has two components, the random Z_t' and determined Z_t'', for which

$$Z_t = Z_t' + Z_t'', \tag{3.148}$$

$$Z_t' = \sum_{i=1}^{p} \varphi_i Z_{t-i} + e_t \tag{3.148a}$$

and

$$Z_t'' = \sum_{j=0}^{q} \gamma_j t^j. \tag{3.148b}$$

The random component Z_t' is the model of random error whose properties are expressed by $g\{\varepsilon\}$. The variable Z_t'' describes the systematic error, corresponding to the function h.

The coefficients φ_i, γ_i may be determined from the experimental data (the values of Z_t over time interval $\langle 1, N\rangle$); this problem will be discussed in Subsection 4.5.1. To show some of the properties of the model, a first-order ($p = 1$ and $r = 1$) difference equation [7] will be considered:

$$Z_t' = \varphi_1 Z_{t-1}' + a_t - \theta_1 a_{t-1}. \tag{3.149}$$

The process Z_t' is stationary if the coefficients are

$$-1 < \varphi_1 < 1, \quad -1 < \theta_1 < 1. \tag{3.150}$$

After multiplying equation (3.149) by Z_t' and taking the mean value, we obtain

$$\sigma_Z^2 = \varphi_1 K_z(1) + K_{za}(0) - \theta_1 K_{za}(-1). \tag{3.151}$$

By multiplying equation (3.149) by Z_{t-1}', a_t, a_{t-1} and calculating subsequent mean values, a system of equations is obtained from which we get

$$\sigma_Z^2 = \frac{1+\theta_1^2-2\varphi_1\theta_1}{1-\varphi_1^2}\,\sigma_a^2 = K_Z(0), \tag{3.152a}$$

$$\varrho_1 = \frac{K_Z(1)}{K_Z(0)} = \frac{(1-\varphi_1\theta_1)(\varphi_1-\theta_1)}{1+\theta_1^2-2\varphi_1\theta_1}, \tag{3.152b}$$

$$\varrho_2 = \frac{K_Z(2)}{K_Z(0)} = \varphi_1 \varrho_1. \tag{3.152c}$$

The correlation coefficients may be calculated directly from the given values of Z_t' for $t = 1, \ldots, N$. Depending on the values of the coefficients in the difference equation the correlation coefficient approaches zero, either aperiodically or with oscillations, with increase of the time shift (Fig. 3.26).[7] The time series may be written as a Fourier series [7] derived from the numerical data or from the model (3.149). The power spectrum derived for the model (3.148a) is

$$S_Z(\omega) = 2\sigma_a^2 \frac{(1-\theta_1 e^{-j\omega} - \ldots - \theta_r e^{-jr\omega})^2}{(1-\varphi_1 e^{-j\omega} - \ldots - \varphi_p e^{-jp\omega})^2}, \tag{3.154}$$

where $\omega = 2\pi/N\Delta T = 2\pi/T_0$ is the minimum pulsation of the spectrum, arising from the observation time T_0 of the variable Z_t'

$$T_0 = N\Delta T. \tag{3.155}$$

[7] For continuous instability functions, the model of the correlation function is usually

$$K_Z(\tau) = \sum_{r=1}^{N} [A_r \exp(\alpha_r \tau)\cos(\omega_r \tau + \varphi_r)], \tag{3.153}$$

similar to the cases shown in Fig. 3.26.

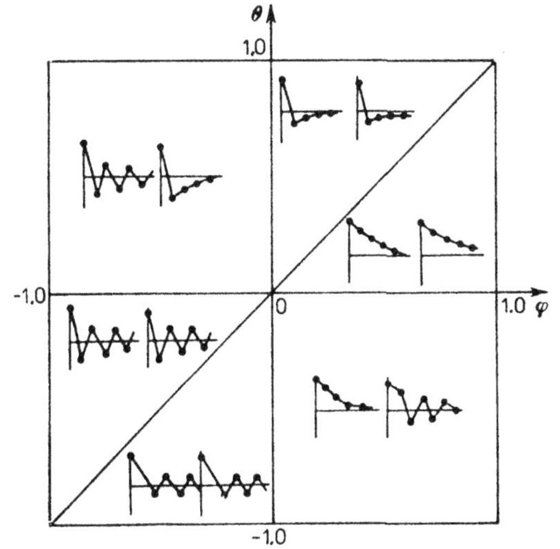

Fig. 3.26. Shape of correlation functions of time series for different values of model coefficients.

The maximum frequency of the spectrum is $\omega_{max} = N\omega/2 = \pi/\Delta T$ for even N values and $(N-1)\,\pi/N\Delta T$ for odd values of N. The constant component of the spectrum is equal to the root mean value of the variable Z_t'. For the applied number of data $N \approx 200$, the ratio $f_{max}/f_{min} = 100$. Therefore, the power spectrum includes a rather narrow range of error frequencies. Widening of the frequency range by increasing the number N is not called for because of the time-consuming and laborious calculations. For this reason, instead of extending the observation period T_0 of the time series Z_t, the variability of Z_t for $t > T_0$ is described by the determined component Z_t''. Here, a significant property is that the coefficients γ_j depend on the coefficients φ_i. This fact is an effect of the spectrum being limited by the frequency f_{min}.

3.4.3 The correlation of errors

Let us briefly summarize the properties of error as expressed by the above model. Two components of error exist. A random component is described by a random process, stationary at time T_0 and correlated in time, but for $t > \tau_{lim}$ the correlation coefficient equals zero. The determined component is described by the functions $g(T)$, $h(T)$. The random component of error referred to the scale of service time T may be treated as white noise. In the scale of measurement time t, correlation of the random component (a coloured noise case) must be considered as well as the variability of the mean value of error $h(T)$ and of the variance $g(T)$.

The time scale has double coordinates t and T, where $t \in T$. Let us now examine the error process as a bidimensional case

$$\{\Delta\} = \{\Delta(x, T)\}. \tag{3.156}$$

The properties of this process are expressed by the *correlation function*

$$\begin{aligned} &K_\Delta(\Delta_1, \Delta_2|t_1, t_2, x_1, x_2, T_1, T_2) \\ &= E(\{\Delta(x_1, T_1)-E[\Delta(x_1, T_1)]\}\{\Delta(x_2, T_2)-E[\Delta(x_2, T_2)]\}), \end{aligned} \tag{3.157}$$

where $t \in T$. The time t is the time for a single (or a few) realizations of the process $\{\Delta\}$ and T is the service time. The value of T is constant within the variability range of the time t. On the other hand, when T is variable, it should be assumed that $t \to \infty$ and that realizations of the error process are not correlated.

The presentation of the properties of correlation function (3.157) becomes a difficult task in the case of a multidimensional process $\{\Delta\}$ [99]. For this reason, only the partial characteristics will be presented here:

$$K_\Delta(\Delta_1, \Delta_2|t_1, t_2) \quad \text{when } x = \text{constant}, T = \text{constant}, \tag{3.158}$$

$$K_\Delta(\Delta_1, \Delta_2|t_1, t_2, x_1, x_2) \quad \text{when } T = \text{constant}, \tag{3.159}$$

$$K_\Delta(\Delta_1, \Delta_2|t_1, t_2, T_1, T_2) \quad \text{when } x = \text{constant}, \tag{3.160}$$

For $x =$ constant and $T =$ constant, the correlation function expresses the dependence of subsequent measurement results obtained at time instants t_1, t_2, where $t_2 - t_1 = \tau$. Then, considering assumptions (3.137), (3.138) and $h(x) =$ constant, we have

$$\begin{aligned} K_\Delta(\Delta_1, \Delta_2|t_1, t_2) &= E(\{\Delta(x, t_1)-E[\Delta(x, t_1)]\}\{\Delta(x, t_2)-E[\Delta(x, t_2)]\}) \\ &= E\{[g(x)\{\varepsilon_{t1}\}+h(x)-h(x)][g(x)\{\varepsilon_{t2}\}+h(x)-h(x)]\} \\ &= E[g^2(x)\{\varepsilon_{t1}\}\{\varepsilon_{t2}\}] = g^2(x)\varrho(\tau). \end{aligned} \tag{3.161}$$

For the realizations of different values of x we obtain

$$\begin{aligned} &K_\Delta(\Delta_1, \Delta_2|t_1, t_2, x_1, x_2) \\ &= E(\{\Delta(x_1, t_1)-E[\Delta(x_1, t_1)]\}\{\Delta(x_2, t_2)-E[\Delta(x_2, t_2)]\}) \\ &= E\{[g(x_1)\{\varepsilon_{t1}\}+h(x_1)-E\{h(x_1)\}][g(x_2)\{\varepsilon_{t2}\}+h(x_2)-E\{h(x_2)\}]\} \\ &= E(g(x_1)g(x_2)\{\varepsilon_{t1}\}\{\varepsilon_{t2}\}+h(x_1)h(x_2)+\bar{h}_x^2+g(x_1)h(x_2)\{\varepsilon_{t1}\}+ \\ &\quad +g(x_2)h(x_1)\{\varepsilon_{t2}\}-\bar{h}_x[g(x_1)\{\varepsilon_{t1}\}+g(x_2)\{\varepsilon_{t2}\}]-\bar{h}_x[h(x_1)+h(x_2)]) \\ &= \underset{x}{E}[g(x_1)g(x_2)\varrho(\tau)]+K_h(x_1, x_2) \\ &= \mu_{2g}(x_1, x_2)\varrho(\tau)+K_h(x_1, x_2). \end{aligned} \tag{3.162}$$

The notation is

$$\bar{h}_x = E[h(x_1)] = E[h(x_2)], \tag{3.162a}$$

$$K_h(x_1, x_2) = E[h(x_1)h(x_2) - \overline{h_x^2}], \tag{3.162b}$$

$$\begin{aligned} E[g(x_1)g(x_2)\{\varepsilon_{t1}\}\{\varepsilon_{t2}\}] &= \underset{x\ t}{EE}[g(x_1)g(x_2)\{\varepsilon_{t1}\}\{\varepsilon_{t2}\}] \\ &= \underset{x}{E}[g(x_1)g(x_2)]\underset{t}{E}[\{\varepsilon_{t1}\}\{\varepsilon_{t2}\}] \\ &= \mu_{2g}(x_1, x_2)\varrho(\tau). \end{aligned} \tag{3.162c}$$

Assuming that

$$\xi = x_2 - x_1, \tag{3.163a}$$

$$K_h(x_1, x_2) = K_h(\xi), \tag{3.163b}$$

$$\mu_{2g}(x_1, x_2) = \mu_{2g}(\xi), \tag{3.163c}$$

we obtain

$$K_\Delta(\Delta_1, \Delta_2 | t_1, t_2, x_1, x_2) = \mu_{2g}(\xi)\varrho(\tau) + K_h(\xi), \tag{3.164}$$

and when $\tau \to \infty$ and $\varrho(\infty) = 0$, then

$$K_\Delta = K_h(\xi). \tag{3.165}$$

Moreover,

$$K_\Delta(0) = \mu_{2g}(0) + K_h(0) = \bar{g}^2 + \sigma_h^2. \tag{3.165a}$$

The above calculations require interpretation. The quantity $K_h(\xi)$ expressed by formula (3.163b) describes the properties of the function $h(x)$, and not of the random variable. It is a so-called *function-correlation function*, the meaning of which is near to that of a multidimensional correlation function [68] or a multiple correlation function [84]. Usually $h(x)$ is determined by a regression method and the derived function $\hat{h}(x)$ is a random variable, as its estimated coefficients depend on the realized random component $g(x)\{\varepsilon\}$. The formal justification for the fact that $K_h(\xi)$ is considered a correlation function may be the form of formula (4.134) derived based on the constraints applied to determine coefficients of $\hat{h}(x)$. The fact that $h(x)$ is random will become obvious if a set of MS of the same type is considered and each MS is treated as an error realization described by an unknown function $h(x)$. The second-order moment (3.163c) of the function $g(x)$ is treated similarly. For example, the function $h(x)$ in the range $Z = x_u - x_l$ is given by the formula

$$h(x) = a_0 + a_1 x. \tag{3.166}$$

Then

$$K_h(\xi) = \tfrac{1}{12} a_1^2 Z^2 = \text{constant}, \tag{3.167}$$

which means that $\varrho(\xi) = 1$. When

$$h(x) = \sin x, \quad \text{then} \quad K_h(\xi) = \cos\xi. \tag{3.168}$$

Generally, $h(x)$ may be expanded in a Fourier series, accepting the interval $\langle x_l, x_u\rangle$ as the period of the fundamental harmonic, i.e. $\omega = 2\pi/(x_u - x_l)$. This series is an approximation of the function $h(x)$:

$$h(x) \approx a_0 + 2\sum_{n=1}^{\infty}(a_n \cos n\omega x + b_n \sin n\omega x). \tag{3.169}$$

Because of the orthogonal properties of the terms of series (3.169), the correlation function calculated for the series is in the form

$$K_h(\xi) = 4\sum_{n=1}^{N}(a_n^2 + b_n^2)\cos n\omega\xi. \tag{3.170}$$

A property of this function is

$$K_h(\xi) \leqslant K_h(0) = \sigma_h^2. \tag{3.171}$$

Final formula (3.164), expressing the correlation of individual realizations of errors, may be interpreted as follows. The particular realizations of errors Δ_j, $j = 1, 2, \ldots, k$, are treated as random variables. If one of the realizations, e.g. Δ_1, is linearly dependent on the others, that is, if

$$\Delta_1 = \sum_{j=2}^{k} \alpha_j \Delta_j, \qquad \alpha_j = \text{constant}, \tag{3.172}$$

then the multiple correlation coefficient will be [84]

$$\varrho_m = 1 - \frac{[\Delta_1 - \sum_{j=1}^{k} \alpha_j \Delta_j]^2}{\operatorname{var}\Delta_1} = 1. \tag{3.173}$$

The relation $\varrho_m < 1$ means that the dependence of Δ_1 on the other variables is stochastic. The value of ϱ_m is a measure of this relationship. As the individual realizations of errors are spaced at the intervals $\xi = x_j - x_i$ and at time intervals $\tau = t_j - t_i$, the correlation function (3.164) is therefore the measure of their interdependence. A large value of time interval spacing implies $\varrho(\tau) \to 0$ and the interdependence is determined by $K_h(\xi)$, because the first component of (3.164) is zero and

$$\varrho = \frac{K_h(\xi)}{\operatorname{var}[\Delta(x, t)]} < 1 \quad \text{when} \quad g \neq 0. \tag{3.174}$$

The distance spacing affects the value of ϱ in different ways, depending on the properties of the function $K_h(\xi)$ (3.170). In the worst case (a linear dependence of $h(x)$), the degree of correlation is constant and independent of ξ. Thus

$$\varrho_h = \frac{\operatorname{var}[h(x)]}{\operatorname{var}[g(x)\{\varepsilon\}] + \operatorname{var}[h(x)]} < 1. \tag{3.175}$$

The coefficient ϱ is a measure of the part of the systematic component in the total error. A small time interval between the realizations of errors leads to

$$|\varrho(t)| > 0, \quad \mu_{2g}(\xi) > 0, \tag{3.176}$$

and the degree of correlation of subsequent error realizations increases and reaches the value

$$\varrho_{ht} = \frac{g^2(x)\varrho(t)+\mathrm{var}[h(x)]}{\mathrm{var}[g(x)\{\varepsilon\}]+\mathrm{var}[h(x)]} > \varrho_h . \tag{3.177}$$

In the case when the argument of the process $(\Delta(x_1, \ldots, x_m, T))\}$ other than $x = x_m$ varies, and the values of the rest are constant, the error correlation may be determined similarly. The coordinate x may be any of $x_1, \ldots, x_m, T$.

The problem of a bidimensional process $\{\Delta(x, T, t)\}$ needs a different approach. As the service time T causes permanent changes in the properties of an MS, and as the investigations (from which N realizations of the process $\{\Delta\}$ were obtained) cannot be repeated, it is assumed that the value of T_j is constant. For each interval $T_j \in T$, the properties of the process $\{\Delta(x, T)\}$ may be determined. The characteristics $g(x)$, $h(x)$, $K_h(\xi)$, etc. written in an analytical form as $f(x, t, \theta_1, \ldots, \theta_q)$ may vary in the subsequent realizations of T_j. By determining the relationship

$$\theta_1(T), \ldots, \theta_q(T) \tag{3.178}$$

this variability may be included, and the properties of the process may be described by characteristics (3.138)–(3.140), (3.164) and by the function

$$f[x, t, \theta_1(T), \ldots, \theta_q(T)]. \tag{3.179}$$

The characteristics of n-dimensional (when $n \geqslant 3$) non-stationary processes may be derived from definitions (3.138)–(3.140), using the same concepts. The number of experiments must be increased adequately, or the model (3.136) must be simplified, for example by assuming g = constant.

3.5 Indentification of Models of Measurement Systems

A mathematical model of an MS may be derived analytically from chemical or physical laws, or may be determined experimentally. In the first case, a so-called *physical* model is obtained, while in the second an MS is considered a "black box", and the model obtained is *statistical.*

A physical model may be determined with different accuracy levels:

(a) the level of independent variables, i.e., of the arguments of the function F (3.8);

(b) the level of coefficients; in the determined function F (3.8), only the values of coefficients b_l (3.72) remain unknown;

(c) the level of the values of coefficients.

When a model as in (c) is obtained, the task of model formation is completed. The model may then be verified experimentally, but this problem remains beyond the field of interest of identification.

The identification of mathematical models is an extensive branch of the theory of automatic control [67, 18]. Because the knowledge of physical phenomena and of the properties of elements is much greater in metrology, a number of identification methods specific for case (a) are not applicable here. Also the purpose of identification and the great accuracy required in metrology make the metrological approach quite different, so that only some methods are suitable. For example, using experimental design as a statistical method is hardly applicable here. In many cases, the range of the identification in metrology, because of the great accuracy required, is like that of fundamental studies in physics. An example may be the studies on the Josephson effect, aimed at applying this phenomenon as a voltage standard.

3.5.1 Determination of the variables of a model

Determination of the variables of a model is not carried out when the characteristics are being determined but is performed in the process of construction of an instrument and when the sources of errors are being investigated. When a physical model cannot be derived, experimental methods must be applied. At a preliminary stage, the hypothesis is accepted that a given construction parameter, an effect or a quantity, is the source of error. Next, an experiment is made in which the hypothesis is tested, and upheld or demolished.

In cases of big projects (in automatic control, in industry and economics), such an experiment is costly or even impossible. Thus, various different heuristic methods are applied (e.g. the Delphic method, the PATTERN method, etc.) which allow limiting of the experiment range.

A common feature of these methods is the use of an heuristic approach. Different kinds of formal treatment are used simultaneously to decrease the probability of erroneous estimation. The aim is to provide a result characterized by objectivity. Of course, making use of the knowledge of many people often improves the result; nevertheless the effect is not objective but an indication of the best choice of direction of further research.

We now look at an example of verification of a hypothesis, although the way in which a working hypothesis was formulated will not be presented. The research by A. Jędrzejewski [43] aimed to identify the sources of errors of a lever-and-tooth

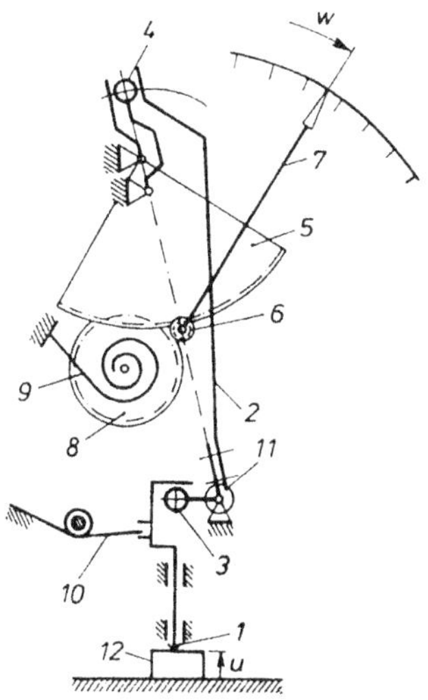

Fig. 3.27. The kinematic diagram of a lever-and-tooth transducer [43]. *1*—plunger tip, *2*—input lever, *3*—intermediate ball, *4*—linearity compensator shaft, *5*—toothed sector, *6*—pinion, *7*—pointer, *8*—gear of the hairspring, *9*, *10*—push springs, *11*—lever bearing, *12*—measured object.

mechanism. The kinematic diagram of the investigated transducer is shown in Fig. 3.27.

The measured dimension u is transferred via the plunger tip onto the measuring organ of the transducer. The position of the tip *1* corresponds to the measured quantity u (dimension of the investigated element). The tip position is transferred by the set of levers *2–5* to pinion *6* activating the pointer. The output quantity w is determined by the pointer position on the scale. The transducer also includes a linearity compensator *4* and spring *10* and hairspring *9* acting as energy accumulators.

The assumed sources of errors were [43]:[8]

(1) strain and slip effects in the process of transferring the position of the tip *1* onto the lever *2*. The characteristic parameter of these sources of errors is the dimension of the ball *3*;

(2) strain of lever *2*—the type of material the lever is made of (its rigidity, mass, etc.) has been included as a parameter;

(3) strain and slip effects in the process of transferring the position by the compensation shaft—the parameter is the diameter of the shaft;

(4) backlash in bearing *11* of input lever *2*;

---

[8] 10 parameters were considered by Jędrzejewski. Here the analysis is simplified.

(5) friction in the pinion bearing, depending on the pinion axle diameter;

(6) energy losses in the transducer; the losses are compensated by the offset tension of hairspring *9*;

(7) friction in the pinion bearing—the parameter is the material of the pinion bearing bush.

The repeatability error of the lever-and-tooth mechanism is a complex metrological property. The error is described by a bimodal probability distribution, arising from two superimposed normal distributions. The first of the distributions characterizes the errors appearing when the measured quantity increases and the second the errors appearing when the measured quantity decreases. The shift of the distributions, shown in Fig. 3.28, is due to the backlash, or hysteresis. The parameters of the distribution are: σ_1, σ_2—the standard deviations of the component distributions, h—measurement hysteresis.

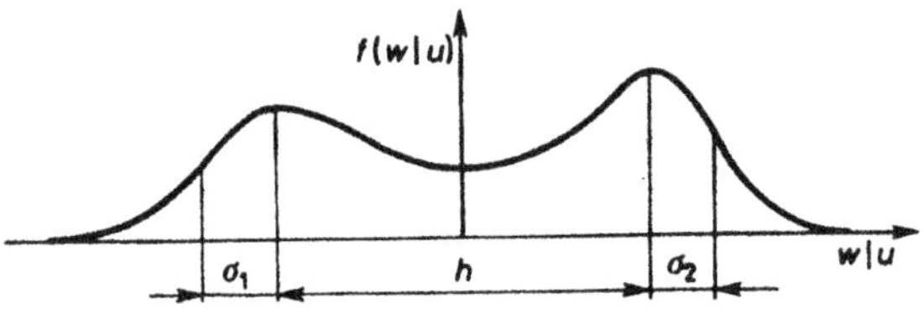

Fig. 3.28. Probability distribution of an error of the lever-and-tooth gauge.

The required model is multidimensional: the number of input quantities is 7, and the number of outputs is 3. The hypothesis was to be proved by experimental studies in which the value of every construction parameter was changed within predetermined limits. The result is the trial space shown in Fig. 3.29. The image of the results is shown in the space coordinates s_1, h_1, s_2, where s_1, s_2 are the estimators of σ_1, σ_2. The investigated model is static:

$$s_1 = F_1(a_1, a_2, \ldots, a_7), \tag{3.180a}$$

$$s_2 = F_2(a_1, a_2, \ldots, a_7). \tag{3.180b}$$

$$h = F_3(a_1, \ldots, a_7). \tag{3.180c}$$

The hypothesis to be verified is that s_1, s_2, h depend on the construction parameters $a_1, \ldots, a_7$, listed above.[9]

[9] Any other parameter describing the repeatability error, e.g. the range, limiting error, variance, etc., may be accepted as a dependent variable. A. Jędrzejewski [43] also applies the information error according to Novitskii's definition [78].

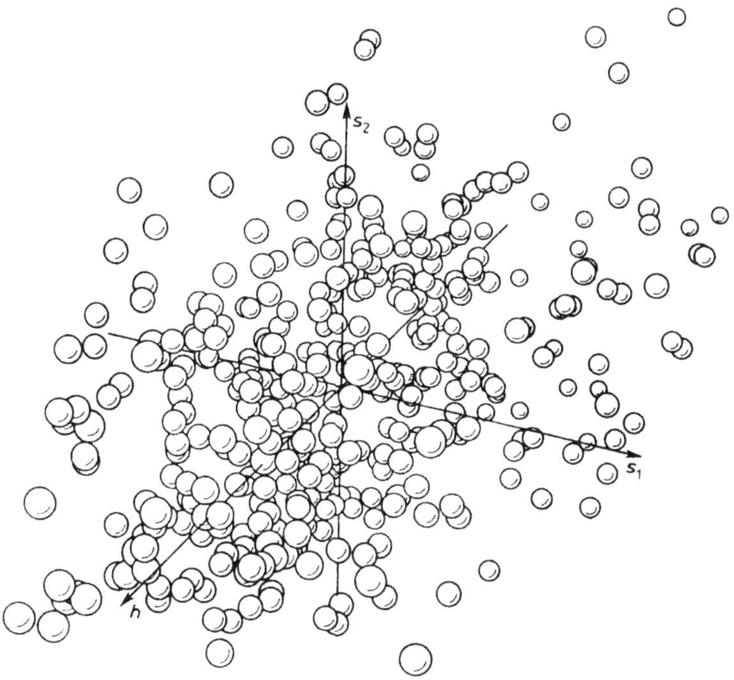

Fig. 3.29. The results of the error measurements of the lever-and-tooth gauge [43].

3.5.2 Correlation Analysis

The *cross-correlation function* is used to test the relations of random variables. The normalized cross-correlation function of the variables U and V with continuous realizations $u(t)$, $v(t)$, $t \in T$, is given by the formula

$$\varrho_{uv}(t_1, t_2) = \frac{R_{uv}(t_1, t_2)}{\sqrt{K_u(t_1, t_2) K_v(t_1, t_2)}}, \tag{3.181}$$

where

$$R_{uv} = E\{[U(t_1) - E[U(t_1)]][V(t_2) - E[V(t_2)]]\} \tag{3.182}$$

is the cross-correlation function and

$$K_u(t_1, t_2) = E\{[U(t_1) - E[U(t_1)]][U(t_2) - E[U(t_2)]]\} \tag{3.183}$$

and $K_v(t_1, t_2)$ are the respective correlation functions of the variables U, V. The property of the random variables stationary in the narrow sense is that the correlation function depends on the distance apart of arguments t_1, t_2 only, i.e., on $\tau = t_2 - t_1$. Thus

$$\varrho_{uv}(t_1, t_2) = \varrho_{uv}(\tau) = \frac{R_{uv}(\tau)}{\sqrt{K_u(\tau) K_v(\tau)}}. \tag{3.184}$$

The values of the normalized cross-correlation function may vary from -1 to $+1$. If the variables are independent, then $\varrho_{uv}(\tau) = 0$. On the other hand, the property $\varrho_{uv}(\tau) = 0$ means that the variables U and V are not correlated (but not necessarily independent). Non-correlated random variables are independent only if their distributions are normal [122]. The value of ϱ being equal to ± 1 means that the relationship is linear, namely:

$$\text{if} \quad \varrho_{uv}(0) = 1, \quad \text{then} \quad U = kV + a, \tag{3.185a}$$

$$\text{if} \quad \varrho_{uv}(\tau_1) = 1, \quad \text{then} \quad U(t+\tau_1) = kV(t) + a, \tag{3.185b}$$

where k and a are constants. If $\varrho = 1$, then $k > 0$; $\varrho = -1$ means that $k < 0$.

The value of the standardized correlation function gives information both on the order and on the kind of relationship. For example, let

$$V = U + Z, \tag{3.186}$$

where Z is an independent random variable with the properties of white noise: $E(Z) = 0$, $E(Z-\bar{Z})^2 = \sigma_z^2$. Then, by substituting (3.186) in (3.182), we obtain

$$R_{uv}(\tau) = K_u(\tau) + K_{uz}(\tau) = K_u(\tau). \tag{3.187}$$

Thus, the correlation coefficient is

$$\varrho_{vu}^2(\tau) = \frac{[K_u(\tau)]^2}{\sigma_u^2 \sigma_v^2} = \varrho_u^2(\tau) \frac{\sigma_u^2}{\sigma_u^2 + \sigma_z^2}. \tag{3.188}$$

The larger the contribution of stochastic phenomena in the relationship $V = f(U)$ (expressed by the random variable Z), the smaller the correlation coefficient. The measure of the contribution is the variances ratio. If the noise significantly exceeds the variations of the variable U, then $\varrho_{vu}^2 \approx \sigma_u^2/\sigma_z^2$, which means that the correlation coefficient approaches zero.

For a simple sample of n elements of each of the variables u_i, v_i, ordered by the index j, the correlation coefficient is

$$R_{uv} = \frac{\sum_{j=1}^{n} (u_j - u_{av})(v_j - v_{av})}{n s_v s_u}, \tag{3.189}$$

where u_{av}, v_{av} are the mean values of the sample:

$$u_{av} = \frac{1}{n} \sum_{j=1}^{n} u_j \quad \text{and} \quad v_{av} = \frac{1}{n} \sum_{j=1}^{n} v_j, \tag{3.190}$$

and the respective sample variances are

$$s_u^2 = \frac{1}{n} \sum_{j=1}^{n} (u_j - u_{av})^2 \quad \text{and} \quad s_v^2 = \frac{1}{n} \sum_{j=1}^{n} (v_j - v_{av})^2. \tag{3.191}$$

When the distributions of the variables U, V are normal, the probability distribution of the sample correlation coefficient may be calculated [22]. The probability density is given by the formula

$$p(r) = \frac{n-2}{\pi}(1-\varrho^2)^{(n-1)/2}(1-r^2)^{(n-4)/2}\int_0^1 \frac{x^{n-2}}{(1-\varrho r x)^{n-1}\sqrt{1-x^2}}\,\mathrm{d}x, \quad (3.192)$$

and the density function varies with the value of ϱ. The quantity r is the value of the random variable R and n is the number of elements in the sample.

For a hypothetical assumption $\varrho = 0$ (the variables are independent), the formula is [22]

$$p(r) = \frac{n-2}{\pi}(1-r^2)^{(n-4)/2}\int_0^1 \frac{x^{n-2}}{\sqrt{1-x^2}}\,\mathrm{d}x = \frac{1}{B[\frac{1}{2}(n-2), \frac{1}{2}]}(1-r^2)^{(n-4)/2}. \quad (3.193)$$

In the formula, $B[\cdot]$ is the beta function. If the distribution is known, the confidence interval may be calculated for the coefficient R derived from the sample. It is convenient to apply a property of this distribution that after substitution of variables

$$t = \frac{R}{\sqrt{1-R^2}}\sqrt{n-2}, \quad (3.194)$$

the random variable t has a Student's distribution with $n-2$ degrees of freedom. The influence of the number of elements in the sample on the confidence of the calculated value of R may be found in the tables of Student's distribution. The

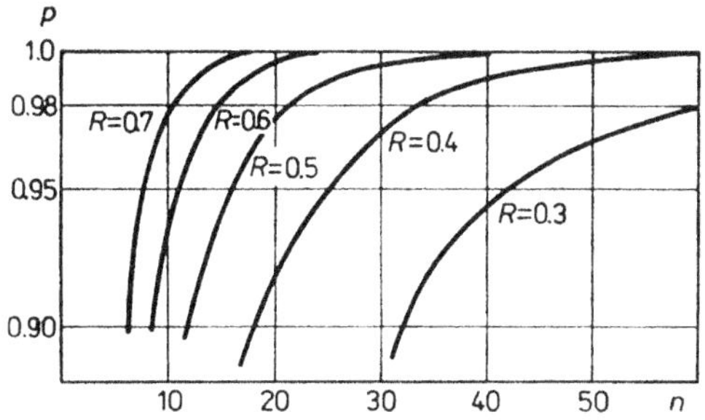

Fig. 3.30. The influence of the number of elements of a sample on the probability of the correlation coefficient for a hypothetical value $\varrho = 0$.

relationship is shown in Fig. 3.30. If the value of $R(n)$ falls into the area of $P < 0.5$, the hypothesis $\varrho = 0$ will be more probable than the opposite one. To ensure that the value of R derived from the sample is certain, in the statistical sense, the sample size should be fairly large.

In the studies of interrelations of phenomena, the relationships are often multidimensional. The measure of the dependence is either the matrix of correlation coefficients

$$R = \begin{bmatrix} 1 & \varrho_{12} & \cdots & \varrho_{1m} \\ \varrho_{21} & 1 & \cdots & \varrho_{2m} \\ \cdots & \cdots & \cdots & \cdots \\ \varrho_{m1} & \varrho_{m2} & \cdots & 1 \end{bmatrix}, \tag{3.195}$$

or the *covariance matrix*

$$M = \begin{bmatrix} \lambda_{11} & \lambda_{12} & \cdots & \lambda_{1m} \\ \lambda_{21} & \lambda_{22} & \cdots & \lambda_{2m} \\ \cdots & \cdots & \cdots & \cdots \\ \lambda_{m1} & \lambda_{m2} & \cdots & \lambda_{mm} \end{bmatrix} \tag{3.196}$$

of m random variables. At the same time, the relations

$$\lambda_{ii} = \operatorname{var}(u_i) = \sigma_i^2, \qquad k, i = 1, 2, \ldots, m. \tag{3.197a}$$

$$\lambda_{ik} = \operatorname{cov}(u_i u_k) = \varrho_{ik}\sigma_i\sigma_k, \qquad i \neq k, \tag{3.197b}$$

are satisfied, i.e., the matrices $\boldsymbol{M}$ and $\boldsymbol{R}$ are equal when the variables U_i are standardized. Both matrices are symmetric and positively determined if no linear relationship of the variables occurs. Since the determinant of the matrix $\boldsymbol{M}$ is

$$\det \boldsymbol{M} = \sigma_1^2\sigma_2^2 \ldots \sigma_m^2 \det \boldsymbol{R} \leqslant \sigma_1^2\sigma_2^2 \ldots \sigma_m^2, \tag{3.198}$$

the greater the variances of the variables, the greater is $\det\boldsymbol{M}$ and the greater is the dispersion of the set of variables in an m-dimensional space. On the other hand, the determinant of $\boldsymbol{R}$ depends on the correlation of the variables. When there is no correlation, the elements $\varrho_{ik} = 0$ for $i \neq k$ and $\det\boldsymbol{R} = 1$. The value $\det\boldsymbol{R} = 0$ means that at least two variables are linearly dependent and the matrix $\boldsymbol{R}$ is singular.

In an m-dimensional space, the matrix $\boldsymbol{R}$ is described by its eigenvector (unit) which generates the axes of an ellipsoid in which the elements of the set of events should be spaced with a given probability. This is illustrated in Fig. 3.31 (for $m = 2$). The major axis of the ellipse forms an angle ϑ with the horizontal axis of the ellipse; then [5]

$$\operatorname{tg} 2\vartheta = \frac{2\varrho\sigma_x\sigma_y}{\sigma_x^2 - \sigma_y^2}. \tag{3.199}$$

In the case of a bidimensional normal distribution, the ellipse equation is [5]

$$\left(\frac{x-\bar{x}}{\sigma_x}\right)^2 - 2\varrho\frac{x-\bar{x}}{\sigma_x}\frac{y-\bar{y}}{\sigma_y} + \left(\frac{y-\bar{y}}{\sigma_y}\right)^2 = 2(1-\varrho^2)k, \tag{3.200}$$

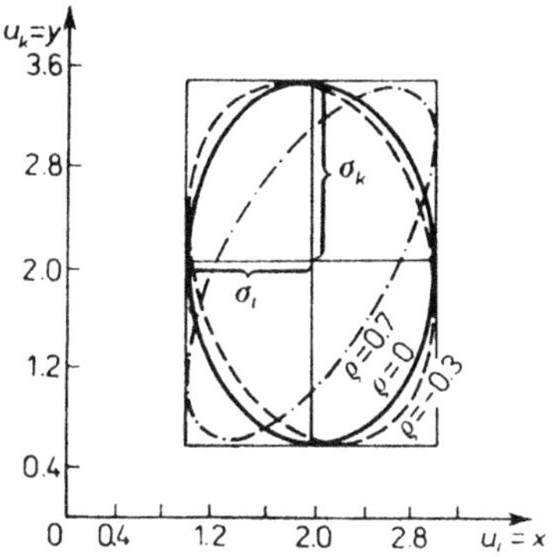

Fig. 3.31. The area of equal probability of realization of elements in a multidimensional space.

where $k = -\ln(2C\pi\sigma_x\sigma_y\sqrt{1-\varrho^2})$ and $C = p(x, y)$, according to the primary assumption. The confidence level over the interval included in the ellipse is

$$P = 1-e^{-k} = \alpha. \tag{3.201}$$

In general, the major components are derived from the equation

$$(\boldsymbol{R}-\eta\boldsymbol{I})\boldsymbol{v} = 0, \tag{3.202}$$

in which $\eta_1, \ldots, \eta_m$ are the eigenvalues of equation (3.202), $\boldsymbol{v}_1, \ldots, \boldsymbol{v}_m$ are the eigenvectors, $\boldsymbol{I}$ is a unit matrix and $\boldsymbol{R}$ is the matrix of correlation coefficients. The values of the major components are

$$\boldsymbol{U}_i = \boldsymbol{v}_i\boldsymbol{W}, \tag{3.203}$$

where $\boldsymbol{W}$ is the vector of primary variables. Also, the relationship

$$\operatorname{var}\det\boldsymbol{U}_i = \eta_i \quad \text{for} \quad i = 1, 2, \ldots, m \tag{3.204}$$

is satisfied. The major components are the new orthogonal and uncorrelated variables, determined as a linear combination of those primary variables which transfer the greatest part of the variance of the experiment. The major components are ordered by proper numbering of the eigenvalues η_i, so that $\eta_1 \geqslant \eta_2 \geqslant \ldots \geqslant \eta_m$. Thus the subsequent major components transfer the greatest variance of the experiment, and allow a reduction of the number of components by neglecting those whose part in the variance is sufficiently small. The eigenvector determines the orientation of the variable in the space of events and the eigenvalue of the vector decides its quantitative part. Let us look again at the example of the analysis of errors of the lever-and-tooth mechanism (by Jedrzejewski [43]). From the experimental results shown in Fig. 3.29, those points were selected for which the influence of a selected construction parameter on the repeatability error had been investigated. The repeatability error in this case is characterized by the hysteresis h and by the standard deviations s_1, s_2 of the ranges of the bimodal distribution.

The backlash in the bearing of the first lever, as an error source, was characterized by the value of backlash varying over the range 6–50 μm. The matrix of the resulting correlation coefficients of random variables h, s_1, s_2 was

$$\boldsymbol{R}_{ww} = \begin{bmatrix} 1 & -0.110 & -0.086 \\ -0.110 & 1 & 0.226 \\ -0.086 & 0.226 & 1 \end{bmatrix}. \tag{3.205}$$

The values of the subsequent eigenvectors were

$$\boldsymbol{v}_1 = 0.429h + 0.649s_1 + 0.629s_2, \tag{3.206a}$$

$$\boldsymbol{v}_2 = 0.899h + 0.235s_1 + 0.370s_2, \tag{3.206b}$$

$$\boldsymbol{v}_3 = 0.092h + 0.724s_1 - 0.684s_2, \tag{3.206c}$$

with eigenvalues η_i 1.2919, 0.9360, 0.7724, respectively.

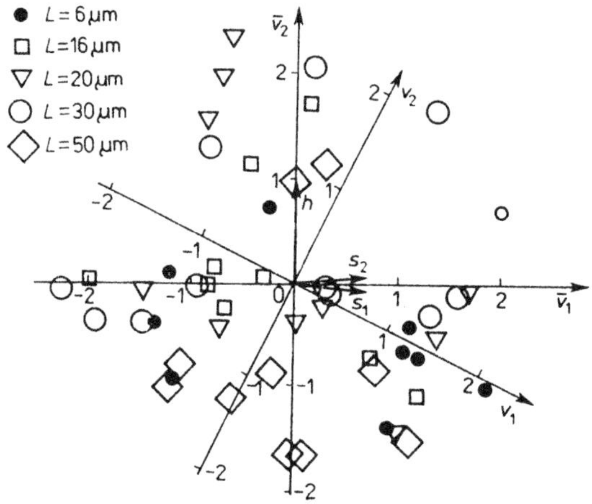

Fig. 3.32. A sample image for a varying backlash L in the bearing [43]; v_1, v_2 are the projections on the primary coordinate system.

The sample image in the coordinates of the first two vectors is shown in Fig. 3.32. It can be clearly seen that the variable v_1 may be identified with the variance of the repeatability error, while v_2 may be identified with hysteresis. The projections of the unit vectors s_1, s_2 are identical, which means here that the error distribution is symmetrical. The image presented shows that the variation of backlash gives a random distribution of experimental results. Therefore, the repeatability error is independent of the backlash in the bearing *11* of the first

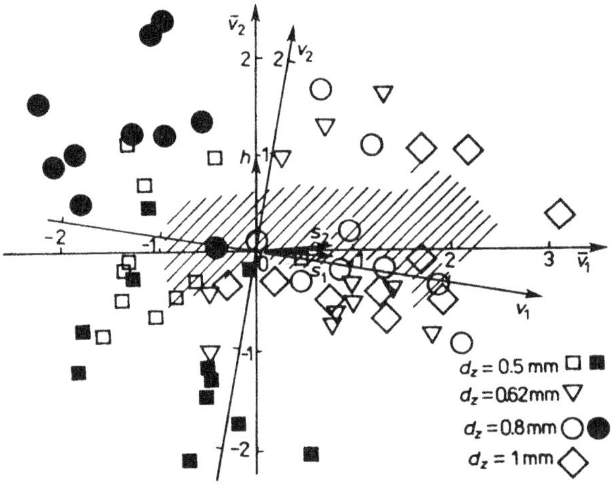

Fig. 3.33. A sample image for the variable pinion axle diameter d_z [43].

lever (see Fig. 3.27). In Fig. 3.33 a similar sample image is shown for the variable diameter of the pinion of the transducer. The diameter change is here a construction parameter, characterizing the errors due to friction in the bearing. Here, the causal connection is clear. A smaller diameter gives a smaller value of dispersion and of hysteresis of the repeatability error. The shaded arrow in the figure shows the direction of the relationship.

The correlation function is used for identification of dynamic properties. The output signal of a linear dynamic element described by the Volterra operator (3.26), which is driven by a stationary signal $x_1(t)$ with correlation function $K_x(\tau)$, is

$$y(t) = \int_0^t k(t-\tau)x(\tau)\mathrm{d}\tau, \tag{3.207}$$

and the correlation function of the output signal is

$$\begin{aligned} K_y(t_1, t_2) &= E\Big\{\Big[\int_0^{t_1} k(t_1-\tau)x(\tau)\mathrm{d}\tau - \bar{y}(t_1)\Big]\Big[\int_0^{t_2} k(t_2-\tau)x(\tau)\mathrm{d}\tau - \bar{y}(t_2)\Big]\Big\} \\ &= \int_0^{t_1}\int_0^{t_2} k(t_1-\tau_1)k(t_2-\tau_2)E[x(\tau_1)x(\tau_2)]\mathrm{d}\tau_1\mathrm{d}\tau_2 + \bar{y}(t_1)\bar{y}(t_2) \\ &= \int_0^{t_1}\int_0^{t_2} k(t_1-\tau_1)k(t_2-\tau_2)K_x(\tau_1, \tau_2)\mathrm{d}\tau_1\mathrm{d}\tau_2. \end{aligned} \tag{3.208}$$

For a first-order inertial element, the kernel of the operator is

$$k(t-\tau) = e^{\lambda(t-\tau)}. \tag{3.209}$$

If

$$K_x(\tau_1, \tau_2) = e^{-a(\tau_2-\tau_1)}, \tag{3.210}$$

then

$$K_y(t_1, t_2) = \frac{1}{\lambda^2-a^2}\,[1-e^{-(\lambda-a)t_1}][1-e^{-(\lambda+a)t_2}], \tag{3.211}$$

or, after substituting $t_1 = t$, $t_2 = t_1+\tau = t+\tau$,

$$K_y(t, \tau) = \frac{1}{\lambda^2-a^2}\,[1-e^{-(\lambda-a)t}][1-e^{-(\lambda+a)t}e^{-(\lambda+a)\tau}]. \tag{3.212}$$

The cross-correlation function for $t_2 > t_1$ may be calculated as

$$\begin{aligned} R_{yx}(t_1, t_2) &= E\left\{\left[\int_0^{t_2} k(t_2-\tau)x(\tau)\mathrm{d}\tau-\bar{y}(t_2)\right][x(t_1)-\bar{x}(t_1)]\right\} \\ &= \int_0^{t_2} k(t_2-\tau)K_x(t_1, t_2)\mathrm{d}\tau-\bar{x}(t_1)\bar{y}(t_2). \end{aligned} \tag{3.213}$$

For the given (3.209), (3.210),

$$R_{yx}(t, \tau) = \int_0^t e^{\lambda(t-\tau)}e^{-a\tau}\mathrm{d}\tau = \frac{1}{\lambda+a}\,e^{\lambda t}[1-e^{-(\lambda+a)t}]. \tag{3.214}$$

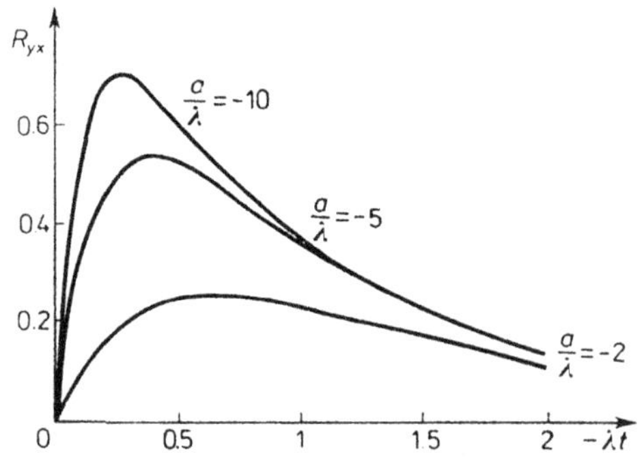

Fig. 3.34. Cross-correlation function of a first-order inertial element.

This function, as shown in Fig. 3.34, has a maximum at the point

$$t_M = \frac{1}{a+\lambda}\ln\frac{a}{\lambda}, \tag{3.215}$$

and the maximum value of the normalized correlation function (3.184) is

$$R_{yx}(t_M) = \frac{a-\lambda}{a}\sqrt{\frac{a-\lambda}{a+\lambda}}\left(\frac{a}{\lambda}\right)^{\lambda/(a+\lambda)}. \tag{3.216}$$

There are some particular cases when the model can be derived in a relatively simple way. The form of the model should then be known a priori and the test signal should be selected appropriately. The form of the cross-correlation function is specific when the test signal is the white noise

$$K_x(\tau) = \sigma_x^2\,\delta(\tau). \tag{3.217}$$

Then

$$R_{yx}(\tau) = E\left\{\left[\int_0^\tau k(t,\tau)\,\delta(t)\sigma_x^2\,\mathrm{d}t\right]\right\} = \sigma_x^2 E[k(\tau)\,\delta(\tau)] = \sigma_x^2 k(\tau), \tag{3.218}$$

i.e., the cross-correlation function is identical with the response of the element to a pulse excitation. This is the property most often applied in the identification of dynamic elements by the stochastic-signal method. The diagram of an identification system with additive noise taken into account is shown in Fig. 3.35. The signal

$$u(t) = y(t)+z(t) \tag{3.219}$$

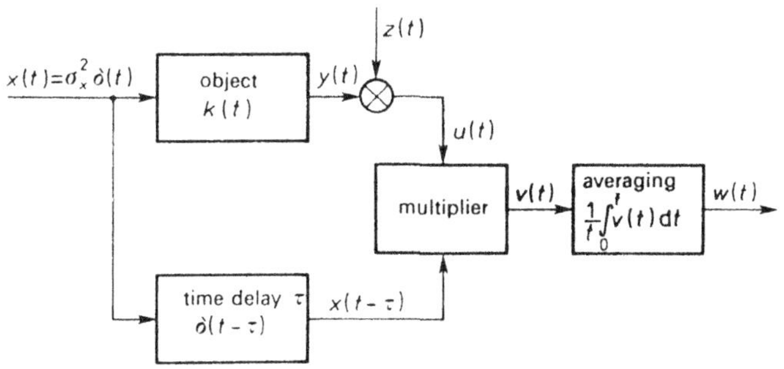

Fig. 3.35. Diagram for determination of an element response to pulse excitation.

is multiplied by the signal $x(t)$ having properties (3.217), delayed for time τ. The processed signal is averaged over the time interval t:

$$w(t) = \frac{1}{t}\int_0^t v(t)\,\mathrm{d}t = \frac{1}{t}\int_0^t u(\tau)x(t-\tau)\,\mathrm{d}\tau = R_{ux}(\tau). \tag{3.220}$$

Since

$$k(\tau) = \frac{1}{\sigma_x^2} R_{yx}(\tau) \neq R_{ux}(\tau), \tag{3.221}$$

the output signal $w(t)$ therefore includes an error. The correlation function of the output signal is

$$K_w(\tau) = E\{[y(t)+z(t)]x(t)[y(t+\tau)+z(t+\tau)]x(t+\tau)\}. \quad (3.222)$$

Considering that $x(t)$ is the white noise (3.217) with variance σ_x^2 and expressing the fourth-order moment by the second-order moments [122, 113],

$$E(x_a x_b x_c x_d) = E(x_a x_b)E(x_c x_d)+E(x_a x_c)E(x_b x_d)+E(x_a x_d)E(x_b x_c), \quad (3.223)$$

an operation which is allowed for variables with normal distribution, we obtain [114]

$$K_w(\tau) = \sigma_x^4[k(\tau)]^2+K_e(\tau) = R_{yx}(\tau)+K_e(\tau), \quad (3.224)$$

where

$$K_e(\tau) = \sigma_x^2\,\delta(\tau)K_z(\tau)+\sigma_x^4 k(t+\tau)k(t-\tau)+\sigma_x^4\,\delta(t)\int_0^\infty k(t)k(t+\tau)\mathrm{d}t. \quad (3.225)$$

Here $K_e(\tau)$ is the correlation function of the error of correlation function measurement, since the first component in formula (3.224) corresponds to formula (3.221). The second and third components in (3.225) may be limited by proper selection of the test signal [114, 62]. The first component of the error arises from the noise signal $z(t)$, with mean value zero. When the signal $w(t)$ is averaged over a time interval t (3.220), the variance of error diminishes t times, i.e.,

$$\mathrm{var}(\Delta_z) = \frac{\sigma_x^2 K_z(0)}{t}, \quad (3.226)$$

and, considering the instrument scaling (3.221) or equation (3.224), we have

$$\mathrm{var}[k(\tau)] = \frac{\mathrm{var}(\Delta_z)}{\sigma_x^4} = \frac{K_z(0)}{\sigma_x^2 t}. \quad (3.227)$$

For a given value of error variance, var $[k(t)]$, the minimum averaging time may be calculated as [114]

$$t_{\min} \geqslant \frac{K_z(0)}{\sigma_x^2}\,\frac{1}{\mathrm{var}[k(\tau)]}. \quad (3.228)$$

3.5.3 Regression analysis

The *regression method* makes it possible to determine quantitative models from experimental results. In the case of two variables, the data for the model are in the form of two corresponding sets of variables X and Y, namely $\{X, Y\}$, where the data are (x_j, y_j) for $j = 1, \ldots, k$. Usually, theoretical analysis makes it possible to answer the question which variable is independent (input variable) and which is dependent (output variable): the assignment is indicated by the symbols X, Y as in the model of MS.

The *regressive model* is the relationship of an independent variable and of a conditional expected value of the dependent variable, i.e.,

$$y = E[y_j|x_j] = m_y(x) = E[Y|X = x]. \tag{3.229}$$

This is called the *first-kind regression*, that is, the regression of the variable Y by the variable X.

Determination of the conditional expected value $y = m(x)$ may be performed point by point if the values of x_j are multiple realizations. Then the point estimation of $m_y(x_j)$ is performed, and the result is a set of points $(x_j, \hat{m}_y(x_j))$ and not a function. In some cases, a small interval may be taken instead of a point x_j, and the mean values over this interval may be calculated. This is a so-called point estimation problem [68]. Also, the confidence intervals resulting from the identified stochastic relationship and from the estimator bias may be determined for individual points of the relationship $\hat{y}_j = m_y(x_j)$. A functional relationship may be determined when some other theoretical data are available, for example, a bidimensional probability distribution $p(x, y)$, yet in this case the experimental data are not necessary.

The task of model determination by the regression method, or so-called *second-kind regression*, is formulated somewhat differently. The experimental results (x_j, y_j), $j = 1, \ldots, k$, are the data and the model is known at the level of coefficients, together with the random component ε which includes all the errors of the model. It is assumed that $E(\varepsilon) = 0$. The required model may either be linear

$$y = \beta x + \alpha + \varepsilon \tag{3.230}$$

or nonlinear

$$y = f(x) + \varepsilon, \tag{3.231}$$

$$f(x) = W_q(x) = b_0 + b_1 x + \ldots + b_q x^q. \tag{3.232}$$

The first case is called *linear regression*, and the second *nonlinear regression*. The model should meet the requirement

$$\sum_{j=1}^{k} [y_j - (\beta x_j + \alpha)]^2 = \sum_{j=1}^{k} \varepsilon_j^2 = \min, \tag{3.233}$$

from which the name *least squares method* (LSM) is taken. The estimators $\hat{b}_0, \hat{b}_1, \ldots, \hat{b}_q$ of the coefficients α, β of the model (or generally, the estimators of the model (3.232)) satisfy the condition of the least distance of the points from the regression curve. The way in which the model is derived will be shown by examples. The method is presented in detail in handbooks of mathematical statistics [16, 68].

Example of the regression method. The model identification concerns an MS for measurement of the carbon content in a heated converter bath (see Fig. 4.2). The measured state coordinate of the object is the carbon content X. It is determined by measurement of the decarburization rate Y (the carbon flux in the form of exhaust gases). On that basis, the value of X is reproduced. Figure 3.36 shows

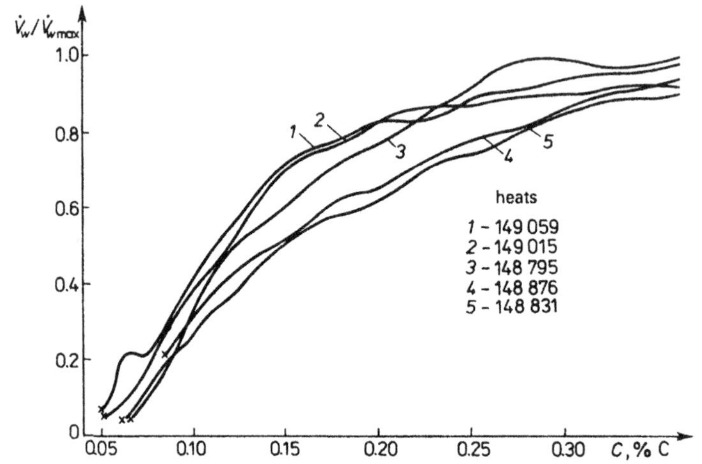

Fig. 3.36. Empirical characteristics of the steel converter heat [140]. C—carbon contents in bath, $\dot{V}_w$—decarburization rate.

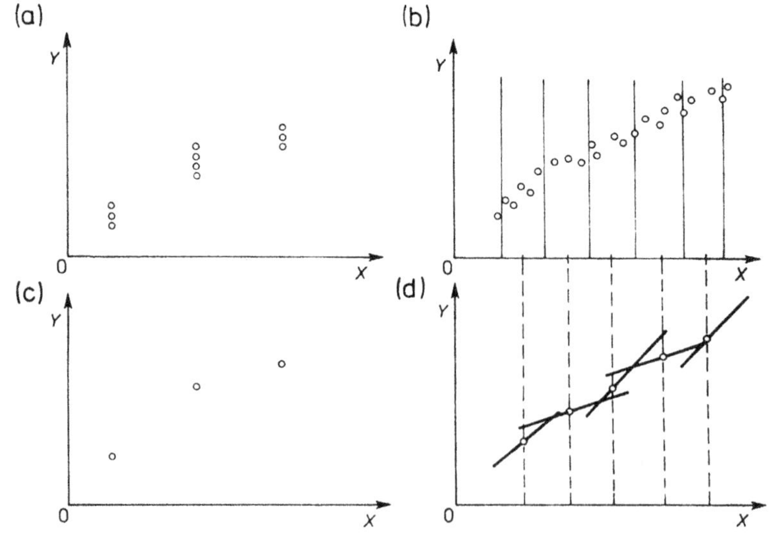

Fig. 3.37. Selection of the measurement points in a first-kind regression method.

the empirical relationship [140]. Each curve is a trajectory at one converter heat, and the results lying on one curve are strongly correlated. The points, each from a different heat, may be used as the independent results. The distribution of the measurement results of the MS model may be as in Figs. 3.37a, b. In the first case the values of the independent variable are identical, while in the second they are different. In the first-kind regression method, point-regression estimates are obtained, as shown in Fig. 3.37c. When the values of an independent variable are different, as in Fig. 3.37b, the domain X is divided into intervals, and the point estimates are calculated for a central point of each interval or for the mean over the interval (as in Fig. 3.37d). The identification of the model of the MS for

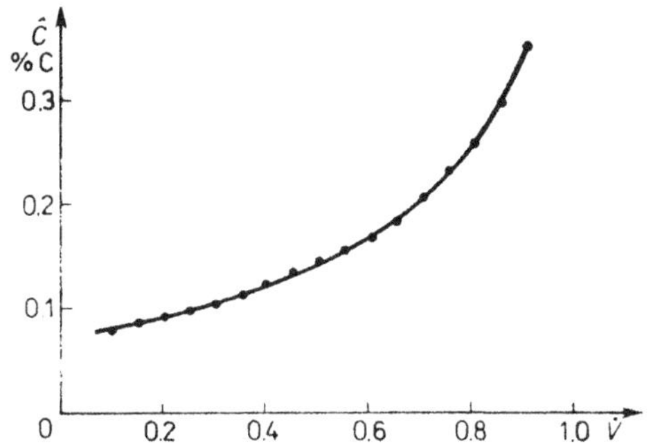

Fig. 3.38. First-kind regression results for the characteristics of an MS for carbon contents measurement in a converter bath [140]. $\dot{V} = \dot{V}_w/\dot{V}_{w\,max}$, $\dot{V}_w$—decarburization rate, $\hat{C}$—estimator of carbon contents in bath.

carbon content measurement gave the results of point estimation of the first kind

$$y_j = \frac{1}{n_j} \sum_{i=1}^{n_j} y_{ij}|_{x=x_j}, \tag{3.234}$$

shown in Fig. 3.38 [140]. The result does not determine the function. At the same points of the independent variable, the confidence interval for the estimator may be calculated from the sample. The estimator of the variance is

$$\hat{\sigma}_j^2 = \frac{1}{n_j-1} \sum_{i=1}^{n_j} (y_{ij}-y_j)^2, \tag{3.235}$$

and the confidence interval at a given confidence level is

$$\Delta_{\lim j} = \pm t_j \hat{\sigma}_j. \tag{3.236}$$

The value of t depends on the distribution of the variable y_j. The first-kind regression method does not produce any analytical form of a model of an MS. Furthermore, the confidence interval is determined at a point and hence the number of measurement points n_j should be considerable if the accuracy of each value of x_j is to be good. In the multidimensional models the number of measurement points increases as $n_j n_i^N$, where n_j is the number of repeated measurements at the point x_j and n_i is the number of points in the domain of each independent variable, N is the number of independent variables (the model is $(N+1)$-dimensional).

A small number of points means worse accuracy of the model, which is thus less useful. The analytical form of the model in the first-kind regression is obtained by applying approximation of the results with any suitable function and by any method. For example, for the data as in Fig. 3.37b, the results over individual intervals may be approximated by straight lines. The result of an approximation like that is shown in Fig. 3.37d.

The second-kind regression method, known as the least squares method (LSM), is the one most widely used. The property of this method is that the calculated estimators of the model coefficients are not biased and are the most effective. The form of the model postulated in the LSM is

$$y = b_0 + \sum_{k=1}^{q} b_k f_k(x) + \varepsilon, \tag{3.237}$$

where $b_0, b_1, \ldots, b_q$ are the model coefficients and $f_1(x), \ldots, f_k(x)$ are any of the unambiguous functions of the argument x, whose form is known a priori. The assumption is that the model is known at the level of coefficients and ε is a random process which models the stochastic (non-functional) relationship of X and Y. The realizations of ε_j are unknown, but their variance is identical: σ_ε^2 = constant[10] and they are not correlated.

A linear form of the model—in relation to the coefficients—implies that the method is the linear regression. The assumption that the form of the model is known at the level of coefficients is essential, since all the deviations are considered random errors. In estimation of the random errors, the number of degrees of freedom is greater and this property is applied. In the LSM it is further postulated that the model (3.237) best describes the experimental results if the value of errors (understood as a quadratic metric) is the smallest one, namely

$$\sum_{j=1}^{nj} \left\{ y_j - \left[b_0 + \sum_{k=1}^{q} b_k f_k(x_j) \right] \right\}^2 = \sum_{j=1}^{nj} \varepsilon_j^2 = \min_{b_0, b_1, \ldots, b_q} . \tag{3.238}$$

This condition is fulfilled by the estimators of the coefficients of the model, i.e., $\hat{b}_0, \hat{b}_1, \ldots, \hat{b}_q$.

[10] The LSM may be applied without the condition σ_ε^2 = constant, but the estimators obtained in this way are not the most effective, as is discussed later.

Let us go back to the example of the MS for determination of carbon content. The function derived in the first-kind regression analysis is nonlinear. From the theoretical analysis, the postulated model is

$$Y = 1-\exp[-k(X-p)], \tag{3.239}$$

where k, p are coefficients. Figure 3.36 confirms this postulate. To obtain a linear function, a function transformation has been applied to the model (3.239). The substitution was

$$U = -\ln(1-Y) \tag{3.240}$$

and the result was

$$U = b_0+b_1X+v \tag{3.241}$$

or

$$X^* = a_0+a_1U+\varepsilon \tag{3.242}$$

if the characteristic is determined as mapping (3.10), inverse to the model. The notation was

$$b_0 = -kp_1, \qquad b_1 = k,$$

$$a_0 = p, \qquad a_1 = 1/k.$$

The results obtained were

$$\hat{a}_0 = 0.0670, \quad \hat{a}_1 = 0.1127 \quad \text{for} \quad X \in \langle 0.08,\ 0.17 \rangle \tag{3.243a}$$

$$\hat{a}_0 = 0.0487, \quad \hat{a}_1 = 0.1317 \quad \text{for} \quad X \in \langle 0.17,\ 0.35 \rangle \tag{3.243b}$$

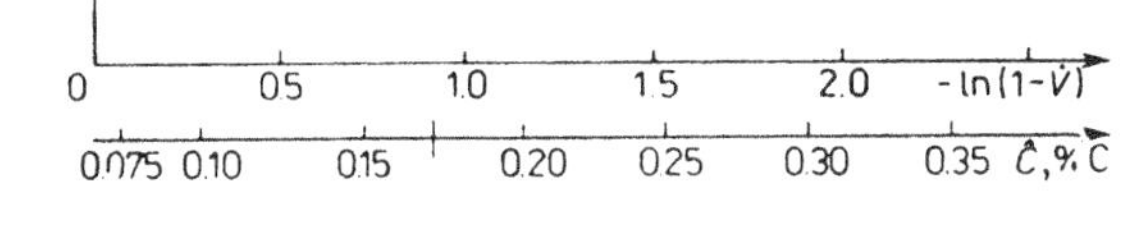

Fig. 3.39. The relationship of the standard deviation of the error of carbon contents vs the measured quantity.

The assumptions were verified by calculating the variance of error ε for the first-kind regression analysis results. The result shown in Fig. 3.39 [140] demonstrates that the estimators obtained for the model coefficients are not the most effective. To obtain the most effective estimators, the model should be transformed into a form for which the variance of the random component ε is constant. When the variance of ε changes within the measurement range $\langle x_l, x_u \rangle$ as

$$\sigma^2 = g^2(x)\sigma_0^2, \quad \sigma_0^2 = \text{constant}, \tag{3.244}$$

while

$$\frac{1}{x_u - x_l}\int_{x_l}^{x_u} g(x)\mathrm{d}x = 1, \tag{3.245}$$

then, after the substitutions

$$v = \frac{y}{g(x)}, \tag{3.246a}$$

$$\varphi_k(x) = \frac{f_k(x)}{g(x)}, \tag{3.246b}$$

$$\varphi_0 = \frac{1}{g(x)}, \tag{3.246c}$$

$$\varepsilon_0 = \frac{\varepsilon}{g(x)}, \tag{3.246d}$$

the model obtained instead of equation (3.237) is

$$v = \sum_{k=0}^{q} a_k \varphi_k(x) + \varepsilon_0. \tag{3.247}$$

Then, the condition $\sum_{j=1}^{n_j} \varepsilon_{0j}^2 = \min$ for the estimation points of the first-kind regression may be written as

$$\sum_{j=1}^{n_j}\left\{\frac{y_j - [b_0 + \sum_{k=1}^{q} b_k f_k(x_j)]}{g(x_j)}\right\}^2$$
$$= \sum_{j=1}^{n_j}\left\{y_j - \left[b_0 + \sum_{k=1}^{q} b_k f_k(x_j)\right]\right\}^2 w_j = \sum_{j=1}^{n_j} w_j \varepsilon_j^2 = \sum_{j=1}^{n_j} \varepsilon_{0j}^2 = \min, \tag{3.248}$$

where

$$w_j = \frac{1}{g^2(x_j)} \tag{3.249}$$

is called the *weight* (or the *weight coefficient*) of a given point. This results from the maximum likelihood function for the normal distribution, namely

$$L = \prod_{j=1}^{n_j} \frac{1}{\sqrt{2\pi}\,\sigma_j} \exp\left[-\frac{(y_j-\hat{y})^2}{2\sigma_j^2}\right]. \tag{3.250}$$

By applying logarithmic and differentiating operations, the likelihood equation is obtained as

$$\frac{dL}{d\hat{y}} - \sum_{j=1}^{n_j} \frac{y_j-\hat{y}}{\sigma_j^2} = 0 \tag{3.251}$$

and its solution is [8]

$$\hat{y} = \frac{\sum_{j=1}^{n_j} \frac{y_j}{\sigma_j^2}}{\sum_{j=1}^{n_j} \frac{1}{\sigma_j^2}} = \frac{\sum_{j=1}^{n_j} w_j y_j}{\sum_{k=1}^{n_j} w_j}. \tag{3.252}$$

The proof for the estimator to be the most effective may be found, for example, in [8]. The Cramer–Rao inequality is applied in the proof. Formula (3.248) is the condition of the so-called *weighted least squares method* (WLSM).

The values of the estimators for the coefficients of the model of an MS for carbon content measurement are different in the WLSM from those in the first-kind regression method (3.238), (3.243) [140]:

$$\hat{a}_0 = 0.0660, \quad \hat{a}_1 = 0.1109 \quad \text{for} \quad X \in \langle 0.08,\ 0.17 \rangle, \tag{3.253a}$$

$$\hat{a}_0 = 0.0542, \quad \hat{a}_1 = 0.1303 \quad \text{for} \quad X \in \langle 0.17,\ 0.35 \rangle. \tag{3.253b}$$

When regression analysis is applied, it is necessary to ensure that the derived *regression function* is significant. The *confidence interval for the regression function* should also be calculated.

To estimate the significance of the regression function, the quotient of two statistics [68] formed from the set of the measurement results is investigated. The first statistic describes the variability of the estimator of the regression function, as in model (3.237). The second describes the random component, i.e., that part of it which is a physical result of the random process $\{\varepsilon\}$ (3.237). The statistics are

$$U_1^2 = \frac{1}{\sigma_1^2} \sum_{j=1}^{N} [\hat{W}_q(x_j) - \bar{W}]^2 = N \frac{s_1^2}{\sigma_1^2}, \tag{3.254}$$

$$U_2^2 = \frac{1}{\sigma_2^2} \sum_{j=1}^{N} [y_j - \hat{W}_q(x_j)]^2 = N \frac{s_2^2}{\sigma_2^2}. \tag{3.255}$$

Both statistics have a chi-square distribution with f_1 and f_2 degrees of freedom, respectively. It is assumed that the distributions of the sets $\hat{W}_q(x_j)\}$ and $\{y_j - \hat{W}_q(x_j)\} = \{\varepsilon_j\}$ are normal with variances σ_1^2 and σ_2^2. With these assumptions, the quotient U_1^2/U_2^2 is a random variable F with F (Snedecor) distribution [22]:

$$F = \frac{U_1^2}{U_2^2}\frac{f_2}{f_1}. \tag{3.256}$$

Both statistics are obtained from the sets of direct measurement results $\{y_j\}$ of the sets of indirect results $\{\hat{W}_q(x_j)\}$ with the same number of elements N. The numbers of degrees of freedom determine the number of independent elements of the statistics and are, respectively,

$f_1 = (q+1)-1 = q$; the estimator $\hat{W}_q(x)$ is described by $q+1$ coefficients and its decrease by 1 is the effect of the constraints expressing $\overline{W}$;

$f_2 = N-q-1$; the set includes N independent results y_j, but $q+1$ constraints are used for estimation of $\hat{W}_q(x)$.

The variable F allows estimation of the likelihood of the hypothesis that the sample variances s_1^2, s_2^2 will differ significantly if it is assumed that $\sigma_1^2 = \sigma_2^2$ [68]. If the value of F is determined from the formula

$$F = \frac{s_1^2}{s_2^2}\frac{N-q-1}{q} = \frac{N-q-1}{q}\,\frac{\sum_{j=1}^{N}[\hat{W}_q(x_j)-\overline{W}]^2}{\sum_{j=1}^{N}[y_j-\hat{W}_q(x_j)]^2} \tag{3.257}$$

and it fulfils the conditions $F > F_{\mathrm{cr}\,\alpha}$ for a given confidence α, then $\sigma_1^2 > \sigma_2^2$, that is, the regression function is significant at the level α. The result $F < F_{\mathrm{cr}\,\alpha}$ means that the experimental results of identification give no confirmation of the relationship $y = f(x) = W_q(x)$ with the random component ε. In such a case, either we should be satisfied with such an estimate or the experiment should be repeated with an increased number of measurements N or with increased range of x. Then another estimation of the significance of the regression curve should be performed.

Instead of applying the variable F, some more elementary figures may be introduced in order to determine and compare the variability of the regression function and the random component [84]. These are:

$R_{\hat{W}y}$ —sample cross-correlation function,

φ^2—convergence coefficient,

which lead to the same estimates as previously obtained.

The components of the uncertainty of the model are:

— estimation uncertainty of the coefficients of the regression function (the model);

— a random component, identified together with the model.

Assuming that σ_ε^2 = constant, the unbiased estimator of the variance of random component ε may be expressed as [68]

$$\hat{\sigma}_\varepsilon^2 = \frac{1}{N-q-1}\sum_{j=1}^{N}[y_j-\hat{W}_q(x_j)]^2. \quad (3.258)$$

The coefficients of the model are calculated from the formula

$$\boldsymbol{b} = \boldsymbol{W}^{\mathrm{T}}(x_j)[\boldsymbol{W}^{\mathrm{T}}(x_j)\boldsymbol{W}(x_j)]^{-1}\boldsymbol{y}, \quad (3.259)$$

where

$$\boldsymbol{W} = [1\ x\ x^2\ \dots\ x^q] \quad (3.260)$$

if $W_q(x)$ is a polynomial, and

$$\boldsymbol{W} = [1\ f_1(x)\ f_2(x)\ \dots f_q(x)] \quad (3.261)$$

in general. The dimension of the matrix $\boldsymbol{W}(x_j)$ is $(q+1)\times N$, namely

$$\boldsymbol{W}^{\mathrm{T}}(x_j) = [\boldsymbol{W}(x_1)\boldsymbol{W}(x_2)\ \dots\ \boldsymbol{W}(x_j)\ \dots\ \boldsymbol{W}(x_N)] \quad (3.262)$$

and

$$\boldsymbol{y}^{\mathrm{T}} = [y_1\ y_2\ \dots\ y_j\ \dots\ y_N]. \quad (3.263)$$

The equation of the model in matrix form is

$$\hat{y} = \hat{W}_q(x) = \hat{\boldsymbol{b}}\hat{\boldsymbol{W}}. \quad (3.264)$$

The source of uncertainty of the model is the random quantities $\hat{\boldsymbol{b}}$. In the regression method, the most probable values of these coefficients are calculated, yet they are not the exact values. Although the value of the estimation error $\Delta\hat{W}$ is constant due to the errors of the coefficients $\hat{\boldsymbol{b}}$, the value of this error is unknown. For a normal ditribution of the random component and for a linear model (in relation to the coefficients), the distribution of the error of the model estimation is normal. Then, the variance of the estimation error at point x_0 is [68]

$$\sigma_{\Delta W}^2 = [\boldsymbol{W}^{\mathrm{T}}(x_0)][\boldsymbol{W}^{\mathrm{T}}(x_j)\boldsymbol{W}(x_j)]^{-1}\boldsymbol{W}(x_0)\hat{\sigma}_\varepsilon^2. \quad (3.265)$$

The variance of the uncertainty of the model characterizes both components. Therefore

$$\sigma^2 = \sigma_\varepsilon^2+\sigma_{\Delta W}^2, \quad (3.266)$$

or

$$\sigma^2 = \{\boldsymbol{W}^{\mathrm{T}}(x_0)[\boldsymbol{W}^{\mathrm{T}}(x_j)\boldsymbol{W}(x_j)]^{-1}\boldsymbol{W}(x_0)+1\}\hat{\sigma}_\varepsilon^2. \quad (3.267)$$

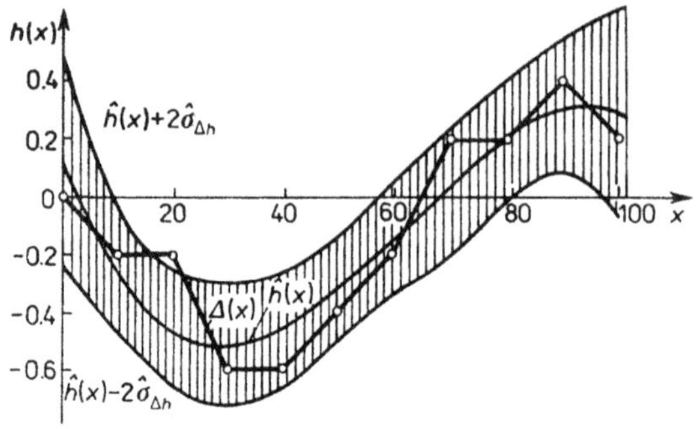

Fig. 3.40. The confidence intervals for the regression function; $h(x) = W_q(x)$—systematic error distribution, $\hat{h}(x)$—estimator, $\pm 2\sigma_{\Delta h}$—confidence interval for the error.

The characteristic feature of the uncertainty of a model is its dependence on the measured quantity x, as shown in Fig. 3.40. It is clearly seen from equation (3.267) that the uncertainty of the model is strongly influenced by the choice of points for identification of the model, described by the *Fisher matrix*

$$\boldsymbol{M} = \boldsymbol{W}^{\mathrm{T}}(x_j)\boldsymbol{W}(x_j). \tag{3.268}$$

The problem of selection of the measurement points is discussed in another chapter. In this author's opinion, a specific feature of the regression method is that the model is assumed to be certain, at the level of coefficients. This fact brings about serious consequences for the estimation of the confidence interval of a regression function. In particular, the variance estimator of the random component (3.258) is determined from the number N of measurement points, diminished by the number of degrees of freedom. The factor $1/(N-q-1)$ gives a favourable estimate of the variance.

In the LSM, a model is the best approximation and the estimate of the model uncertainty is different.

3.5.4 The least squares method

The least squares method had been developed by Legendre and Gauss before the regression method was introduced. The LSM is applied as an independent method of approximation of the experimental results to an arbitrary model. In particular, the model may be sought out by a trial-and-error method. The best model is the one for which the sum of the squares of its error is least. For this reason, in the computational iterative procedures applied here, the coefficients calculated previously should remain unchanged. Most often orthonormal polynomials or series are applied in the search for the model. These are the Legendre and Chebyshev

TABLE 3.2

Orthogonal polynomials [18, 23].

---

1. Chebyshev polynomials [6]

$P_n^{(0)}(x) = 1$, $P_n^{(1)}(x) = \lambda_n^{(1)}x$, $P_n^{(2)}(x) = \lambda_n^{(2)}[x^2 - \frac{1}{12}(n^2-1)]$,

$P_n^{(3)}(x) = \lambda_n^{(3)}[x^3 - \frac{1}{2}(3n^2-7)x]$,

$P_n^{(4)}(x) = \lambda_n^{(4)}[x^4 - \frac{1}{14}(3n^2-13)x^2 + \frac{3}{560}(n^2-1)(n^2-9)]$,

$P_n^{(5)}(x) = \lambda_n^{(5)}[x^5 - \frac{5}{18}(n^2-7)x^3 + \frac{1}{1008}(15n^4-230n^2+407)x]$,

$P_n^{(6)}(x) = \lambda_n^{(6)}[x^6 - \frac{5}{44}(3n^2-31)x^4 + \frac{1}{176}(5n^4-110n^2+329)x^2 - \frac{1}{14784}(n^2-1)(n^2-9)(n^2-25)]$.

The domain is $0 < x < \infty$.

$i = 0, 1, 2, \ldots, m$—the order of the polynomial.

$\lambda_n^{(i)}$—the smallest positive number for which all the values of $P_n^{(i)}(x_t)$ are integer.

$(x_t = t - \frac{1}{2}(n+1)$; $t = 1, 2, \ldots, n$; $x_t - x_{t-1} = \text{const.})$.

---

2. *Legendre polynomials*

$$P_n(x) = \frac{1}{2^n n!}\frac{d^n}{dx^n}(x^2-1)^n.$$

The domain is $-1 \leqslant x \leqslant 1$.

Recursive formula:

$nP_n(x) = (2n-1)xP_{n-1}(x) - (n-1)P_{n-2}(x)$,

$P_0(x) = 1$, $P_1(x) = x$, $P_2(x) = \frac{3}{2}x^2 - \frac{1}{2}$, $P_3(x) = \frac{5}{2}x^3 - \frac{3}{2}x$,

$P_4(x) = \frac{35}{8}x^4 - \frac{15}{4}x^2 + \frac{3}{8}$.

---

3. *Laguerre polynomials*

$$L_n(x) = e^x \frac{d^n}{dx^n}(x^n e^{-x}).$$

The domain is $0 \leqslant x < \infty$.

Recursive formula:

$L_n(x) = (2n-1-x)L_{n-1}(x) - (n-1)^2 L_{n-2}(x)$.

---

4. *Hermite polynomials*

$$H_n(x) = (-1)^n e^{x^2} \frac{d^n}{dx^n}(e^{-x^2}).$$

The domain is $-\infty < x < \infty$.

Recursive formula:

$H_n(x) = 2xH_{n-1}(x) - 2(n-1)H_{n-2}(x)$.

---

5. *Walsh functions*

$$\tau_0(t) = 1 \quad \text{for} \quad 0 < t < 1,$$

$$\tau_1(t) = \begin{cases} 1 & \text{for} \quad 0 < t < \frac{1}{2}, \\ -1 & \text{for} \quad \frac{1}{2} < t < 1. \end{cases}$$

$$\tau_2^{(1)}(t) = \begin{cases} 1 & \text{for} \quad 0 < t < \frac{1}{4}, \frac{3}{4} < t < 1, \\ -1 & \text{for} \quad \frac{1}{2} < t < \frac{3}{4}, \end{cases}$$

$$\tau_2^{(2)}(t) = \begin{cases} 1 & \text{for} \quad 0 < t < \frac{1}{4}, \frac{1}{2} < t < \frac{3}{4}, \\ -1 & \text{for} \quad \frac{1}{4} < t < \frac{1}{2}, \frac{3}{4} < t < 1, \end{cases}$$

....

$$\tau_{m+1}^{(2k-1)}(t) = \begin{cases} \tau_m^{(k)}(2t) & \text{for} \quad 0 < t < \frac{1}{2}, \\ (-1)^{k+1}\tau_m^{(k)}(2t-1) & \text{for} \quad \frac{1}{2} < t < 1, \end{cases}$$

$$\tau_{m+1}^{(2k}(t) = \begin{cases} \tau_m^{(k)}(2t) & \text{for} \quad 0 < t < \frac{1}{2}, \\ (-1)^{k}\tau_m^{(k)}(2t-1) & \text{for} \quad \frac{1}{2} < t < 1, \end{cases}$$

for $m = 1, 2, 3, \ldots$, $k = 1, 2, 3, \ldots, 2^{m-1}$.

The domain is $0 < t < 1$.

polynomials, as well as the Laguerre, Chebyshev, Hermite, Walsh and Haare functions, as presented in Table 3.2 [23].

For example, the measurement results of the decarburization rate as a function of time were fitted to the Gram polynomial [108]:

$$p_j(s, 2n) = \sum_{k=0}^{j} (-1)^{k+1} \frac{(j+k)^{2k}(n+s)^k}{2(k!)^2 n^k}, \tag{3.269}$$

where n are the standardized values of an independent variable with uniformly spaced points $-n, -n+1, \ldots, 0, 1, \ldots, n$; $s = (x-x_0)/d-n$ is a current variable; x_0 is the starting point of the independent variable, d is the sampling interval (d = constant) and j is the degree of the polynomial. The model is the polynomial

$$y_q(s) = \sum_{j=0}^{q} b_j p_j(s, 2n). \tag{3.270}$$

The result of identification and fitting is the following sequence of models, with errors M [56]:

$$\begin{aligned}
b_0 &= 3.4710\times 10^{-3}, & M_0 &= 3.3399\times 10^{-8},\\
b_1 &= -2.7952\times 10^{-4}, & M_1 &= 1.6241\times 10^{-9},\\
b_2 &= 2.9040\times 10^{-5}, & M_2 &= 1.4443\times 10^{-9},\\
b_3 &= -1.8534\times 10^{-5}, & M_3 &= 1.4256\times 10^{-9},\\
b_4 &= -3.7595\times 10^{-5}, & M_4 &= 6.5602\times 10^{-10},\\
b_5 &= 7.4549\times 10^{-7}, & M_5 &= 7.2828\times 10^{-10}.
\end{aligned}$$

The error was calculated from the formula

$$M_j = \frac{1}{2n-q} \sum_{s=-n}^{n} \Big[y_s - \sum_{j=0}^{q} b_j p_j(s, 2n)\Big]^2. \tag{3.271}$$

The best approximation is in this case the fourth-degree polynomial. The idea of the LSM was presented in the preceding section. Now, the most important properties of the method will be presented, using vector notation.

The model of the phenomenon investigated is the equation

$$Y = A^{\mathrm{T}} X + \varepsilon, \tag{3.272}$$

where Y is the column vector whose components y_j ($j = 1, \ldots, N$) are the realizations of the dependent variable; X is the row vector of independent variables $X_0, X_1, \ldots, X_q$, each having $j = 1, \ldots, N$ realizations, thus X is the matrix whose dimension is $(q+1)\times N$ and elements are x_{kj};[11] A is the row vector of the

[11] The variables x_k may be independent; the model in this case is linear and multidimensional. The symbol x_k may also denote the function $f_k(x)$ in which case the model is one-dimensional and nonlinear. Usually the value of x_0 is constant and equal to 1.

coefficients of the model, whose components are $a_0, a_1, \ldots, a_q$ (in general a_k, $k = 0, 1, \ldots, q$); ε is the column vector of the random component describing the errors of the model referred to the output. The components ε_j of this vector meet the following conditions: (1) $E(\varepsilon_j) = 0$—the errors are random; (2) $E(\varepsilon_i \varepsilon_j) = 0$ for $i \neq j$—individual errors are independent; (3) $E(\varepsilon_i \varepsilon_j) = \sigma_\varepsilon^2$ for $i = j$—the variance of error is constant (though unknown).

The coefficients of the model should fulfil the condition

$$M = \boldsymbol{\varepsilon}^{\mathrm{T}}\boldsymbol{\varepsilon} = (Y - A^{\mathrm{T}}X)^{\mathrm{T}}(Y - A^{\mathrm{T}}X) = \min_{A}. \tag{3.27)}$$

which, differentiated by A, takes the form

$$2X^{\mathrm{T}}(A^{\mathrm{T}}X - Y) = 0, \tag{3.274}$$

and thus

$$\hat{A} = X(X^{\mathrm{T}}X)^{-1}Y^{\mathrm{T}}. \tag{3.275}$$

After the model is derived, the errors may by calculated as

$$\varepsilon = Y - \hat{A}^{\mathrm{T}}X = Y - X(X^{\mathrm{T}}X)^{-1}X^{\mathrm{T}}Y, \tag{3.276}$$

and then the results may be corrected

$$Y_{\mathrm{corr}} = Y - \varepsilon = X(X^{\mathrm{T}}X)^{-1}X^{\mathrm{T}}Y. \tag{3.277}$$

There are no profound grounds for such correction, although some of the monographs do mention it [68, 8]. The problem is similar when the accuracy of the measurements is not constant. In this case, instead of condition (3) being fulfilled, the covariance matrix

$$C_y = C_\varepsilon = \begin{bmatrix} \sigma_1^2 & & & 0 \\ & \sigma_2^2 & & \\ & & \ddots & \\ 0 & & & \sigma_N^2 \end{bmatrix}. \tag{3.278}$$

should be known. This is the case of the WLSM, and the coefficients of the model are then calculated as

$$\hat{A} = X(X^{\mathrm{T}}GX)^{-1}GY^{\mathrm{T}}, \tag{3.279}$$

where $G = C^{-1}$. Formula (3.279) was obtained from the condition

$$M = \boldsymbol{\varepsilon}^{\mathrm{T}}G\boldsymbol{\varepsilon} = (Y - A^{\mathrm{T}}X)G(Y - A^{\mathrm{T}}X) = \min_{A}. \tag{3.280}$$

The coefficients of the matrix G are the weights of particular measurements $w_j = 1/\sigma_j^2$. The property of the LSM and WLSM is that the estimators produced are unbiased and also—when the distribution of errors is normal—the most effective.

Like the other statistical methods, the LSM and the WLSM do not allow detection of the systematic errors. In a case when the distribution of errors is normal, the fact that the assumptions are not satisfied may be detected by checking the sum of squares M (3.273). When the distribution of the random component ε of the model is normal, then $M = \sum_{j=1}^{N} \varepsilon_j^2$ has the chi-square distribution with $N-(q+1)$ degrees of freedom. The expected value of M is then

$$E(M) = N-(q+1). \tag{3.281}$$

If the value of M exceeds the value of $\chi_{\alpha}^2(N-q-1)$ for a given confidence level α, errors in the assumptions can usually be found. The reason may be an improper model, the error having a distribution other than normal, the variance σ_ε^2 not being equal in all the measurements, or an error in the calculations.

3.5.5 Other methods

The common use of computers has led to the creation of a number of new identification methods [18, 114, 67]. The idea usually is to determine those values of coefficients which would satisfy adequate optimality criteria. For example, the following requirements may be formulated for the errors:

$$\begin{aligned} &\text{when } |\varepsilon| \leqslant \delta, \quad \text{then} \quad F(\varepsilon) = 1, \\ &\text{when } |\varepsilon| > \delta, \quad \text{then} \quad F(\varepsilon) = 0, \end{aligned} \tag{3.282}$$

and

$$M = \sum_{j=1}^{N} F(\varepsilon_j) = \max_{A}. \tag{3.283}$$

The error ε is defined as in the LSM, i.e.,

$$\varepsilon = y - f(x), \tag{3.284}$$

where x, y are the realizations and $j = 1, \ldots, N$.

The function f and the coefficients may be chosen arbitrarily, and the model obtained will be the best of the models investigated, in the sence of criterion (3.283). Criterion (3.283) is less convenient for analytical calculations, but is relatively simple for numerical processing. In the example of the MS for decarburization rate measurement, the linear model (3.242) was investigated, applying criterion (3.283).

The results are shown in Fig. 3.41 [140]. In principle, when a model is being identified, the values of coefficients are selected either with a trial-and-error method by scanning the domain, by generation of random numbers (the Monte-Carlo

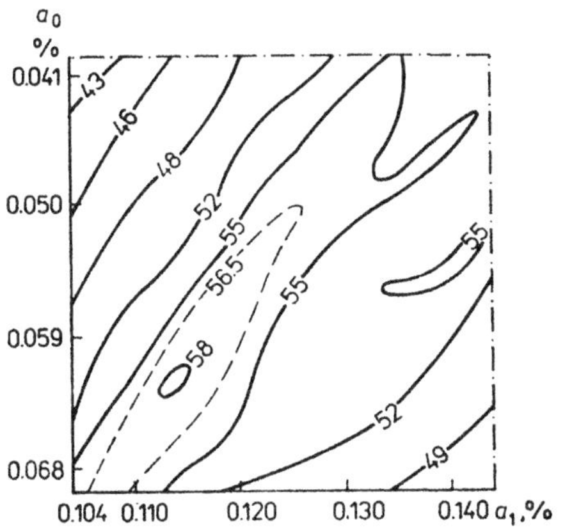

Fig. 3.41. Experimental results for the MS for carbon contents measurement [140]; a_1, a_0—parameters of model (3.242).

method), by applying experimental design techniques or by minimum-finding methods. The advantage of these methods is that commercially available standard computer programs may be used.

One of the well known methods is the stochastic approximation method [18, 113], in which the expected value of the squares of errors is minimized

$$E\{[y-f(x)]^2\} = \min_{A}. \tag{3.285}$$

Advantages of this method are the recursive form of the algorithm and the small number of computations required.

Chapter 4

The Theory of Error

4.1 Images of Reality in Measurement Systems. Principles of Image Production

In Chapter 1 we postulated the axiom that an image of reality is always evoked with some error. In the theory of error, one basic *definition of error* is applied. The error may be expressed by different concepts.

(i) An error is the difference between a state of a given quantity and the real state of this quantity.

(ii) If a value of a quantity is x and the accepted value is x^*, the difference $x^* - x$ is called an error.

(iii) An error is the difference between a value of a quantity and the correct value of this quantity.

Note that the above definitions deal with error itself and not with a measurement error; this was intentional. In mental experiments each of the above definitions may be applied, although the second one is the best.

When images of reality are produced in a process of measurement, two domains are involved: reality and abstraction. In the first definition, the error is expressed in the domain of reality, and in the third, in the domain of abstraction. To a first approximation, both definitions express the same error in adequate domains.

The notion of error is not connected with the mapping of a state of a quantity into a value. It expresses the fact that a distance between two states (or values) exists. The cause of this difference need not be known.

The errors expressed according to the first and the third definitions may be combined. Let us denote: x—a state of a quantity; x_{real}—a real state of the quantity; x^*—a value of the quantity; x^*_{corr}—correct value of the quantity. Now we have

$$\Delta_1 = x - x_{real}. \tag{4.1}$$

and

$$\Delta_3 = x^* - x^*_{corr}. \tag{4.2}$$

There is a mapping (1.35) of the state x into the value x^*:

$$x^* = f^*(x), \tag{4.3}$$

which is one-to-one mapping at the point x and within some neighbourhood of this point. Let x_{real} belong to the neighbourhood of x. Considering (4.3) and (1.39), the relationship

$$\Delta_3 = x^* - x^*_{\text{corr}} = f^*(x) - f^*(x_{\text{real}}) = f^*(x) - f^*(x - \Delta_1) = f^*(\Delta_1) \tag{4.4}$$

is obtained which is true for interval and quotient scales.

The symbol f^* denotes the postulated mapping. Physically, the mapping is of point $\Delta_1 \Rightarrow$ set $\{\Delta_3\}$ and is an ambiguous mapping of a point in the real domain into a set in an abstract domain, similar to the case of mapping (1.30) of a state into a value.

It has been assumed above that the mapping of a real state is the *true value* of a quantity. This assimption causes the most confusion in the theory of error, since the real state of a quantity remains unknown and likewise the correct value. As it cannot be proved that

$$x^*_{\text{corr}} = f^*(x_{\text{real}}), \tag{4.5}$$

both notions x_{real} and x^*_{corr} form the definition of error.

Measurement errors may also be expressed in the domain of states or in the domain of images. The value of a quantity obtained as the output of an MS model (as in Fig. 3.1b) is expressed by formula (3.6) and the correct value is expressed by formula (4.3). The error in the domain of values is then

$$\Delta^* = b_\Delta = x^* - f^*(x) = \varphi[F(x)] - f^*(x). \tag{4.6}$$

The measurement result expressed in the domain of states is obtained from the mapping inverse to (4.3), namely

$$x^{**} = f^{*-1}(x^*), \tag{4.7}$$

while the correct state is x. The error is the difference

$$\Delta = a \;\; = x^{**} - x = f^{*-1}\{\varphi[F(x)]\} - x. \tag{4.8}$$

It may be proved that correspondence with (4.4) occurs in this case, since

$$f^*(a_\Delta) = f^* f^{*-1} \varphi F - f^* = \varphi F - f^* = b_\Delta \tag{4.9a}$$

or

$$f^*(\Delta) = \Delta^*. \tag{4.9b}$$

Subsequently the shortened notation

$$\Delta = x^* - x = \varphi[F(x)] - x, \tag{4.10}$$

will be used, where x^* is a correct result and x is a correct state (or correct value) of a measured quantity. If it is of no importance, it will be not stated in which domain the error is expressed. The above symbols will be used as follows:

x—state (value) at the input of an MS;

x^*—value (state) at the output of an MS;

The property of an MS arising from condition (3.6c)

$$x^* = \varphi[F(x)] = I_0(x) = f_x(x) \tag{4.11a}$$

is that the operator

$$I_0(x) = x^* \tag{4.11b}$$

is a unitary operator which maps the real domain into the abstract domain. This property gives grounds for application of the shortened notation of error (4.10). If the domain of I_0 is the abscissa axis and the range is the ordinate axis, the mapping is a straight line passing through the origin of coordinates and sloping an angle of 45°. This means that the error of an MS may be defined as the difference of a reading (output) of an MS and the state of a given quantity at the input.

Because of the stochastic character of the model F (3.126)

$$y = F(x_1, \ldots, x_{m-1}, x, T, a_1, \ldots, a_k, z),$$

there is no procedure φ which satisfies condition (3.6d) and thus also condition (4.11a). The calibration procedure φ is determined, for example, from model (3.128)

$$\varphi = F_0^{-1}(x_{10}, \ldots, x_{(m-1)0}, x, T_0, a_{10}, \ldots, a_{k0})$$

in which the random effects have been neglected. Thus

$$\varphi[F_0(x)] = I_0(x) = x. \tag{4.12}$$

The performance of a real MS is described by the operator

$$I(x) = \varphi[F(x)] = x^*, \tag{4.13}$$

which differs from the unitary operator I_0 by the value of error

$$\Delta = I - I_0 = x^* - x. \tag{4.14}$$

In the case of a linear calibration procedure we have

$$\Delta = \varphi[F(x)] - \varphi[F_0(x)] = \varphi[\Delta F(x)], \tag{4.15}$$

which means that the error is caused by simplification of the model.

In Chapter 3 (on modelling) ways of formulating mathematical models of MSs were presented, with internal and external sources of errors considered. Each source of error appears in a model of an MS as a random variable or a ran-

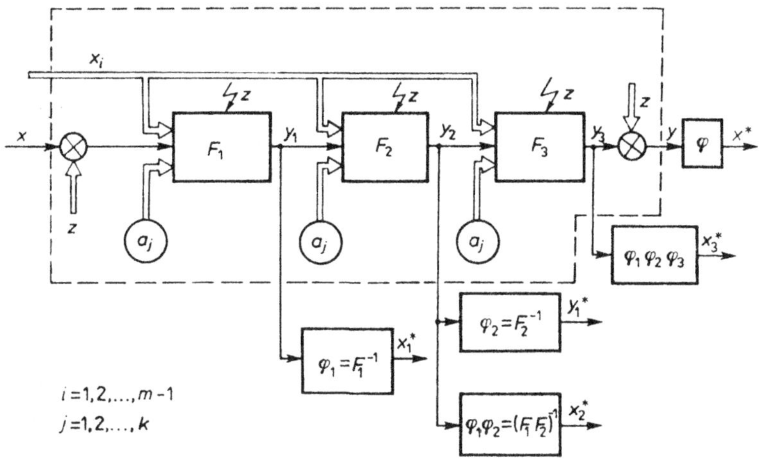

Fig. 4.1. Location of the internal error sources in an MS.

dom process with known characteristics, as an input quantity to one of the conversion stages. This fact is shown in Fig. 4.1. Thus, the input quantities for the first stage, apart from the measured quantity $x = x_m$, may be the influencing quantities $x_1, \ldots, x_{m-1}$, the construction parameters $a_1, \ldots, a_k$ and some unidentified disturbances z. As a result of a physical conversion, a physical quantity y_1 is obtained at the output of the first stage. The quantity is generally a non-stationary, multidimensional random process $\{y_1(x_1, \ldots, x_m, T)\}$ [99]. From y_1, the image of the state of the measured quantity x_1^* may now be derived by applying procedure φ_1, because

$$\varphi_1(y_1) = x_1^*. \tag{4.16}$$

The difference

$$x_1^* - x_1 = \Delta_1 \tag{4.17}$$

is an error introduced by the first conversion stage. The mathematical models of the next conversion stages are constructed similarly. The application of a proper procedure $\varphi = \varphi_1\varphi_2\ldots\varphi_n$ makes the determination of the measured quantity x_n^* from the value of y_n possible.

The mathematical model of a whole MS is obtained by performing the mathematical operations $F_1, F_2, \ldots$ describing the transformation of the quantities x_i $(i = 1, 2, \ldots, m)$ and a_j $(j = 1, 2, \ldots, k)$ into a random process $\{y\}$. Also here, the application of a proper procedure $\varphi(y) = x^*$ makes it possible to determine the value x^* corresponding to the state x at the input of the MS.

4.2 Optimization of Calibration Procedures

The purpose of optimization of calibration procedures is to adapt them to measurement conditions and to diminish the calibration errors. The formulation of the optimization criterion concerns a set of errors (3.134) expressed by formula (3.130). The unattainable condition $\Delta = 0$ may only be a guide in the selection of φ. A calibration procedure is an operator or a function

$$\varphi = \varphi(y, x_1, \ldots, x_{m-1}, T, c_1, \ldots, c_l) \tag{4.18}$$

where the arguments $y, x_1, \ldots, x_{m-1}, T$ are measurable and the coefficients $c_1, \ldots, c_l$ are controllable and measurable. The coefficients are controlled when the procedure is fixed (and the MS is adjusted). The structure of the calibration operator is established on the basis of just an idea, using only some general guidelines. These are:

(1) A *simplified model of an MS* is considered in reference conditions and according to the notation $\varphi = F_0^{-1}$.

(2) The basis is an *extended model of an MS* (3.12); the extension includes the influencing quantities and then $\varphi = F_0^{-1}$.

(3) The influence of random components may be limited by *averaging* $\varphi = [E(F)]^{-1}$.

(4) Averaging may be performed with weighted elements of the set X. In this case, $\varphi = [E\{F[x, p(x)]\}]^{-1}$. The weight $p(x)$ may be, for example, a frequency of realization of the element x or may be established on the basis of specific requirements.

(5) The *prediction* of the operator changes *in time* T is introduced. In this, case $\varphi = [F+\Delta F_T]^{-1}$, where ΔF is a predicted change of the operator F.

Usually, the decision on the choice of a calibration procedure is conditioned by the possibility of realizing this procedure in a given MS, by the complexity of the MS, by the importance of the measurements, and the cost of calibration, with reference to other solutions.

The *figure of merit*, called an *objective function* or a *loss function*, is formulated in order to select the parameters of the procedure (with l degrees of freedom), to compare different procedures and to choose the best one.

The objective function takes into account the planned designation of an MS [101, 113]; the intended application may have several criteria. Most frequently the objective function is the distance of the elements of the set $\{\Delta\}$ in a space normalized by one of metrics (1.48)–(1.51), and the distances are properly weighted. At the same time, the set X is assumed and for this reason the figure of merit and the condition of selection of the *procedure parameters* $c_1, c_2, \ldots, c_l$ is in the form

$$\min_{c_1 \ldots c_l} J(\Delta/X) = \min_{c_1 \ldots c_l} J\{\varphi[F(x_1, \ldots, x_m, a_1, \ldots, a_k, T, z), x_1, \ldots, x_{m-1}, T_1, c_1, \ldots, c_l] - x\}|_X. \quad (4.19)$$

The problem of optimization of a calibration procedure may be treated as an optimum control. The task formulation is shown in Fig. 5.4. The problem will be not discussed here, since a wide bibliography is available [18, 116].

For practical reasons, the simplest procedures are usually applied. Some of the known procedures are compiled below; the classification and the names are proposed by the present author.

1. *Static procedures*

(a) linear

$$x^* = cy, \quad (4.20a)$$

(b) nonlinear

$$x^* = \varphi(y), \quad (4.20b)$$

(c) linear, with correction of supplementary errors,

$$x^* = cy + \sum_{i=1}^{m-1} c_i x_i, \quad (4.20c)$$

(d) nonlinear, with correction of supplementary errors,

$$x^* = \varphi(y, x_1, \ldots, x_{m-1}). \quad (4.20d)$$

2. *Dynamic procedures*

(a) first-order linear

$$x^* = y + T\frac{\Delta y}{\Delta t}, \quad (4.21a)$$

(b) linear

$$x^* = \mathscr{F}^{-1}[\mathscr{F}(y)H(j\omega)]. \quad (4.21b)$$

3. *Averaging procedures*

(a) simple averaging

$$x^* = c\frac{1}{N}\sum_{j=1}^{N} y_j, \qquad x^* = cE(y), \quad (4.22a)$$

(b) weighted averaging

$$x^* = c\sum_{j=1}^{N} w_j y_j, \qquad x^* = cE[w(y)y]. \quad (4.22b)$$

4. *Predictive procedures*

(a) prediction of the changes of a mathematical model

$$x^* = \varphi(y - \check{\Delta}F), \quad \check{\Delta}F = \check{F}(T) - F_0(T_0), \tag{4.23a}$$

(b) prediction of the delay time

$$x^*(t) = \varphi[y(t-t_m) + \check{\Delta}y(t_m | t-t_m)], \tag{4.23b}$$

(c) prediction with filtering (Kalman, Wiener or digital filters, etc.) [116, 77].

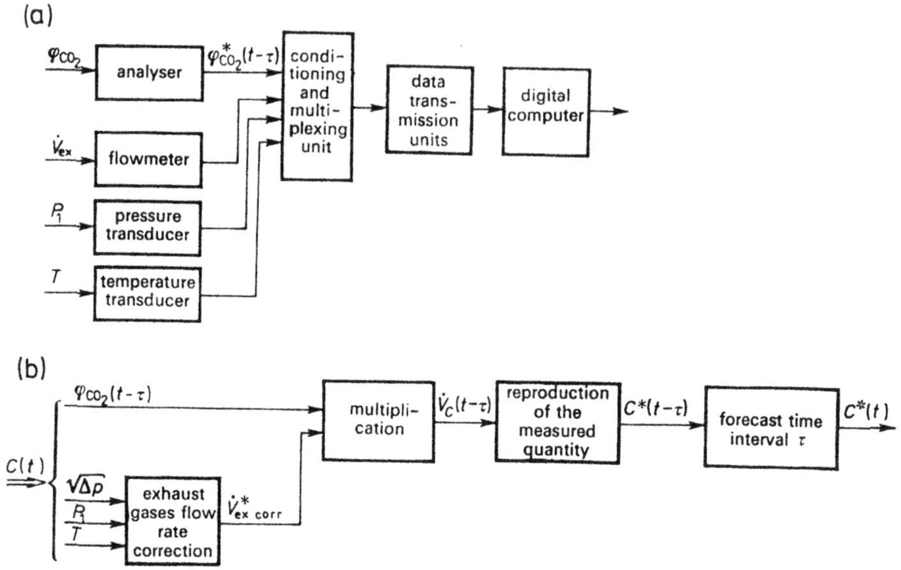

Fig. 4.2. Simplified diagram of an MS for carbon contents measurement in a converter bath: (a) a structural diagram of the MS, (b) processing flowchart. $X = C(t)$—measured carbon contents in the bath unmeasurable directly, $Y = \dot{V}_C(t) = \varphi_{CO_2} \dot{V}_{ex\,corr}$—decarburization rate ($\dot{V}_C = dC/dt$), $C^*(t-\tau)$, $C^*(t) = X$—measurement results, φ_{CO_2}—CO_2 concentration in exhaust gases, $\dot{V}_{ex}$—flow rate of exhaust gases, Δp—differential pressure at the orifice plate, P_1—pressure, T—temperature.

For example, in the MS for carbon content measurement in the converter bath shown in Fig. 4.2 [136, 140] and discussed in Chapter 3, the procedure applied was dynamic with correction of supplementary errors, with averaging and prediction of a delay time. The analyser of CO_2 concentration in the exhaust gases was described by the model

$$K(s) = e^{-s\tau_m} \frac{k}{1+sT}, \tag{4.24}$$

where τ_m is the delay time and T is a time constant. The supplementary temperature and pressure errors in the flowmeter were corrected.

Measurement procedures are also applied in control systems [77, 116]. An example of application of the Kalman–Bucy filter is the MS for temperature measurement and control, described in [38], [165].

4.3 Definitions and Classification of Errors

In order to make the definition simple, the errors are defined under certain idealized conditions. At the same time, the definition should allow distinguishing of one kind of error from another in situations when all kinds of errors appear. The definition of error should suffice for either its theoretical or empirical determination. If the errors are determined in an empirical way, the number of degrees of freedom in the form of controls of input quantities is limited. This fact will be considered here.

Errors are classified with respect to:

(1) the character of the error—errors are systematic or random;

(2) the measurement conditions—the error in reference conditions is called fundamental; if the conditions differ from the reference situation, a supplementary error occurs;

(3) the character of the measured and input quantities—the errors are static and dynamic;

(4) the physical causes of errors, i.e., so-called sources of errors—a number of types follow this classification: calibration error, instability error, supplementary error, dynamic error, quantization error, sampling error, counting error, etc. [91].

The basis for all the definitions of errors is the elementary definition (4.2) that an error is a difference between the value of a quantity and the correct value of this quantity. Furthermore, the error of readings of an MS is regarded as a multidimensional non-stationary random process (3.134):

$$\{\Delta\} = \{\Delta(x_1, \ldots, x_m, T)\}, \tag{4.25}$$

where T is the service time of the MS.

The definitions of systematic and random errors are elementary notions in metrology. A *systematic error* is an error whose value is constant or changes by a known rule when the measurement of some value of the measured quantity is performed repeatedly under the same conditions.[1] Similarly, a *random error* may be defined in an elementary way [97]. The effect is the definition

---

[1] Recommendation 1 (CI–1981) of the CIPM [178] suggests that errors be divided into two categories A and B, and claims that the notion of systematic error is unnecessary. This author is convinced that the recommendation is not correct and may concern only standards [174].

$$\{\Delta_r\} = \{\Delta\} - \Delta_s, \tag{4.26}$$

where Δ_s is a systematic error and Δ_r is a random error.

For the errors understood as a random process, a relatively wide class of processes may be distinguished in which the component errors may be expressed as a stationary random process (3.136). This may be done by applying the functions $h(x_1, \ldots, x_m, T)$, $g(x_1, \ldots, x_m, T)$, i.e.,

$$\{\Delta_k\} = g(x_1, \ldots, x_m, T)\{\varepsilon_k\} + h(x_1, \ldots, x_m, T). \tag{4.27}$$

The random process $\{\varepsilon\}$ is stationary, with its mean value equal to zero and unit variance (3.137); k denotes the number of realizations of the process.

Since the value of a systematic error is constant or changes by a known rule, the definition

$$\Delta_s = E\{\Delta_k\} = h(x_1, \ldots, x_m, T) \tag{4.28}$$

determines the value of this error and the rule by which it changes. The averaging operation is performed on the realizations of processes $\{\Delta\}$ and $\{\varepsilon\}$ (3.139).

A systematic error is the expected value of error when the same value of a measured quantity is measured under the same conditions. An important condition for the determination of the value of systematic error is to know the values of the influencing quantities $x_1, \ldots, x_{m-1}$, as well as the value of the measured quantity $x = x_m$ and the time T since the last calibration. This definition is also applicable to the time functions $x(t)$. Each waveform is one realization.

The physical sense of the definition of systematic error does not change as compared with the classical definition. The errors with random sources and with the same sign or a probability distribution which is not symmetrical at the zero point are easily and clearly classifiable. For example, the coincidence error occurring in Geiger counters is always negative; the background error in isotope measurements is always positive. The expected values of these errors differ from zero and each of these values is a systematic error.

The definition of a random error remains unchanged though the properties change. The definitions of random (4.26) and systematic (4.28) errors imply that the expected value of a random error is zero [145], because

$$E\{\Delta_r\} = E\{\Delta\} - \Delta_s = 0.$$

This property is not always in agreement with the physical sense of random errors.

In the previously accepted model of error (4.27), a random error is described by a function g and a stationary process $\{\varepsilon\}$

$$\{\Delta_r\} = g(x_1, \ldots, x_m, T)\{\varepsilon\}. \tag{4.29}$$

The definitions of systematic and random errors assume that the measurement time T_m is short enough for the functions g and h to be constant. The interpretation of the influence of the service time T of an MS on its errors will be given in the discussion of the fundamental error. In the same place, the dynamic properties of error, i.e., the properties of the process $\{\varepsilon\}$, will be discussed.

A *fundamental error* of an MS is the error in reference conditions. In reference conditions, the values of influencing quantities x_i, $i = 1, 2, \ldots, m-1$, remain constant and equal to x_{i0}. The fundamental error is thus a bidimensional non-stationary random process

$$\{\Delta_0\} = \{\Delta_0(x, T)\}. \tag{4.30}$$

The index 0 denotes the reference conditions.

For the previously discussed class of non-stationary random processes (4.27) which may be reduced to stationary processes, we obtain

$$\{\Delta_0\} = g_0(x, T)\{\varepsilon_0\}+h_0(x, T), \tag{4.31}$$

where g_0, h_0 are the functions of x and T, and $\{\varepsilon_0\}$ is a stationary random process with its mean value equal to zero and unit variance.

The systematic component of the fundamental error is called the *calibration error* [97]

$$\Delta_c = E\{\Delta_0\} = h_0(x, T), \tag{4.32}$$

and the random component is called the *instability error*

$$\{\Delta_{\text{in}}(x, T)\} = \{\Delta_0(x, T)\}-E\{\Delta_0(x, T)\} = \{\Delta_0\}-\Delta_c = g_0(x, T)\{\varepsilon_0\}, \tag{4.33}$$

and

$$\operatorname{var}\{\Delta_0\} = \operatorname{var}\{\Delta_{\text{in}}\} = g_0^2(x, T)E\{\varepsilon_0^2\} = g_0^2(x, T). \tag{4.33a}$$

The notion of calibration error is classical, and definition (4.32) introduces mathematical precision only. The definition gives an interpretation of the experimental results of error determination. A quite different approach is applied when the error is theoretically calculated from the data of a standard, from a calibration method and its conditions. This problem is discussed in a wider context in [91]. The limiting values of calibration error are determined theoretically, and in the calibration process only one (or a few) value of this error is realized. The value is constant, though unknown. By proper selection of the calibration procedure, the calibration error may be further reduced, either by averaging or by smoothing the values of errors of several standards. An additional component of the calibration error appears when a characteristic being fixed is approximated by a model which is more convenient or simple. This problem is presented in Chapter 5.

The instability error is the notion introduced and promoted by this author [91, 88]. The substance of this treatment is the fact that random errors appear in every measurement, while the reference conditions provide the most convenient measurement conditions, reducing the random errors to a minimum. The instability error is thus a characteristic property of every specimen instrument and its technical condition at time T. The sources and mechanism of creation of an instability error are known; they have been presented in Chapter 3. The value of an instability error may be determined theoretically or empirically, but the empirical way [89] is more accurate and usually simpler. As a matter of fact, only the limiting properties of the instability error can be determined theoretically.

A *supplementary error*, according to the classical qualification, is due to the measurement conditions differing from the reference situation. Consequently, regarding this qualification and (4.27) and (4.30), the definition is

$$\{\Delta_{\text{sup}}\} = \{\Delta\} - \{\Delta_0\} = \{\Delta_{\text{sup}}(x_1, \dots, x_m, T)\}. \tag{4.34}$$

A supplementary error as a difference of two random processes is a multidimensional non-stationary random process. The arguments of the process are the influencing quantities $x_1, \dots, x_{m-1}$, the measured quantity x and the service time T.

To make the notion more simple, a distinction is made between the errors caused by individual influencing quantities which are independent error sources. This does not mean that an additive concatenation of errors may be applied here, for this form is exact only for on MS which is linear with respect to the input quantities. The components of error are calculated as a total differential of the mathematical model (3.8)

$$\mathrm{d}y = \sum_{i=1}^{m-1} \frac{\partial F}{\partial x_i} \mathrm{d}x_i = \sum_{i=1}^{m-1} k_i \tilde{x}_i \quad \text{for } i = 1, \dots, m-1. \tag{4.35}$$

When $\partial F/\partial x_i$ = constant, $x_i = x_i - x_{i0}$ or when $k_i \neq$ constant, we obtain by integrating from the reference point to $(x_1, \dots, x_{m-1})$

$$y - y_0 = \sum_{i=1}^{m-1} \int_{x_{i0}}^{x_i} \left(\frac{\partial F}{\partial x_i}\right) \mathrm{d}x_i = \Delta y_{\text{sup}}. \tag{4.36}$$

To calculate the error, a slightly different definition of supplementary error must be applied. According to this definition, a supplementary error is the error resulting from the calibration procedure applied, which is simplified, in comparison with a *natural procedure*.[2] The simplification means neglecting the fact that

[2] The nature of a supplementary error is systematic and for this reason a procedure with correction of the supplementary errors was called natural. In the meaning of this definition, a supplementary error is an error of method [91].

the measurement conditions are different from the reference conditions. The reasons for this simplification are practical: the influencing quantities need not be measured and the instrument construction does not have to be developed. According to this qualification, for a linear calibration procedure φ_0, applying formula (4.3) gives

$$\begin{aligned} \Delta_{sup} &= \varphi_0(y)-\varphi(y) = \varphi_0(y_0+\Delta y_{sup})-\varphi(y) \\ &= \varphi_0(y_0)+\varphi_0(\Delta y_{sup})-\varphi(y) = \varphi_0(\Delta y_{sup}) = \Delta x^*_{sup}, \end{aligned} \tag{4.37a}$$

since $\varphi_0(y_0) = x^*$. For a nonlinear procedure, we still have $\Delta_{sup} = \varphi_0(y)-\varphi(y)$.

When the supplementary errors are determined empirically, it is more convenient to apply model (4.37). According to formulae (4.27) and (4.32), the supplementary error

$$\begin{aligned} \{\Delta_{sup}\} = {} & [g(x_1, \ldots, x_{m-1}, x, T)\{\varepsilon\}+h(x_1, \ldots, x_{m-1}, x, T)]- \\ & -[g_0(x, T)\{\varepsilon_0\}+h_0(x, T)] \end{aligned} \tag{4.37b}$$

has two components: a *systematic component*

$$\Delta_{sups} = h(x_1, \ldots, x_{m-1}, x, T)]-h_0(x, T) = h_{sup}(x_1, \ldots, x_{m-1}, x, T) \tag{4.38}$$

and a *random component*

$$\{\Delta_{sup\,r}\} = g(x_1, \ldots, x_{m-1}, x, T)\{\varepsilon\}-g_0(x, T)\{\varepsilon_0\}. \tag{4.39a}$$

After substituting (3.135) and (3.136), the variance of the supplementary error is

$$\mathrm{var}\{\Delta_{sup}\} = E[(g\{\varepsilon\}-g_0\{\varepsilon_0\})^2] = g^2+g_0^2-2g_0 g\varrho(\tau) = g_{sup}^2. \tag{4.39b}$$

The value of the *correlation coefficient* ϱ varies from 0 to 1, depending on the time interval of the realizations of the processes $\{\varepsilon\}$ and $\{\varepsilon_0\}$ expressed by the value τ. This fact means that in case of a time-correlated process $\{\varepsilon\}$, the methodology of measurement of a supplementary error considerably affects the results. For example, a supplementary voltage error may be measured by small intervals of the error realizations. In this case, $\varrho(0) \approx 1$ and

$$g_{sup} = g-g_0. \tag{4.40a}$$

When the time interval is large enough, $\varrho(\infty) = 0$ and

$$g_{sup} = \sqrt{g^2+g_0^2}. \tag{4.40b}$$

The above interpretation agrees with (3.164). Thus, a supplementary temperature error cannot be measured in small time intervals [97] and the case that usually occurs here is (4.40b). The time scale is relative and determined by the properties of the process $\{\varepsilon\}$. On the other hand, the accuracy of determining a systematic component of supplementary error is dependent on the variance of the error. Thus, the measurement conditions must be defined precisely. To characterize a systematic component of a supplementary error, the function $h(x_1, \ldots, x_{m-1},$

$x, T)$ is developed into a Taylor series. By rejecting the terms with second- and higher-order derivatives, we obtain

$$\Delta_{\text{sups}} = \sum_{i=1}^{(1/2)(m-1)(m+2)} k_i \tilde{x}_i, \tag{4.41}$$

with the notations as in Table 4.1.

TABLE 4.1

Notations in formulae (4.41), (4.44)

i	$\tilde{x}_i$	k_i	q_i
$1, \dots, m-1$	$x_i - x_{i0}$	$\frac{dh}{dx_i}$	$\frac{dg}{dx_i}$
$m, \dots, 2(m-1)$	$(x_i - x_{i0})^2$	$\frac{1}{2}\frac{d^2h}{dx_i^2}$	$\frac{1}{2}\frac{d^2g}{dx_i^2}$
$2m-1, \dots, \frac{1}{2}(m-1)\times$ $\times(m+2)$	$(x_i - x_{i0})(x_j - x_{j0})$	$\frac{d^2h}{dx_i dx_j}$	$\frac{d^2g}{dx_i dx_j}$

Each sensitivity coefficient depends on the measured quantity x and on the time T. The values of k_i are determined in the reference conditions, i.e.,

$$k_i = k_i(x_{10}, \dots, x_{(m-1)0}, x, T) = k_{i0}(x, T). \tag{4.42}$$

The function (4.42) of x is usually approximated by a polynomial

$$k_{i0}(x, T) = b_{i0}(T) + b_{i1}(T)(x - x_0) + b_{i2}(T)(x - x_0)^2 + \dots \tag{4.43}$$

The first term describes a supplementary systematic error of zero ($x = x_0$), while the second describes the systematic error of sensitivity. It may also be convenient to apply model (3.12).

A random supplementary error may be presented in a form similar to formula (4.41), namely

$$\{\Delta_{\text{supr}}\} = \left[\sum_{i=1}^{(1/2)(m-1)(m+2)} q_i \tilde{x}_i\right] \{\varepsilon\}, \tag{4.44}$$

and every component of $q_{i0}(x, T)$ may be approximated by the polynomial as in (4.43).

In measurement conditions, the influencing quantities are random variables or random processes, and because of the limited period of observation, they may be non-stationary. In this case, g and h (4.27) are functions of the random variables x_i, $i = 1, \dots, m-1$. For example, the properties of power supply voltage and of atmospheric pressure are shown in Figs. 4.3 and 4.4 [64].

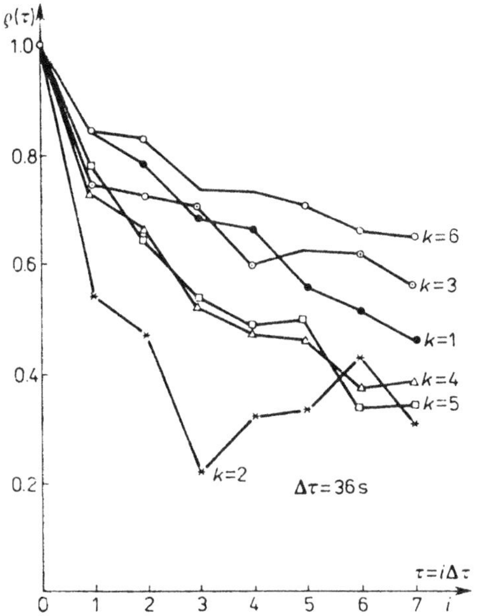

Fig. 4.3. Averaged correlation coefficient of the random component of power supply voltage in May and June 1969, 8–11 a.m. [64]; *k*—selected realizations.

The above errors of an MS always appear together, though they have been presented separately. To show clearly the physical causes of individual errors, the basic mathematical model of an MS will be considered once again. The conversion equation (3.126)

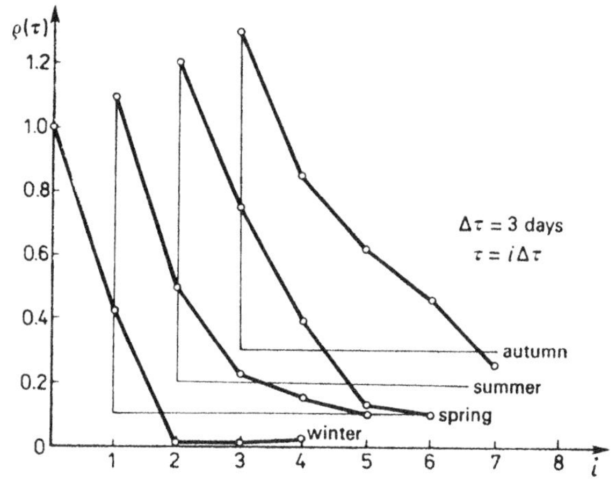

Fig. 4.4. Averaged correlation coefficient of atmospheric pressure. Data gathered in the Chorzów Weather Station, 1962–68 [64].

$$y = F(x_1, \ldots, x_{m-1}, x, T, a_1, \ldots, a_k, z)$$

is developed in a Taylor series at the point $(x_{10}, \ldots, x_{(m-1)0}, T_0, a_0, \ldots, a_{k0})$. The higher-order terms and the component $(\partial F/\partial T)\Delta T$ are neglected (the last component is negligible since the service time is usually not present in a mathematical model in explicit form). Considering (3.112) we obtain

$$y = F_0(x_{10}, \ldots, x_{(m-1)0}, x, T_0, a_0, \ldots, a_{k0})+$$

$$+\sum_{j=1}^{k} \frac{\partial F}{\partial a_j}\Delta a_{j0} + \left[\sum_{j=1}^{k} \frac{\partial F}{\partial a_j} \sum_{i=1}^{m-1} \frac{\partial a_j}{\partial x_i}\Delta x_i + \sum_{i=1}^{m-1} \frac{\partial F}{\partial x_i}\Delta x_i\right]+$$

$$+\left[\sum_{j=1}^{k} \frac{\partial F}{\partial a_j}\frac{\partial a_j}{\partial T}\Delta T + \sum_{j=1}^{k} \frac{\partial F}{\partial a_j}\Delta a_{j\,\mathrm{in}} + \frac{\partial F}{\partial z} z\right]. \tag{4.45}$$

Writing

$$y_0 = F_0(x_{10}, \ldots, x_{(m-1)0}, x, T_0, a_0, \ldots, a_{k0}), \tag{4.45a}$$

$$\Delta y_0 = \sum_{j=1}^{k} \frac{\partial F}{\partial a_j}\Delta a_{j0}, \tag{4.45b}$$

$$\Delta y_{\mathrm{sup}} = \sum_{j=1}^{k} \frac{\partial F}{\partial a_j} \sum_{i=1}^{m-1} \frac{\partial a_j}{\partial x_i}\Delta x_i + \sum_{i=1}^{m-1} \frac{\partial F}{\partial x_i}\Delta x_i, \tag{4.45c}$$

$$\Delta y_{\mathrm{in}} = \sum_{j=1}^{k} \frac{\partial F}{\partial a_j}\frac{\partial a_j}{\partial T}\Delta T + \sum_{j=1}^{k} \frac{\partial F}{\partial a_j}\Delta a_{j\,\mathrm{in}} + \frac{\partial F}{\partial z} z, \tag{4.45d}$$

we get

$$y = y_0 + \Delta y_0 + \Delta y_{\mathrm{sup}} + \Delta y_{\mathrm{in}}. \tag{4.46}$$

The errors in formulae (4.45) have been grouped so that the causes of the errors may be indicated:

Δy_0—(further called calibration error) is the effect of deviations of the real values of construction parameters from their nominal values;

Δy_{sup}—a supplementary error caused by deviations of the influencing quantities from their reference values; when the conditions are equal to the reference ($\Delta x_i = 0$), this error does not occur;

Δy_{in}—an instability error expressing the ageing processes of the elements $((\partial a_j/\partial T)\Delta T)$, the fluctuations and oscillations of the parameters $(\Delta a_{j\,\mathrm{in}})$ and a random influence of the environment.

According to a classical interpretation, the applied calibration procedure is $\varphi_0 = F_0^{-1}$. Then, the formulae

$$x^* = \varphi_0(y_0) = x, \tag{4.47}$$
$$x^* = \varphi_0(y_0+\Delta y_0) = x+\Delta_c, \tag{4.47a}$$
$$x^* = \varphi_0(y_0+\Delta y_0+\Delta y_{sup}) = x+\Delta_c+\Delta_{sup}, \tag{4.47b}$$
$$x^* = \varphi_0(y_0+\Delta y_0+\Delta y_{in}) = x+\Delta_c+\Delta_{in} \tag{4.47c}$$

define respectively

— a correct value of the measured quantity,
— a calibration error,
— a supplementary error,
— an instability error.

If the calibration procedure φ_0 is differentiable, the definitions of errors may be replaced by the approximations:

$$\Delta_c = \frac{\partial\varphi_0}{\partial y}\Delta y_0 = h_0(x), \tag{4.48a}$$

$$\Delta_{sup} = \frac{\partial\varphi_0}{\partial y}\Delta y_{sup} = h_{sup}(x_1, \ldots, x_{m-1}, x), \tag{4.48b}$$

$$\Delta_{in} = \frac{\partial\varphi_0}{\partial y}\Delta y_{in} = h_{in}(x, T)+g(x_1, \ldots, x_m)\{\varepsilon\}. \tag{4.48c}$$

According to the recommendation given in Section 4.2, the applied procedure φ should allow elimination of the calibration errors, supplementary errors and the systematic component of the instability error. The effect of such a procedure is

$$x^* = \varphi(y) = \varphi(y_0+\Delta y_0+\Delta y_{sup}+\Delta y_{in}) = x+g(x_1, \ldots, x_m, T)\{\varepsilon\}. \tag{4.49a}$$

The averaging procedures

$$\varphi_1 = E[\varphi(y)] \quad \text{or} \quad \varphi_2 = \varphi[E(y)], \tag{4.49b}$$

which decrease the variance g^2 of a random component of error, are applied as well.

Practically, for various different reasons, a simplified calibration procedure φ_3 is applied

$$x^* = \varphi_3(y) = x+h(x_1, \ldots, x_m, T)+g(x_1, \ldots, x_m, T)\{\varepsilon\}. \tag{4.49c}$$

Some of the errors may be determined theoretically, when the conversion equation and the calibration procedure are known with enough precision.

A *dynamic error* will be defined in two ways.

(i) A dynamic error is the error caused by the dynamic properties of an MS diverging from the ideal. The ideal dynamic properties are regarded as the kernel of the Volterra operator (3.26) in the form

$$k(t, \tau) = k\delta(t-\tau), \tag{4.50}$$

or as a transmittance $K(s)$ = constant, independent of its argument s.

(ii) A dynamic error is the error caused by applying a static calibration procedure, while the natural procedure[3] is dynamic.

The first definition implies

$$y_{ideal} = y_{corr} = F[x(t)] = kx(t). \tag{4.51}$$

By applying definition (4.2) of an error of a quantity y, we have

$$\Delta_{dyn} = y(t) - y_{corr}(t) = \int_0^t k(t, \tau)x(\tau)d\tau - kx(t). \tag{4.52}$$

For a linear and static calibration procedure,

$$x^* = cy = \frac{1}{k_0}y \tag{4.53}$$

and after applying (4.52)

$$\Delta^*_{dyn} = \frac{1}{k_0}y(t) - x(t) = \frac{1}{k_0}\int_0^t k(t, \tau)x(\tau)d\tau - x(t). \tag{4.54}$$

For an MS with ideal properties (4.50), the dynamic error is zero.

In the second definition, the dynamic procedure $h(t, \tau)$[4] should be applied, for which

$$y_{corr}(t) = \int_0^t h(t, \tau)y(\tau)d\tau = x(t). \tag{4.55}$$

Application of procedure (4.53) creates the error

$$\Delta^*_{dyn} = \frac{1}{k_0}y(t) - \int_0^t h(t, \tau)y(\tau)d\tau = \frac{1}{k_0}y(t) - x(t). \tag{4.56}$$

The difference between formulae (4.54) and (4.56) is that in (4.54) the argument is an unknown input quantity x, while in (4.56) the argument is an observed output quantity. In both cases it is necessary to know the dynamic properties (in formula (4.56) in order to determine $h(t, \tau)$).

The procedure $h(t)$ may be determined [23] by assuming $x(t) = \delta(t)$. Then, by substituting expression (3.26) in formula (4.55), we obtain

$$\int_0^t h(\tau)\left[\int_0^\tau k(\theta)x(t, \theta)d\theta\right]d\tau = \int_0^t h(\tau)k(t, \tau)d\tau = \delta(t) \tag{4.57}$$

---

[3] The natural procedure is a dynamic procedure $\varphi = F^{-1}$, where F is the dynamic model of the MS.

[4] $h(t, \tau)$ is a *conjugated kernel of the operator*, satisfying equation (4.57) [23].

and thus[5]

$$h(t) = \mathcal{F}^{-1}[H(j\omega)] = \mathcal{F}^{-1}\left[\frac{1}{K(j\omega)}\right] = \mathcal{F}^{-1}\left[\frac{1}{\mathcal{F}[k(t)]}\right]. \tag{4.58}$$

This procedure is realized, for example, with the fast Fourier transform FFT [83, 77]. The instrumentation of the procedure is a correction circuit as in Fig. 4.5. However, not every procedure is physically realizable.

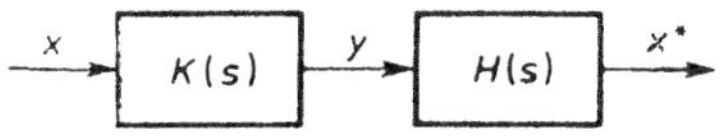

Fig. 4.5. A system performing a dynamic procedure of reproduction of a measured quantity.

When the procedure $h_1(t)$ is applied instead of $h(t)$ (4.55), the error, according to definition (ii), is

$$\Delta^*_{dyn} = \int_0^t h_1(\tau)y(t,\tau)d\tau - \int_0^t h(\tau)y(t,\tau)d\tau$$
$$= \int_0^t [h_1(\tau)-h(\tau)]y(t,\tau)d\tau = \int_0^t \Delta h(\tau)y(t,\tau)d\tau. \tag{4.59}$$

Application of definition (ii), together with formula (4.59), separates unambiguously a dynamic error from a static error, in the presence of an instability error. If the instability error is $z(t)$, then

$$y(t) = y_0(t)+z(t), \tag{4.60}$$

and

$$\Delta^*_{dyn} = \int_0^t \Delta h(\tau)y_0(t,\tau)d\tau + \int_0^t \Delta h(\tau)z(t,\tau)d\tau, \tag{4.61}$$

i.e., the error, being an effect of $z(t)$, is not a part of the dynamic error if $\Delta h = h_1 - h = 0$. However, application of the first definition (4.54) leads to

$$\Delta^*_{dyn} = \int_0^t h(\tau)y_0(t,\tau)d\tau + \int_0^t h(\tau)z(t,\tau)d\tau - x(t) = \int_0^t h(\tau)z(t,\tau)d\tau, \tag{4.62}$$

because $\int_0^t h(\tau)y_0(t,\tau)d\tau = x(t)$. This means that the instability error is included in the dynamic error. For this reason, definition (i) also includes the condition that the static errors are neglected [91, 82]. The problem of the dynamic procedures

[5] The Fourier transform $\mathcal{F}$ of equation (4.57) gives the result $H(j\omega)K(j\omega) = 1$. The symbol $\mathcal{F}$ denotes the Fourier transform and $\mathcal{F}^{-1}$ is the inverse transform. In (4.57) the property of convolution commutation is applied.

reproducing the state $x(t)$ from $y(t)$ is considered at length within the theory of automatic control [116, 77]. One of the most frequently applied methods is filtration in different variants of Kalman filters [116, 113].

We now look at a simple example. The operator form of the model of a magnetoelectric meter is

$$y(s) = \frac{k}{As^2+Bs+1}x(s). \tag{4.63}$$

The calibration procedure is static and is expressed by (4.53). Then, according to (4.54),

$$\Delta^*_{\mathrm{dyn}}(s) = \frac{1}{k}\frac{k}{As^2+Bs+1}x(s)-x(s) = -\frac{(As+B)s}{As^2+Bs+1}x(s). \tag{4.64}$$

Application of formula (4.56) gives the same result

$$\Delta^*_{\mathrm{dyn}}(s) = \frac{1}{k}y(s)-\frac{As^2+Bs+1}{k}y(s) = -\frac{(As+B)s}{k}y(s) = -\frac{(As+B)s}{As^2+Bs+1}x(s), \tag{4.65}$$

when $H(s) = As^2+Bs+1$.

There is still the problem of the nature of a dynamic error. The mean values of error calculated from (4.54) and (4.56) are the same when $E[x(t)] = x(t)$, $E[y(t)] = y(t)$ (i.e., when $x(t)$ and $y(t)$ are determined quantities), as the reproduction procedures are functions of y. Therefore, the nature of the dynamic error is systematic. When $x(t)$ or $y(t)$ is a random variable, the dynamic error also has a random component. Considering that in (4.60) $E[y(t)] = y_0(t)$, the systematic component is expressed by the first component in (4.61), while the random component is expressed by the second. The random component may be modelled by formula (3.29), namely

$$K_{\Delta^*}(\tau) = \int_0^{t_2}\int_0^{t_1} \Delta h(t_1, \tau_1)\Delta h(t_2, \tau_2)K_z(\tau_1, \tau_2)\mathrm{d}\tau_1\mathrm{d}\tau_2, \tag{4.66}$$

when $\tau = t_2-t_1$.

The Parseval theorem is used to express the error in the frequency domain [23].

4.4 The Metrological Characteristic of Measurement Systems and Standards

A metrological characteristic is understood to be the complete data concerning the errors of an MS, presented in proper order, according to conventional rules. The metrological characteristics should be known before an MS is applied, since it is impossible to determine the uncertainty of a measurement result without knowledge of these data (which would mean that the MS cannot be usefully applied).

There is an essentially practical significance attached to the methods of expressing the metrological characteristic. Both the way in which the characteristics are determined and the methodology of determining the measurement result depend on the type of characteristic.

The variety of measurement systems and of their uses is the reason why metrological characteristics cannot be expressed in a uniform way. Various simplifications, necessary for practical purposes, are applied. The principles of simplification have not been given a sufficient theoretical basis. The effect of this is an impression that the problem of metrological characteristics still has no satisfactory solution. Some standard recommendations were issued by OIML [70], based on the GOST standard [27].

The problem is looked at in a different way below and the physical sense of some of the characteristics is discussed in the next section. Ways of using metrological characteristics are presented in the next chapter, on measurement procedures. The ways in which the metrological characteristics are expressed are here called *classes*. The model of error as a non-stationary multidimensional random process

$$\{\Delta\} = g(x_1, \ldots, x_m, T)\{\varepsilon\} + h(x_1, \ldots, x_m, T) \tag{4.67}$$

is accepted as the most complete one.

Class 1 *of metrological characteristics of MS*

The data are:

(1) a probability distribution of a stationary process $\{\varepsilon\}$ and a correlation coefficient $\varrho_\varepsilon(\tau)$ of the process:

(2) estimators of the functions g, h and their confidence intervals;

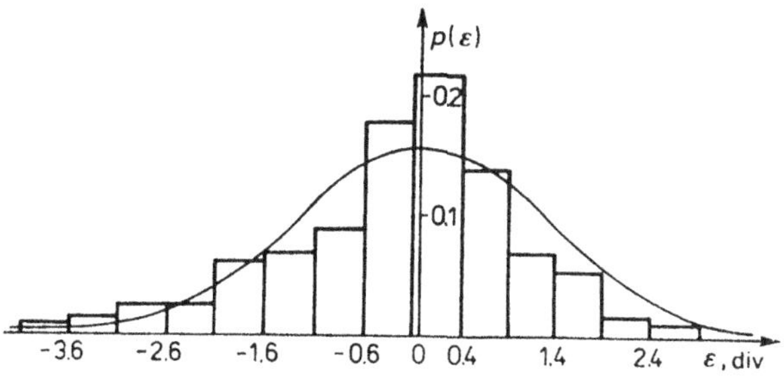

Fig. 4.6. The probability distribution of the instability error of an oxygen analyser type KTM-7.

(3) predictions of the functions g and h and their confidence intervals (in particular the predictions $\check{g}_0(T)$ and $\check{h}_0(T)$ in reference conditions).

These characteristics are applicable both to standards and unique MS. This way of expressing the characteristics is shown for the example of the KTM-7 thermomagnetic oxygen analyser [24, 71]. The distribution of the process $\{\varepsilon\}$ is shown in Fig. 4.6. The distribution is normal, with variance $g_0^2 = \sigma^2 = 1.58$. The time correlation is described by the correlation coefficient

$$\varrho(\tau) = e^{-0.23\tau} \quad (\tau \text{ in hours}). \tag{4.68}$$

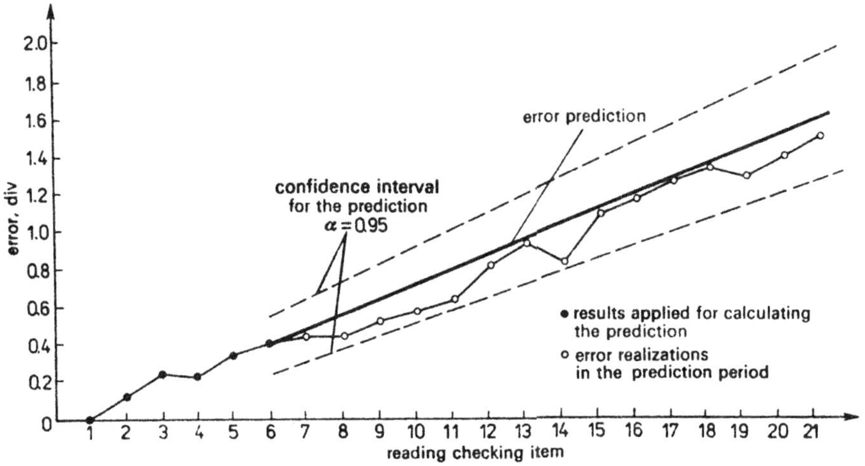

Fig. 4.7. Instability error changes and its prediction.

For this result, which is rather unexpected, no physical interpretation has been found. In Fig. 4.7 the time changes of the systematic error at point $x = 0$ are shown. The prediction $\check{h}_0(T)$ and the confidence intervals for the prediction are indicated. Different models $\hat{h}_0(T)$ are obtained depending on the method applied. In model (4.31),

$$\hat{h}_0(T) = 0.05T + 0.0001T^2, \tag{4.69}$$

but in model (3.147) the coefficient γ_2 is not significant, and the form of the model is [80]

$$\Delta_{0t} = 0.89\Delta_{0t-1} + 0.014T + e_t, \tag{4.70}$$

where e_t is a white noise with variance $\sigma_e^2 = 0.81$ (which means that $\theta_1, \theta_2, \ldots, \theta_r = 0$). The value of coefficient b_1 (= 0.05 in equation (4.69)), as calculated from model (4.70), is $b_1 = 0.064$. The value of the random error is constant, i.e., $g_0(T) = \text{constant}$.

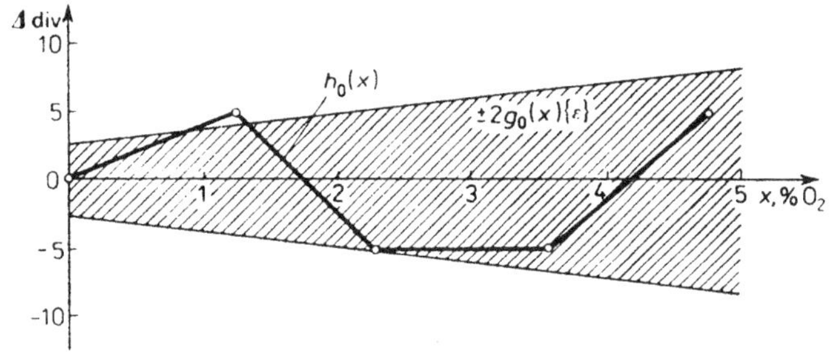

Fig. 4.8. Distributions of the systematic and random errors of the oxygen analyser KTM-7 in reference conditions.

The distributions of $h_0(x)$ ang $g_0(x)$ are shown in Fig. 4.8. In this case the coefficients of the model $h_0(x) = W_q(x)$ are insignificant. The reason is that the analysers are calibrated individually. The distribution of the random error is

$$\hat{g}_0(x) = b_0+b_1x = 1.3+0.8x. \tag{4.71}$$

The value of the coefficient $b_1 \neq 0$ is the result of temperature control in the thermostat of the analyser and the correlation coefficient of the respective component of the process $\{\varepsilon\}$ is a cosine function with a period equal to the period of the thermostat cycle. Since the characteristic of the analyser is nearly linear $y_0 = k_0x+x_0$, the supplementary errors describe two parameters [91]: A_i—the dependence of the zero level from the influencing quantities x_i, and δ_i—the influence of the quantities x_i on the sensitivity, according to formula (3.12)

$$k = k_0\left[\frac{p}{p_0}\right]^{\delta_p}\left[\frac{I}{I_0}\right]^{\delta_I}\left[\frac{T}{T_0}\right]^{\delta_T}, \tag{4.72}$$

where p, I, T are, respectively, pressure, supply current and temperature. The properties of the analyser are

$$A_p = -0.4\,\text{div/Tr}, \quad A_I = -3\,\text{div/div}, \quad A_T = 0, \tag{4.73a}$$

$$\delta_p = 2.1\%/\%, \quad \delta_I = 2.3\%/\%, \quad \delta_T = -0.43\%/^\circ\text{C}. \tag{4.73b}$$

The sensitivity k_0 of the analyser in the conditions of the experiment was 200 div/1% O_2. The analyser's service time without adjustment is determined from the predicted value of the error. According to the prediction given by model (4.69), for an allowable error $\Delta_{all} = \pm 20$ div, the service time is 220 hours, while according to model (4.70), it is 240 hours.[6]

The problems of the service time and the calibration techniques will be discussed in Chapter 5.

---

[6] During a three months test of two analysers, the errors never exceeded the range $\pm 0.16\%$ O_2 [71].

Class 2 of metrological characteristics

The data are:

(1) a probability distribution of the process $\{\varepsilon\}$ and its correlation function $\varrho_\varepsilon(\tau)$;

(2) estimators of the functions g_0 and h_0 and their confidence intervals;

(3) characteristics of the supplementary errors $h_i(x, x_i)$ and $g_i(x, x_i)$ for $i = 1, \ldots, m-1$;

(4) the service time without adjustment of an MS, to satisfy the condition

$$|\Delta_0(T)| \leqslant \Delta_{\mathrm{all}} \quad \text{when} \quad T \in \langle 0, T_E \rangle. \tag{4.74}$$

As compared with the Class 1, Class 2 is simplified. The simplification lies in the evaluation of the future metrological properties in relation to the time when the characteristic was determined. Usually the service time without calibration is determined by the producer of an MS, according to his experience, and the grounds for the decision are not known precisely. The user assumes that the characteristic conforms to the data at the moment $T = T_0$, which means the uncertainty is not determined. The user may only check the readings of the MS; usually only $\{\Delta_0(x_j, T_0)\}$ is being determined. Since the characteristic of the set $\{\Delta_0(x_j, T_0)\}$ is known, a proper hypothesis is formulated and then verified by a proper statistical test. The hypothesis may concern the equality of mean values, of variances, of error distributions for a given $x = x_j$, etc. The tests are: the test of goodness of fit, the Wilcoxon tests, F test, Kolmogorov–Smirnov's test, etc. [22, 9, 8]. Usually, from $h_0(x_j, T) = h_0(x_j, T_0)$, one concludes that the characteristic of a random component did not change. The reasons for such a conclusion are as follows. In the model of error (3.147), the coefficients φ_i, γ_j are generally the coefficients b_l (3.72) which depend on the construction parameters $a_1, \ldots, a_k$. The degree of correlation of the coefficients may be estimated from formula (4.110). If only the stability of a systematic error is to be investigated, the method for correlated quantities known from reliability theory is applied. Without a detailed analysis of the sources of errors, the degree of correlation remains unknown, but may be assumed. The interpretation and the way of determining the service time T_E are presented in Chapter 5.

Class 3 of metrological characteristics

The metrological characteristics include;

(1) a systematic error distribution in reference conditions $\hat{h}_0(x)$;

(2) a random error model in the form of $g_0(x)$, and information on the type of distribution of this error;

(3) a characteristic of the systematic supplementary errors $h_i(x, x_i)$ for $i = 1, \ldots, m-1$;

(4) the service time, meeting condition (4.74).

The simplification in Class 3 is that the characteristics of systematic and random errors are not given precisely and that the supplementary random errors are neglected. The systematic error is still given as an individual property of each instrument (e.g. as in Fig. 4.9), but the confidence interval for this error is not

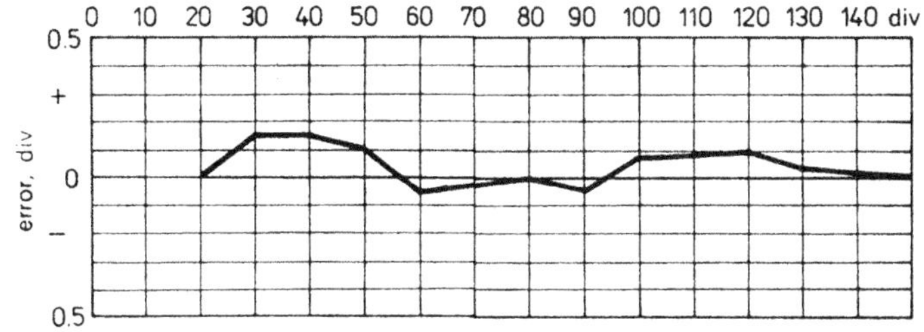

Fig. 4.9. Distribution of the systematic error of an electrodynamic ammeter.

quoted. The uncertainty of the systematic error and the random error are described conjointly by the limiting value of error, either constant or dependent on x. Since the systematic error is given, it may be assumed that the limiting error is random. The distribution of the random error determines the function g or $g(x)$, according to the relationship

$$g(x)t = \Delta_{\text{lim}}(x), \tag{4.75}$$

where t is a half of the confidence interval, expressed as the multiplication factor of the standard deviation g. The value of t is not known exactly, neither is the value of the confidence level. A putative value is at least 3, in order to create a margin for an instability error in the service time.

Class 4 of metrological characteristics

The MS data include:

(1) the limiting distributions of systematic and random errors in the reference conditions, $h_{\text{llm}}(x)$, $q_{\text{llm}}(x)$;

(2) the limiting distributions of the systematic supplementary errors $h_{\text{lim}\,i}$ for $i = 1, \ldots, m-1$;

(3) an approximate value of the service time of an MS without adjustment.

In comparison with the previous cases, this class involves a significant qualitative change. The characteristics concern the type of MS and not the individual properties of a given specimen. The error distributions of the instruments of a given type may be different and may remain unknown. It is only known that the following conditions are met:

$$|\hat{h}_0(x)| \leqslant h_{\lim}(x), \tag{4.76a}$$

$$\hat{g}_0(x) \leqslant g_{\lim}(x). \tag{4.76b}$$

It must be stressed that the conditions are met with an unknown probability depending on the production technology, on the way in which the tolerance of the construction parameters were determined, on the kinds of product tests performed and on the margin of the instability error in the service time. This class of characteristics is often presented implicitly. For example, a nonlinearity error (or, in other words, linearity) below 0.5% is a characteristic of systematic error. A drift below 0.2% per day or month is a margin for an instability error, or the instability error characterized as a separate component of error.

Class 5 of metrological characteristics

The characteristics include:

(1) a limiting distribution of total error in reference conditions $\Delta_{\text{all}}(x)$;

(2) the limiting distributions of supplementary errors.

This class includes the simplest characteristics. The systematic and random errors are not distinguished, but instead are treated conjointly. Exemplary distributions of an allowable error are given in Fig. 4.10.

Each specimen of an istrument may have different distributions and different proportions of systematic and random errors, fulfilling the condition

$$|\Delta_0(x)| \leqslant \Delta_{\text{all}}(x). \tag{4.77}$$

The way in which the value of an allowable error is calculated is not determined precisely. Thus there are no grounds for treating this value as a confidence interval. When the allowable error is expressed by an accuracy class, i.e., by a number from a specified sequence of numbers, in most cases an unknown margin of the rounding error arises. The methodology of checking the readings (indications) during product tests creates a basis for a more complete estimation of the errors of an MS, as is presented below.

The problem of the metrological characteristics is really important, but has yet to be solved satisfactorily. Knowledge of probability theory and mathematical statistics has become widespread, and disparities have arisen between the development of the theory of measurement and the practical level of measurements performed. The difficulty lies in the choice which must be made between the complexity of the characteristics and the effort necessary for their determination on the one hand, and on the other, the required simplicity due to the universal nature of measurements and a rational level of calibration and maintenance costs that must be borne in the service time. The automation of measurements and the

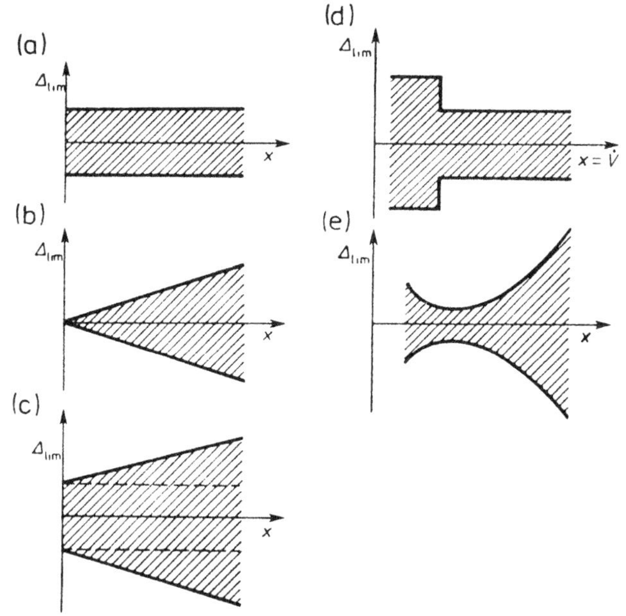

Fig. 4.10. Exemplary limiting error distributions: (a) analog electric meters, (b) thermocouple, strain gauge, (c) digital voltmeter, resistance thermometer, (d) water meter, (e) orifice plate.

great potential of digital techniques applied in measurements make the cost barrier relatively lower.

The effects of the most advanced process of standardization of metrological characteristics are presented in the GOST 8.009-84 Standard [27]. In 1972, work on an earlier version of this standard (GOST 8.009-72) was in progress, although not yet completed. The concepts and characteristics covered by the standard have since been modified in subsequent documents, such as the Comecon Recommendations 53–78 (1978) [74] and the III Recommendations of OIML [70] (in 1981–83). However, they have never been published as an official paper of the OIML.[7]

The present standard, GOST 8.009-84, is similar to the recommendations of the OIML. The standard itself is 15 pages long plus 27 pages of appendices, while the methodological instructions cover more than 100 pages. The instructions include interpretations of the models and the methods of determining the characteristics, as well as the criteria for selecting these characteristics.

The GOST standard determines the metrological characteristics of types of measuring instruments or of the transducers which are elements of measurement

[7] OIML—Organisation International de Metrologie Legal.

systems. The characteristics presented in the standard may not be used to describe a particular specimen of an instrument, checking post, standard instrument or standard.

The characteristics (in particular the random and systematic components of error) are random variables for the set of instruments of a given type.

The following characteristics of errors are recommended by the GOST 8.009-84 Standard.

1. The characteristics of systematic errors:

(a) value of a systematic error Δ_s;

(b) expected value $E(\Delta_s)$ and mean square deviation $\sigma(\Delta_s)$.

It should be borne in mind that these are distributions with independent variable x.

2. The characteristics of random errors:

(a) mean square deviation of a random error $\sigma(\Delta_r)$;

(b) as in (a), together with the standardized correlation function of a random error $r_\Delta(\tau)$ or the spectral density $S_\Delta(\omega)$;

(c) variance H of the output hysteresis error; H is one half of the hysteresis interval, and the distribution of the hysteresis error is uniform over the interval $2H$.

3. Complete or partial dynamic characteristics of those instruments which may be considered linear elements. The characteristics are well known; the partial characteristics are the functionals or the parameters of respective complete characteristics.

The dynamic characteristics of a given kind of elements should also include the tolerances of the parameters; standardization is thus detailed.

The standard includes some details about cases when particular kind of errors are considered. It should be noted that it is recommended to neglect the supplementary errors if their sum is small enough, namely

$$\left(\sum_{i=1}^{m-1} \Delta_{i\,\max}\right) < 0.17\Delta_{\max}. \tag{4.78}$$

If a maximum value of a dynamic error exceeds

$$\Delta_{\text{dyn}} > 0.17\Delta_{\max}, \tag{4.79}$$

the error is also regarded in some non-conventional way. The methodological instructions give the interpretation of an error of a measuring instrument as a non-stationary random process with the following components:

$$\Delta = \Delta_s + \Delta_{\text{rlf}} + \Delta_{\text{rhf}} + \Delta_{\text{rq}}, \tag{4.80}$$

where Δ_s or $E(\Delta_s)$ is a systematic error whose time changes make the process non-stationary, Δ_{rlf} is a low-frequency random component of error, and is described by the correlation function, Δ_{rhf} is a high-frequency random component of error; it is considered a white noise, Δ_{rq} represents other components of random error (e.g. quantization error).

The above interpretation is similar to the general model of error (3.136).[8] The standards in other countries [149, 169] do not include the definitions of metrological characteristics. The DIN 1319 Standard, Parts 3 and 4, partially follows the Recommendation CI-1981 of the CIPM [178]. The concept of systematic error is understood differently and random errors are not divided into two classes A and B explicitly.

Novitskii [78] has proposed an original way of simplifying some of the metrological characteristics. His idea is that the probability distributions of errors of instruments of one type are of the same kind. For example, in the case of analog electrical meters, the distribution is normal, for digital instruments it is uniform, etc. Novitskii's proposal was to characterize the errors by substitute uniform distributions with parameters such that the entropies of true and substitute distributions are equal. The relationship of the parameter of a uniform distribution (probability value $p = 1/2\Delta$) and the standard deviation σ of the true distribution may be obtained from the comparison of entropies. The relationship is called the *entropy coefficient* K:

$$K = \frac{\Delta}{\sigma}. \tag{4.81}$$

The coefficient K may be calculated for different distributions; its value varies from about 0.5 up to 2.07 for a normal distribution (see Fig. 4.11b). If Novitskii's thesis is true, the coefficient K would be characteristic for a type of MS whose individual properties were described by Δ. Then, instead of two parameters (a confidence interval and a confidence level), one parameter only could be used, since the interval 2Δ gives $\alpha = 1$ in the case of a uniform distribution.

The above thesis had to be proved experimentally. Novitskii experimented with:

(1) analog electrical meters—19 samples;

(2) electronic instruments for measurements of non-electrical quantities—21 samples;

(3) digital meters—7 samples.

---

[8] The GOST 8.009-72 Standard also included other characteristics, such as the intensity of exceeding the allowed value (a treshold) by the systematic component:

$$\lambda_s = \frac{1}{P_s}\frac{\mathrm{d}P_s}{\mathrm{d}T},$$

where P_s is the probability that the systematic component remains within the allowed limits and T is time. This characteristic was not included in the new Standard.

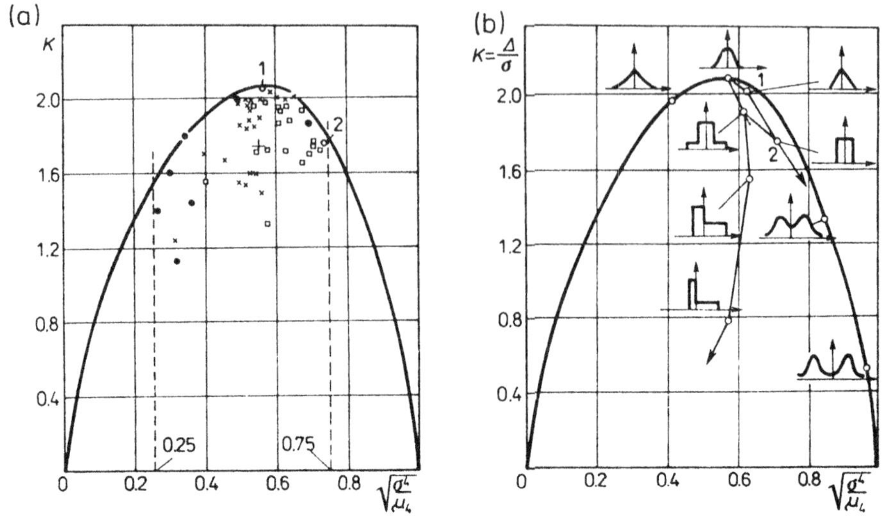

Fig. 4.11. Dispersion of the probability distributions of the instruments investigated by Novitskii [78]: (a) experimental results, (b) location of various distributions in a selected system of coordinates; μ_4—fourth-order central moment, σ—standard deviation, $\times$—the electric meters of non-electric quantities, □—electromechanical meters, ●—digital meters, +—mean value of K from the results obtained in all the experiments, *1*—normal distribution, *2*—uniform distribution.

The results show that the error distributions are not identical for every type of instrument (see Fig. 4.11a). As may be seen, the dispersion of the values of K is considerable: for a given type, it is $\pm(0.2\text{–}0.4)$ [78]. This result means that the idea must be rejected, for the metrological properties of a particular type of instrument cannot be described by one kind of probability distribution. Therefore, the distribution of errors is not a specific property of a type of MS. Moreover, it is well known that the distributions change over the service time of an MS because of contamination, ageing and wearing. Thus, the entropy coefficient cannot be introduced as an individual property of an MS. The scope of experiments necessary to determine the probability distribution of errors would be much too great in this case.

4.5 Metrological Characteristics. Methodology of their Determination and Interpretation

4.5.1 Determination of Class 1 Characteristics

The characterization of an error process $\{\Delta_0(x, T)\}$ is a somewhat complex task because of the fact that the random process is bidimensional and non-stationary. While the model of a characteristic of an MS may be a linear equation, the errors should usually be modelled with a higher-order polynomial or another nonlinear function—this follows from the models, e.g. (3.36), (3.42). The model of an error may be

$$\{\Delta_0(x, T)\} = \{b_0\} + \{b_1\}x + \{b_2\}x^2 + \ldots + \{b_q\}x^q, \tag{4.82}$$

where each of the random processes $\{b_0\}, \ldots, \{b_q\}$ is non-stationary and may usually be reduced to model (3.136):

$$\{b_q\} = g_q\{\varepsilon_q\} + h_q. \tag{4.83}$$

Then, instead of (4.32), we may write

$$\Delta_c = h_0(x, T) = h_0(T) + h_1(T)x + h_2(T)x^2 + \ldots + h_q(T)x^q, \tag{4.84}$$

and instead of (4.33):

$$\{\Delta_{in}(x, T)\} = g_0(x, T)\{\varepsilon\} = g_0(T)\{\varepsilon_0\} + g_1(T)\{\varepsilon_1\}x + g_2(T)\{\varepsilon_2\}x^2 + \ldots + g_q(T)\{\varepsilon_q\}x^q. \tag{4.85}$$

To describe the errors, $q+1$ pairs of the service time functions must be determined. To do this, it is necessary to make measurements at least at $n = q+1$ points of the measurement range, i.e., at points $x = x_1, \ldots, x = x_n$. When $n = q+1$, the number of measurements is equal to the number of unknowns and a certain number of equations should be solved. If $n > q+1$, the equations should be solved by the least squares method. This problem is more complicated, because:

(a) the quantities being determined are time functions $g(T)$, $h(T)$ and not the constant coefficients g, h. The task is thus two-dimensional;

(b) the processes $\{\varepsilon_r\}$ and $\{\varepsilon_s\}$ may be cross-correlated and in this case not only $q+1$ functions $g(T)$, $h(T)$, but also $\binom{q+1}{2}$ cross-correlation functions should be determined.

For the case of $n = q+1$ equations and unknowns, the number will be: n for the function $h(T)$, n for the function $g(T)$, and $\binom{n}{2}$ for the cross-correlation coefficients $\varrho_{rs}(\tau)$, which together make $N = 2n + \binom{n}{2}$.

The computational procedure will be presented in detail for the simplest case $q = 1$,[9] i.e., the error may be expressed by the following two non-stationary random processes:

$$\{\Delta(x, T)\} = \{\tilde{k}(T)\}x + \{\tilde{x}_0(T)\}, \tag{4.86}$$

[9] This model is in many cases insufficient. In an MS with linear characteristics, the phenomena described by the error components b_2x^2, b_3x^3, ... influence the output quantity of an MS about 100 times less strongly than the phenomena affecting the zero and the sensitivity. Our discussion concerns the variability of these phenomena. The variability of b_2, b_3, ... will be taken into consideration only if it is many times greater.

which characterize the sensitivity error and the zero error [91]. Referring to model (4.82), we have $b_0 = \tilde{x}_0$ and $b_1 = \tilde{k}$. Each of the component processes (4.82) may be described by model (4.83):

$$\{\tilde{k}\} = g_k(T)\{\varepsilon_k\}+h_k(T), \tag{4.87a}$$

$$\{\tilde{x}_0\} = g_0(T)\{\varepsilon_0\}+h_0(T). \tag{4.87b}$$

Since model (4.86) has two degrees of freedom, the instability error may be investigated only for two values of x: x_1 = constant and x_2 = constant. The numerical results may be analytically described by model (4.83), namely

$$\{\Delta(x_1, T)\} = g_1(T)\{\varepsilon_1\}+h_1(T), \tag{4.88a}$$

$$\{\Delta(x_2, T)\} = g_2(T)\{\varepsilon_2\}+h_2(T). \tag{4.88b}$$

According to the assumption of model (4.28), we have

$$E\{\Delta(x_1, T)\} = \hat{h}_1(T); \quad E\{\Delta(x_2, T)\} = \hat{h}_2(T). \tag{4.89}$$

By applying the same definition to (4.86) and substituting (4.87) in (4.86), we obtain

$$E\{\Delta(x_1, T)\} = \hat{h}_k x_1+\hat{h}_0, \quad E\{\Delta(x_2, T)\} = \hat{h}_k x_2+\hat{h}_0. \tag{4.90}$$

Hence,

$$\hat{h}_k x_1+h_0 = h_1, \tag{4.91a}$$

$$\hat{h}_k x_2+\hat{h}_0 = \hat{h}_2. \tag{4.91b}$$

From the above system of equations we may calculate

$$\hat{h}_0(T) = \frac{x_2\hat{h}_1(T)-x_1\hat{h}_2(T)}{x_2-x_1}, \tag{4.92a}$$

$$\hat{h}_k(T) = \frac{\hat{h}_2(T)-\hat{h}_1(T)}{x_2-x_1}. \tag{4.92b}$$

The random components of error may be presented in a similar way, based on the processes

$$g(x_1, T)\{\varepsilon_1\} = \{\Delta(x_1, T)\}-\hat{h}_1(T) \quad \text{for} \quad x = x_1, \tag{4.93a}$$

$$g(x_2, T)\{\varepsilon_2\} = \{\Delta(x_2, T)\}-\hat{h}_2(T) \quad \text{for} \quad x = x_2, \tag{4.93b}$$

obtained from estimation of the functions $\hat{h}_1$, $\hat{h}_2$. The processes $\{\varepsilon\}$ are stationary, with mean values equal to zero (an effect of condition (4.89)). It is also assumed that

$$g(x_1, T) = g(x_1, T+\tau) = g_1 = \text{constant}, \tag{4.94a}$$

$$g(x_2, T) = g(x_2, T+\tau) = g_2 = \text{constant}. \tag{4.94b}$$

The correlation functions are

$$K_1(\tau) = E[g_1\{\varepsilon_T\}g_1\{\varepsilon_{T+\tau}\}] = \hat{g}_1^2\varrho_1(\tau), \tag{4.95a}$$

$$K_2(\tau) = E[g_2\{\varepsilon_T\}g_2\{\varepsilon_{T+\tau}\}] = \hat{g}_2^2\varrho_2(\tau), \tag{4.95b}$$

$$R_{12}(\tau) = E[g_1\{\varepsilon_T\}g_2\{\varepsilon_{T+\tau}\}] = \hat{g}_1\hat{g}_2\varrho_{12}(\tau), \tag{4.95c}$$

and may be calculated from the results (4.93).

By applying formulae (4.86), (4.87), we may obtain a relationship with either x_1 or x_2:

$$g(x, T)\{\varepsilon\} = g_k\{\varepsilon_k\}x + g_0\{\varepsilon_0\}, \tag{4.96}$$

and

$$\begin{aligned} K(\tau) &= E[(g_k\{\varepsilon_{k,T}\}x + g_0\{\varepsilon_{0,T}\})(g_k\{\varepsilon_{k,T+\tau}\}x + g_0\{\varepsilon_{0,T+\tau}\})] \\ &= \hat{g}_k^2x^2\hat{\varrho}_k(\tau) + 2\hat{g}_k\hat{g}_0x\hat{\varrho}_{k0}(\tau) + \hat{g}_0^2\hat{\varrho}_0(\tau). \end{aligned} \tag{4.97}$$

A comparison of formulae (4.95) and (4.97) rewritten with x_1, x_2 and including the cross-correlation functions gives a system of equations:

$$x_1^2\hat{g}_k^2\hat{\varrho}_k(\tau) + 2x_1\hat{g}_k\hat{g}_0\hat{\varrho}_{k0}(\tau) + \hat{g}_0^2\hat{\varrho}_0(\tau) = \hat{g}_1^2\hat{\varrho}_1(\tau), \tag{4.98a}$$

$$x_2^2\hat{g}_k^2\hat{\varrho}_k(\tau) + 2x_2\hat{g}_k\hat{g}_0\hat{\varrho}_{k0}(\tau) + \hat{g}_0^2\hat{\varrho}_0(\tau) = \hat{g}_2^2\hat{\varrho}_2(\tau), \tag{4.98b}$$

$$x_1x_2\hat{g}_k^2\hat{\varrho}_k(\tau) + (x_1+x_2)\hat{g}_k\hat{g}_0\hat{\varrho}_{k0}(\tau) + \hat{g}_0^2\hat{\varrho}_0(\tau) = \hat{g}_1\hat{g}_2\hat{\varrho}_{12}(\tau). \tag{4.98c}$$

The solution of this system is

$$\hat{g}_0^2\hat{\varrho}_0(\tau) = \frac{x_1^2\hat{g}_2^2\hat{\varrho}_2(\tau) + x_2^2\hat{g}_1^2\hat{\varrho}_1(\tau) - 2x_1x_2\hat{g}_1\hat{g}_2\hat{\varrho}_{12}(\tau)}{(x_2-x_1)^2}, \tag{4.99a}$$

$$\hat{g}_k^2\hat{\varrho}_k(\tau) = \frac{\hat{g}_2^2\hat{\varrho}_2(\tau) + \hat{g}_1^2\hat{\varrho}_1(\tau) - 2\hat{g}_1\hat{g}_2\hat{\varrho}_{12}(\tau)}{(x_2-x_1)^2}, \tag{4.99b}$$

$$\hat{g}_k\hat{g}_0\hat{\varrho}_{k0}(\tau) = -\frac{x_1\hat{g}_2^2\hat{\varrho}_2(\tau) + x_2\hat{g}_1^2\hat{\varrho}_1(\tau) - (x_1+x_2)\hat{g}_1\hat{g}_2\hat{\varrho}_{12}(\tau)}{(x_2-x_1)^2}. \tag{4.99c}$$

In some cases the above formulae take a much more simple form. Thus when $x_1 = 0$, instead of (4.92) we obtain

$$\hat{h}_0(T) = \hat{h}_1(T), \tag{4.100a}$$

$$\hat{h}_k(T) = \frac{\hat{h}_2(T) - \hat{h}_1(T)}{x_2}, \tag{4.100b}$$

and instead of (4.99)

$$\hat{g}_0^2\hat{\varrho}_0(\tau) = \hat{g}_1^2\hat{\varrho}_1(\tau), \tag{4.101a}$$

$$\hat{g}_k^2\hat{\varrho}_k(\tau) = \frac{\hat{g}_2^2\hat{\varrho}_2(\tau) + \hat{g}_1^2\hat{\varrho}_1(\tau) - 2\hat{g}_1\hat{g}_2\hat{\varrho}_{12}(\tau)}{x_2^2}, \tag{4.101b}$$

$$\hat{g}_k\hat{g}_0\hat{\varrho}_{k0}(\tau) = \frac{\hat{g}_1\hat{g}_2\hat{\varrho}_{12}(\tau) - \hat{g}_1^2\hat{\varrho}_1(\tau)}{x_2}. \tag{4.101c}$$

When $\tau = 0$, the equations are further simplified. The correlation coefficients are all one and equations (4.101) become

$$\hat{g}_0^2 = \hat{g}_1^2, \tag{4.102a}$$

$$x_2^2\hat{g}_k^2 = \hat{g}_1^2+\hat{g}_2^2-2\hat{g}_1\hat{g}_2\hat{\varrho}_{12}, \tag{4.102b}$$

$$x_2\hat{g}_0\hat{g}_k\hat{\varrho}_{k0} = \hat{g}_1\hat{g}_2\hat{\varrho}_{12}-\hat{g}_1^2. \tag{4.102c}$$

The above formulae are the basis for determination of the function $g(T)$ for $T \gg \tau$. The functions $\hat{g}_1(T)$ and $\hat{g}_2(T)$ may be approximated by polynomials, and then $\hat{g}_0, \hat{g}_k$ are obtained in this form as well. The bases for finding these functions are the results at instants $T_0, T_1, T_2, \ldots$, i.e., as for $\hat{h}_0, \hat{h}_k$. Note the fact that lack of correlation of the processes $\{\tilde{k}\}$ and $\{\tilde{x}_0\}$ in equation (4.102c) implies the condition

$$\hat{\varrho}_{12} = \frac{\hat{g}_1}{\hat{g}_2}, \tag{4.103}$$

which, considering (4.102b), takes the form

$$x_2^2\hat{g}_k^2 = \hat{g}_2^2-\hat{g}_1^2. \tag{4.104}$$

When the correlation is strong ($\hat{\varrho}_{12} = 1$), we obtain instead of (4.102b)

$$x_2^2\hat{g}_k^2 = (\hat{g}_2-\hat{g}_1)^2. \tag{4.105}$$

The physical interpretation of the above formulae is simple and may thus provide grounds for the methodology of determination of the presented characteristics of errors.

To determine the characteristics, the experiments should be performed according to the following programme. For constant values of x_1 and x_2, the error should be measured with a sampling interval Δt. The result is a sequence of error realizations:

$$\Delta_{1ij} \quad \text{for } x = x_1, \quad i = 1, 3, 5, 7, \ldots, N-1, \quad j = 0, 1, 2, \ldots, M,$$

$$\Delta_{2ij} \quad \text{for } x = x_2, \quad i = 2, 4, 6, 8, \ldots, N, \quad j = 0, 1, 2, \ldots, M.$$

The index i is the sample number at the jth realization of a non-stationary process. At the time $N\Delta t$ when the process is realized, $g(T)$ should be constant. The same condition for $h(T)$ is not necessary, and depends only on the calculation technique. The index j is the number of current realizations of the processes $\{\Delta_1\}$, $\{\Delta_2\}$. For subsequent realizations it should be $g(T_{j+1})-g(T_j) \neq 0$ or $h(T_{j+1})-$ $-h(T_j) \neq 0$, and $T_{j+1}-T_j > N\Delta t$. The duration of the experiment is

$$T = \sum_{i=1}^{M} T_j.$$

The minimum sample size is $N \geqslant 200$, $M \geqslant 5$, depending on the properties of an MS. The value of the sampling interval Δt is limited by the switching techniques and by the dynamic properties of the MS. The duration of the experiment should be sufficient part of the service time of an MS, to provide a worthwhile application of the MS. The characteristics may be calculated from the above formulae, following the schemes shown in Fig. 4.12. Similar recommendations were presented by Olszewska [80], who applied model (3.147) to describe errors using some different method of identification of the coefficients of the model.

As compared with the model of error (3.147) introduced in Chapter 3, the scope of the properties of the errors of an MS considered here is wider. Model

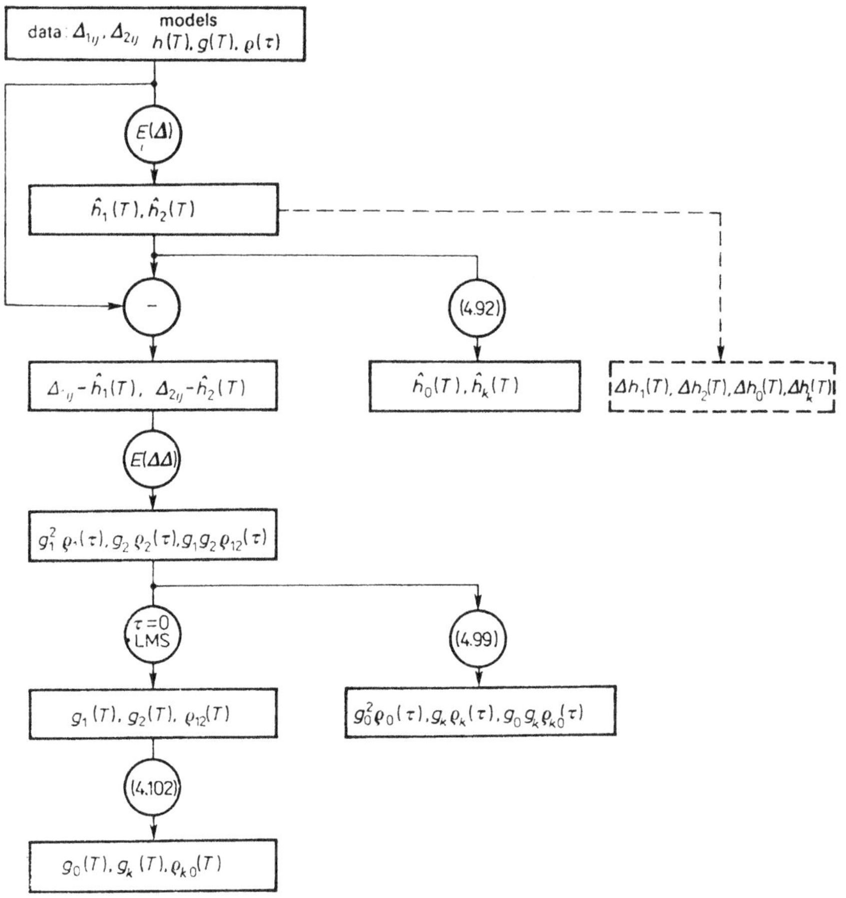

Fig. 4.12. Flowchart for calculating the characteristics of systematic and random errors expressed by two random processes.

(3.147) is equivalent to model (4.88), where $h(T)$ corresponds to the component $\Sigma\gamma_j t^j$ (3.148b)[10] and $g\varrho(\tau)$ corresponds to (3.148a). In model (4.88), which includes two processes for x_1, x_2, their cross-correlation was considered as well as the time variability of their properties, described by the functions $g(T)$ and $\varrho_{12}(T)$.

The correlation of the processes $\{\tilde{k}\}$ and $\{\tilde{x}_0\}$ included in the above models may be given theoretical grounds, based on the mathematical model of an MS. Referring to model (4.86), the sensitivity k and zero point x_0 are generally the coefficients b_l in model (3.72), dependent on the construction parameters $a_1, \ldots$ $\ldots, a_k$. Let us consider a situation when the numbers k_r and k_s of the construction parameters appear in formula (3.72) for $l = 0, 1, 2, \ldots, q$ and the formulae express any two coefficients b_r, b_s (e.g. $b_r = k$, $b_s = x_0$). At the same time, let $n > 0$ parameters appear in the formulae describing b_r and b_s. The coefficients b_r, b_s are then correlated and the correlation coefficients may be calculated as described below. According to the interpretation (3.112),

$$a_j = a_{j0} + \Delta\tilde{a}_j. \tag{4.106}$$

The total differential (4.37) applied to formula (3.72) expresses

$$\Delta\tilde{b} = \sum_{j=1}^{k} c_j \Delta\tilde{a}_j = u. \tag{4.107}$$

Then, after denoting

$$\Delta\tilde{a}_j = z_j \quad \text{or} \quad v_j \quad \text{or} \quad w_j, \tag{4.108}$$

we may write

$$u_r = \sum_{j=1}^{n} c_{rj} z_j + \sum_{j=n+1}^{k_r} c_{rj} v_j, \tag{4.109a}$$

$$u_s = \sum_{j=1}^{n} c_{sj} z_j + \sum_{j=n+1}^{k_s} c_{sj} w_j. \tag{4.109b}$$

The symbols z_j denote the variables included in the coefficients b_r and b_s. When the mean values of particular variables equal zero and the variables z, v, w are not correlated, the correlation coefficient is

$$\varrho_{rs} = \frac{E(u_r u_s)}{\sigma_r \sigma_s} = \frac{\sum_{j=1}^{n} c_{rj} c_{sj} z_j^2}{\sqrt{\sum_{j=1}^{n} c_{rj}^2 z_j^2 + \sum_{j=n+1}^{k_r} c_{rj}^2 v_j^2}\sqrt{\sum_{j=1}^{n} c_{sj}^2 z_j^2 + \sum_{j=n+1}^{k_s} c_{sj}^2 w_j^2}}. \tag{4.110}$$

---

[10] The equality $\hat{\gamma}_j = \hat{b}_q$ for $j = q$ is not valid here, since the coefficients of model (3.147) are determined simultaneously and thus $\hat{b}_q = f(\hat{\gamma}_j, \hat{\varphi}_i)$.

The value of ϱ_{rs} depends on the specific data. To make the relationship clearer, let us assume the equal sensitivities and variances of individual variables. Then [84]

$$\varrho_{rs} = \frac{n}{\sqrt{k_r k_s}}. \tag{4.111}$$

As an example, simplified models of gas analysers will be considered. The analysers are thermoconductometric and thermomagnetic [110]. In a thermoconductometric analyser (cathetometer), the second stage of conversion is described by the following heat balance equation (see Fig. 4.13):

$$Q = Q_C + Q_R, \tag{4.112}$$

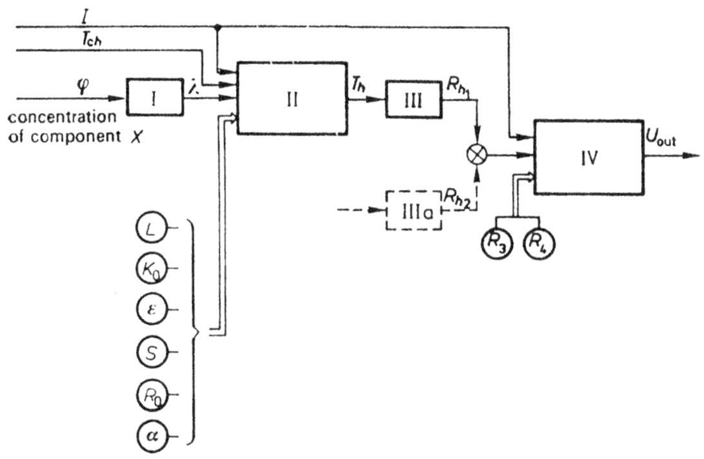

Fig. 4.13. The structural diagram of a thermoconductometric analyser (cathetometer). L—heater length, K_0—chamber form factor, ε—radiation rate, S—heater surface area, R_0—heater resistance at 0°C, α—temperature coefficient of resistance, R_3, R_4—resistances of the fixed resistors of the bridge.

where the components are:

$Q = I^2 R_0(1+\alpha T_h)$ power supplied to the heater, (4.112a)

$Q_C = LK_0 \lambda(T_h - T_{ch})$ heat losses by conduction, (4.112b)

$Q_R = \sigma\varepsilon S\left[\left(\frac{T_h}{100}\right)^4 - \left(\frac{T_{ch}}{100}\right)^4\right]$ heat losses by radiation. (4.112c)

The influencing quantities are the heater current I and the chamber temperature T_{ch}. The output quantity of this conversion stage is the heater temperature T_h. The measured quantity is λ, σ is the Boltzmann constant and the other quantities are the construction parameters listed in Table 4.2 and explained in Fig.

TABLE 4.2

Notation of the construction parameters and sensitivities to these parameters. The factor M was not calculated, as it is reduced in formula (4.110)

j	1	2	3	4	5	6
$\Delta\tilde{a}_j$	$\frac{dL}{L}$	$\frac{dK_0}{K_0}$	$\frac{d\varepsilon}{\varepsilon}$	$\frac{dS}{S}$	$\frac{dR_0}{R_0}$	$\frac{d\alpha}{\alpha}$
C_{jx_0}	1	$\frac{Q_C}{Q}$	$\frac{Q_R}{Q}$	$\frac{Q_R}{Q}$	-1	$-\frac{\alpha T_h}{1+\alpha T_h}$
	1	0.9	0.1	0.1	-1	-0.5
C_{jk}	$\frac{Q_C}{Q}$	$\frac{Q_C}{Q}$	0	0	0	0
	0.9	0.9	0	0	0	0

4.13. Disregarding the conversion stages I, III, IV,[11] the changes of the zero point may be expressed as[12]

$$dT_h = M\left[\frac{dL}{L}+\frac{Q_C}{Q}\frac{dK_0}{K_0}+\frac{Q_R}{Q}\left(\frac{d\varepsilon}{\varepsilon}+\frac{dS}{S}\right)-\frac{dR_0}{R_0}-\frac{\alpha T_h}{1+\alpha T_h}\frac{d\alpha}{\alpha}\right]. \tag{4.113a}$$

The respective sensitivity changes of the second stage are described by the formula

$$d\left(\lambda\frac{dT}{d\lambda}\right) = M\frac{Q_C}{Q}\left[\frac{dL}{L}+\frac{dK_0}{K_0}\right]. \tag{4.113b}$$

The value of the correlation coefficient, calculated from formula (4.110) using the data from Table 4.2, is

$$\varrho_{kx_0} \approx 0.8.$$

The heat balance equation for thermomagnetic analysers is [87]

$$Q = Q_C+Q_R+Q_T+Q_M,$$

where $Q_T = LK\lambda(T_h-T_{ch})(\mathrm{GrPr})^a$—heat losses by thermal convection, $Q_M = LB\lambda(T_h-T_{ch})(H\,\mathrm{grad}\,H)^b(T_h+T_{ch})\varphi c(\mathrm{GrPr})^b/\varrho_0 T_{ch}^2$—heat losses by thermomagnetic convection, a, b, c,—constant coefficients, λ—thermal conductivity of the gas, ϱ_0—density, Gr—Grasshoff number, Pr—Prandtl number, φ—oxygen concentration (measured quantity), H—magnetic field strength, grad H—its

[11] This means that some of the construction parameters (e.g. the fixed resistors of the bridge) are disregarded.

[12] Two or four heaters are mounted in an analyser; this has not been taken into account in the discussion.

gradient, other symbols as in (4.112a-c). λ, ϱ_0, Gr, Pr are considered influencing quantites.

Denoting the quantities as in Table 4.3 and after calculations applying formula (4.110), the cross-correlation coefficient of the zero and sensitivity is $\varrho_{kx0} \approx 0.3$.

TABLE 4.3

A list of the construction parameters of a thermomagnetic oxygen analyser and the sensitivities to these parameters for a second conversion stage and a single heater

j	1	2	3	4	5	6	7	8	9	10
$\Delta\tilde{a}_j$	$\frac{dL}{L}$	$\frac{dK_0}{K_0}$	$\frac{d\varepsilon}{\varepsilon}$	$\frac{dS}{S}$	$\frac{dR_0}{R_0}$	$\frac{d\alpha}{\alpha}$	$\frac{dK}{K}$	$\frac{dB}{B}$	$\frac{dH}{H}$	$\frac{d\,\text{grad}\,H}{\text{grad}\,H}$
C_{jx_0}	1	$\frac{Q_C}{Q}$	$\frac{Q_R}{Q}$	$\frac{Q_R}{Q}$	-1	$-\frac{\alpha T_h}{1+\alpha T_h}$	$\frac{Q_T}{Q}$	$\frac{Q_M}{Q}$	$\frac{Q_M}{Q}$	$\frac{Q_M}{Q}$
	1	0.75	0.15	0.15	-1	-0.15	0.07	0.03	0.03	0.03
C_{jk}	$\frac{Q_M}{Q}$	0	0	0	0	0	0	$\frac{Q_M}{Q}$	$\frac{Q_M}{Q}$	$\frac{Q_M}{Q}$
	0.03	0	0	0	0	0	0	0.03	0.03	0.03

In the above calculations no proof was provided that the potential sources of errors were the real sources and the assumption was made that the standard deviations of the random variables $\Delta\tilde{a}_j$ ($j = 1, \ldots, k$) are equal (1% for all the variables). Thus, the calculations are theoretical and their aim is to show that the components of errors may be correlated and the degree of correlation may vary. To make the calculations precise, the real sources of the instability error should be determined empirically together with their statistical characteristics (at least their variance). The output quantity sensitivity to individual sources of errors and the correlation coefficients arise from the mathematical model, either derived theoretically or empirically.

In a particular instrument, the correlation between zero point output signal and sensitivity output signal may be quite different. For example, in a thermoconductometric analyser the only error source may be the change of heater position in the chamber (resulting from vibrations, transient dynamic forces when switching on the current, etc.), causing a change of K_0. Thus, the correlation coefficient is $\varrho = 1$. If the only source of error is a change in the properties of the material of a heater (i.e. the change of R_0, α), the correlation coefficients for the zero and sensitivity errors will equal 0. This last assessment, however, is not precise, since Table 4.2 concerns only conversion stages I and II.

The cause of correlation of zero and sensitivity errors is the variability of influencing quantities, because both components of supplementary error are deterministically dependent on the same influencing quantities. This fact is clearly seen from the mathematical model of characteristic (3.12).

The error of the standards used to determine a characteristic is another source of correlation of the coefficients of the model, analysed by Allisy [143]. For a standard resistance thermometer, the characteristic is

$$\frac{R}{R_0} = 1+A\theta+B\theta^2, \tag{4.114}$$

where θ is the temperature and R_0, A, B are constant coefficients. The value of R_0 is determined at the water triple point and the values of A and B at the freezing points of tin and zinc.

The errors of the coefficients A, B are expressed by the equations

$$\mathrm{d}A = \frac{\partial A}{\partial \theta_1}\mathrm{d}\theta_1 + \frac{\partial A}{\partial \theta_2}\mathrm{d}\theta_2 = \lambda_{A1}\mathrm{d}\theta_1 + \lambda_{A2}\mathrm{d}\theta_2, \tag{4.115a}$$

$$\mathrm{d}B = \frac{\partial B}{\partial \theta_1}\mathrm{d}\theta_1 + \frac{\partial B}{\partial \theta_2}\mathrm{d}\theta_2 = \lambda_{B1}\mathrm{d}\theta_1 + \lambda_{B2}\mathrm{d}\theta_2. \tag{4.115b}$$

If

$$E(\mathrm{d}\theta_1) = E(\mathrm{d}\theta_2) = 0, \quad \operatorname{var}\theta_1 = \operatorname{var}\theta_2 = \sigma^2, \quad \operatorname{cov}(\theta_1, \theta_2) = \varrho\sigma^2,$$

then the variances as well as the correlation coefficients of the errors of determining A and B may be calculated:

$$\sigma_A^2 = \sigma^2(\lambda_{A1}^2 + \lambda_{A2}^2 + 2\varrho\lambda_{A1}\lambda_{A2}), \tag{4.116a}$$

$$\sigma_B^2 = \sigma^2(\lambda_{B1}^2 + \lambda_{B2}^2 + 2\varrho\lambda_{B1}\lambda_{B2}), \tag{4.116b}$$

$$\sigma_{AB} = \sigma^2[\lambda_{A1}\lambda_{B1} + \lambda_{A2}\lambda_{B2} + \varrho(\lambda_{A1}\lambda_{B2} + \lambda_{A2}\lambda_{B1})]. \tag{4.116c}$$

Allisy [143] used the above formulae to calculate the variances for specific numerical data. The results are presented in Table 4.4. The final result is interesting. The correlation of the errors ΔA and ΔB, described by the coefficient ϱ_{AB}, is

TABLE 4.4

The effects of correlation of the coefficients A and B [143].

ϱ	$(\sigma_A/\sigma)\times 10^5$	$(\sigma_B/\sigma)\times 10^7$	ϱ_{AB}
−1	4.61	1.30	−1
−0.5	4.19	1.14	−0.992
0	3.72	0.96	−0.980
0.5	3.19	0.74	−0.965
1	2.55	0.41	−1

very strong and practically independent of the value of ϱ [143]. On the other hand, the influence of the correlation of errors ($\varrho = 0$, $\varrho = 0.5$) on the estimate of the limiting error of the thermometer characteristic reaches as much as 20%.

4.5.2 Theoretical Determination of the Limiting Distributions of Errors

Limiting distributions are used to describe the properties of a type of MS. The reason for the dispersion of individual characteristics is the random realizations of construction parameters and the variability of technological operations. In the production process, the form of the metrological characteristics may be influenced by the use of elements whose parameters fall in a finite tolerance range, by testing and correction of the properties of individual units, and by proper selection of the calibration procedures. The result of these operations is verified in the product tests.

The way in which the limiting distribution is calculated is the factor which decides whether the metrological properties are determined with proper precision or in great excess (according to information theory, the latter case is also a disadvantage). Because of a large number of degrees of freedom and the complex physical situation, the task is not easy.

The example of a magnetoelectric meter given below demonstrates a method of theoretical determination of the limiting distribution. When the only source of error is improper balancing of the measurement organ, the distribution of the error is

$$h_0(x) = -\frac{mg}{D} R \sin\beta \sin(k_1 x + \varphi_R - \varphi_\beta), \tag{4.117}$$

where the symbols in the formula are as explained in Fig. 3.25. The repeatability of the parameters m, g, D (mass of the measuring element, acceleration of gravity, spring rigidity) is good in every meter specimen. The values of R and φ_R determine the position of the centre of mass of the measuring element, referred to the axis. The angle φ_β determines the direction in which the axis is deflected from the perpendicular. The angle β is usually regarded as an influencing quantity; an example may be the case of a meter mounted in a car.

The following probability distributions of the construction parameters are assumed (the parameters in the production process are random variables):

radius R—Rayleigh distribution

$$p_1(R) = \frac{R}{\sigma_1^2} e^{-R^2/2\sigma_1^2}, \qquad R \geq 0, \tag{4.117a}$$

angle β—truncated normal distribution

$$p_2(\beta) = \frac{2}{\sqrt{2\pi}\sigma_2} e^{-\beta^2/2\sigma_2^2}, \qquad \beta \geq 0, \tag{4.117b}$$

angle φ_R—uniform distribution

$$p_3(\varphi_R) = \frac{1}{2\pi}, \quad \varphi_R \in \langle 0, 2\pi \rangle, \tag{4.117c}$$

angle φ_β—uniform distribution

$$p_4(\varphi_\beta) = \frac{1}{2\pi}, \quad \varphi_\beta \in \langle 0, 2\pi \rangle, \tag{4.117d}$$

angle $\varphi = k_1 x$—uniform distribution over the measurement range, e.g.

$$p_5(\varphi) = \frac{2}{\pi} = \frac{1}{Z}, \quad \varphi \in \langle 0, \pi/2 \rangle. \tag{4.117e}$$

The distribution $h_0(x)_{\lim}$ is limiting if the value of the error $h_0(x)$ does not exceed $h_0(x)_{\lim}$ when the random variables are realizations from the set of parameters R, β, φ_0 allowed by the production technology. The limiting value of the error may be calculated in two ways. The simplified method involves simple calculations, but the estimates of $h_0(x)_{\lim}$ are not very accurate: they are excessive and the value of the excess depends on the number of parameters (the more parameters, the greater the excess). The examples are intended to show the value of the excess. In the simplified method, the limiting value of the error is calculated from the limiting values of particular parameters. It is assumed at the time that the previously assumed limiting values of the parameters are not exceeded (i.e., element checking has been performed, and the elements with parameters outside the limits have been rejected). Thus, for the meter discussed, when the distributions of the construction parameters are (4.117), the limiting values of these parameters at a confidence level 0.99 are

$$R_{\lim} = 2.81\sigma_1, \quad \beta_{\lim} = 2.58\sigma_2, \quad \sin(k_1 x + \varphi_0) = \pm 1. \tag{4.118}$$

Then

$$h_0(x)_{\lim} = \pm \frac{mg}{D} R_{\lim} \sin \beta_{\lim} \tag{4.119}$$

and denoting

$$u_0 = \frac{mg}{D}, \quad b = u_0 \sigma_1 \sigma_2, \tag{4.120}$$

we obtain

$$h_0(x)_{\lim} = u_0 \cdot 2.81\sigma_1 \cdot 2.58\sigma_2 = 7.25b. \tag{4.121}$$

The distribution of limiting error as a function of the measured quantity x is hidden in the factor $\sin(k_1 x + \varphi_0)$ (4.118). The limiting values of the factor are ± 1, but the error is a part of a sine curve with the domain $k_1 x \in \langle 0, \pi/2 \rangle$, depending on the random realization of φ_0 in a particular meter.

For different meters, the limiting distributions of errors

$$h_0(x)_{\lim} = 7.25b\sin(k_1 x+\varphi_0)$$

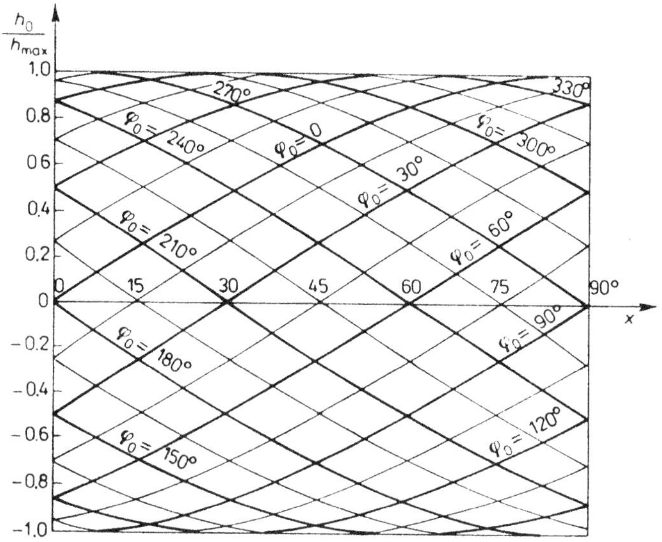

Fig. 4.14. Distributions of the systematic error of a magnetoelectric meter for different realizations of $\varphi_0 = 0$, 15, 30, 45, ..., 345°.

are shown in Fig. 4.14. The probability distributions of the error (by geometrical interpretation) are shown in Figs. 4.15a, b for different random values of the angle φ_0. Examples of real error distributions are shown in Figs. 4.15c, d. As may be seen, the error distributions of these meters are similar to the model distributions.

Some comment on the distributions shown in Fig. 4.14 is necessary.

(1) Because of the limited (90° deflection) measurement range of the meter (4.117e), the limiting values of the factor $\sin(k_1 x+\varphi_0)$ may be not ± 1, as stated in (4.118). In the worst case (angle $\varphi_0 = 45°$ or 225°), the limiting values are $\pm\sqrt{2}/2 = \pm 0.707$. In the best case ($\varphi_0 = 135°$ or 315°), the error varies from 0.707 to 1 (or -0.707 to -1).

(2) The error discussed has a particular cause, but its major part is eliminated in the calibration process. This effect must be considered in all calculations of that type. In the case of $\varphi_0 = 45°$ and a tangent approximation at the centre point of the measurement range (Fig. 3.4), we obtain the limiting values of the error which was not eliminated in the calibration process as ± 0.3. In the chord approximation (Fig. 3.4), the limiting values are smaller. For $\varphi_0 = 135°$ and

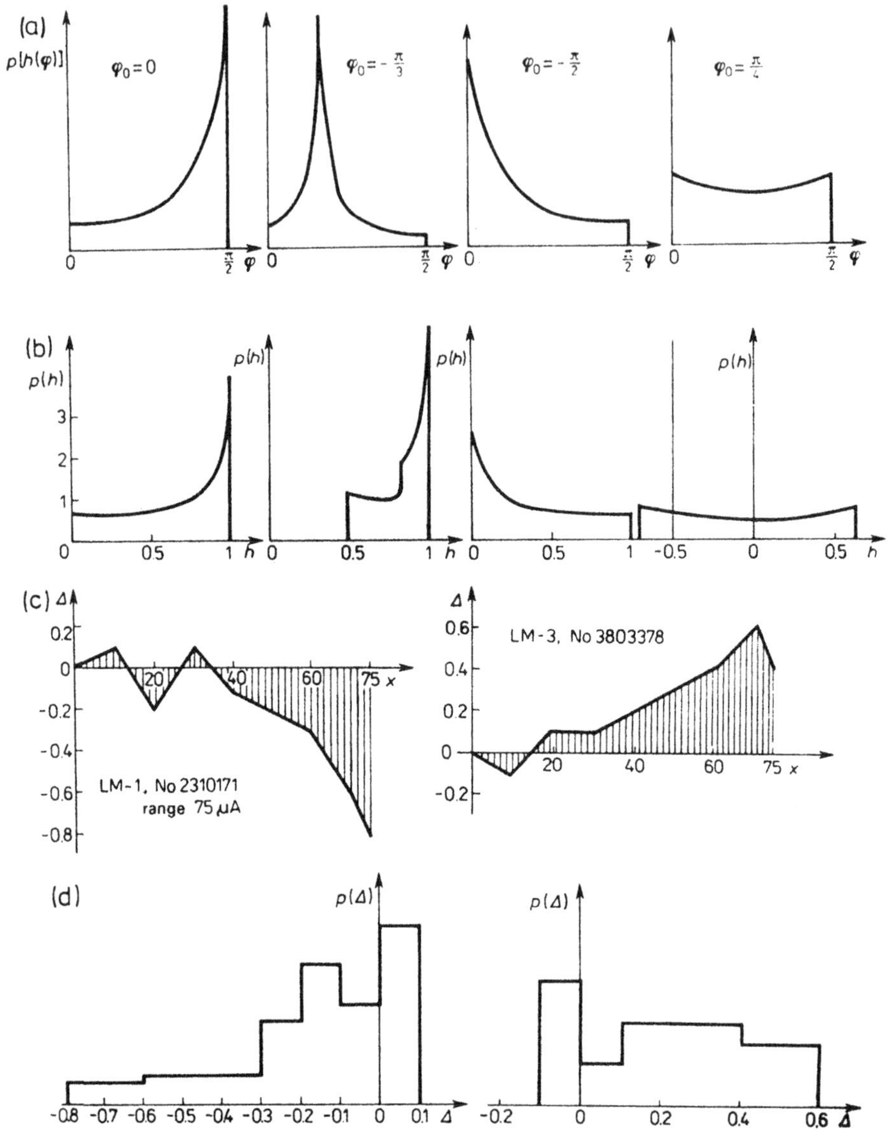

Fig. 4.15. The probability distribution of the systematic error (by geometric interpretation) of a magnetoelectric meter for different values of φ_0 (range of φ is $\pi/2$): (a) theoretical distribution of an argument φ, (b) theoretical distribution of an argument h, (c) error distribution of particular meters, (d) probability distributions of the errors as in (c).

a meter scale shift of -0.853 occurring during the calibration process, the limits of error are ± 0.15.

Accurate values of the limiting error are calculated after the probability distribution $p(h)$ is determined from the error model and from the distribution of parameters (4.117). The formulae from Table 3.1 should be used in calculations. For example, assuming $\sin\beta \approx \beta$ and after substituting the distributions of R and β, the distribution of the variable

$$z = R\beta \tag{4.122}$$

calculated using the formulae from Table 3.1 is (see Fig. 4.16)

$$p_6(z) = \int_{-\infty}^{\infty} \frac{1}{R} p_1(R) p_2\left(\frac{z}{R}\right) \mathrm{d}R = \int_0^{\infty} \frac{1}{R} \frac{R}{\sigma_1^2} \mathrm{e}^{-R^2/2\sigma_1^2} \frac{2}{\sqrt{2\pi}\sigma_2} \mathrm{e}^{-(1/2\sigma_2^2)(z^2/R^2)} \mathrm{d}R$$

$$= \frac{2}{\sqrt{2\pi\sigma_1^2}\sigma_2} \int_0^{\infty} \mathrm{e}^{-(a^2R^2+c^2/R^2)} \mathrm{d}R = \frac{2}{\sqrt{2\pi\sigma_1^2}\sigma_2} \frac{\sqrt{\pi}}{2a} \mathrm{e}^{-2ac} = \frac{u_0}{b} \mathrm{e}^{-u_0 z/b}. \tag{4.123}$$

Taking into account (4.122) and (4.120) we obtain instead of (4.116)

$$h_0(x) = u_0 z \sin(\varphi + \varphi_0), \tag{4.124}$$

where $\varphi = k_1 x$ and $\varphi_0 = \varphi_R - \varphi_\beta$. As the probability distributions of the angles φ_R, φ_β are uniform, the distribution of the angle φ_0 is also uniform in the domain

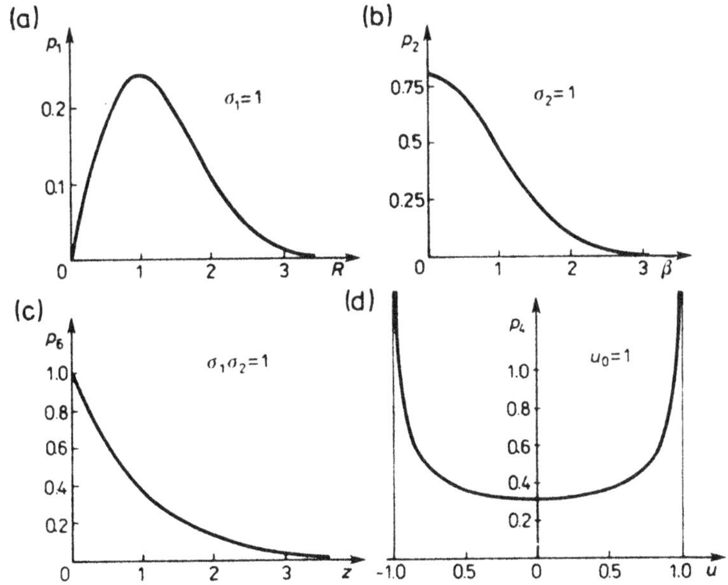

Fig. 4.16. The probability distributions of the construction parameters of a magnetoelectric meter: (a) distance of the mass centre from the axis, (b) deflection angle of the axis from the perpendicular, (c) random variable $z = R\beta$, (d) function $u = \sin\varphi$.

2π, as shown in Fig. 4.17. The limiting distribution of the error $h_0(x)$ is determined by the amplitude, since the values ± 1 are the limiting values of the factor $\sin(\varphi+\varphi_0)$, as in (4.118).

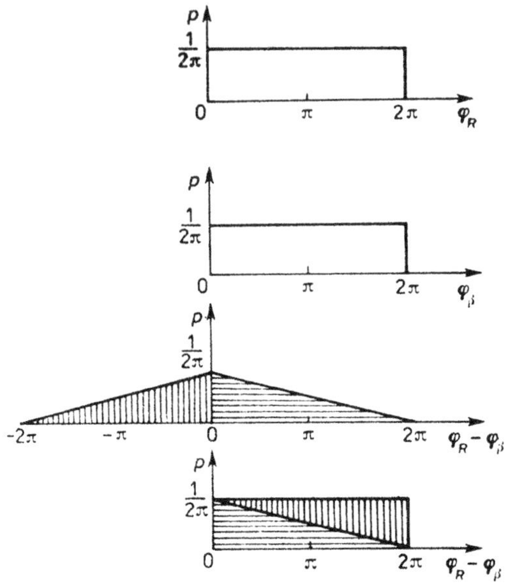

Fig. 4.17. The distribution of the random variable $\varphi_0 = \varphi_R - \varphi_\beta$.

For the confidence level α, the limiting value of error calculated from (4.123) fulfils the condition

$$\int_0^{z_{\lim\alpha}} p_6(z)\,dz = \int_0^{h_{\lim\alpha}} p_6\left(\frac{h}{u_0}\right)dz = \alpha, \tag{4.125}$$

and after substituting (4.123)

$$\int_0^{h_{\lim\alpha}} \frac{1}{b}e^{-h/b}\,dh = 1-e^{-h_{\lim\alpha}/b} = \alpha. \tag{4.126}$$

Hence

$$h_{\lim\alpha} = 4.61b \quad \text{for} \quad \alpha = 0.99 \tag{4.127a}$$

and

$$h_{\lim\alpha} = 3.00b \quad \text{for} \quad \alpha = 0.95. \tag{4.127b}$$

Comparison of the results (4.121) and (4.127) indicates that exact estimation of the limiting error gives a value more than 1.5 times smaller, and thus more advantageous. The smaller the limiting value, the lower is the probability that the limiting value will not be exceeded (lower warrant of the instrument properties). Nevertheless, the process of determination of the limiting error is more complicated. In a magnetoelectric meter, some ten sources of systematic error may be distinguished, and the influence of the components $\sin x$, $\sin 2x$, $x \sin x$, etc., as well as the components with at least seven construction parameters realized randomly, is also strong [58].

A simpler example is a resistance thermometer whose mathematical model is

$$R = R_0(1+ax), \tag{4.128}$$

where R_0 is the resistance at 0°C, a is the temperature coefficient of resistance, x is the measured temperature. Let the values of the construction parameters R_0, a be limited by the values $\pm t_r$, $\pm t_a$ as follows:[13]

$$t_r = t\frac{\sigma_r}{R_0}, \quad t_a = t\frac{\sigma_a}{a}. \tag{4.129}$$

A rough estimate of the limiting distribution of error is given by the formula

$$\pm h_{\mathrm{lim}} = R_0[(1+ax)t_r + axt_a + axt_r t_a] \approx R_0[(1+ax)t_r + axt_a]. \tag{4.130}$$

For exact estimation, an approximate model of thermometer error may be used:

$$\Delta R = (1+ax)\Delta R_0 + R_0 x \Delta a, \tag{4.131a}$$

and the variance of the error is

$$\sigma^2 = R_0^2[(1+ax)^2\sigma_r^2 + (ax)^2\sigma_a^2]. \tag{4.131b}$$

Assuming a normal distribution of the parameters R_0 and a, denoting the confidence level α as $2t$ and applying the symbols in (4.129), we obtain

$$\pm h_{\mathrm{lim}\alpha} = R_0 \sqrt{(1+ax)^2 t_r^2 + (ax)^2 t_a^2}. \tag{4.131c}$$

Comparison of both estimators (4.130) and (4.131c) of the limiting error is shown in Fig. 4.18. For $x = 0$, the limits are identical, but the limiting distributions are different. The limiting distribution $h_{\mathrm{lim}\alpha}$ depends on the assumed confidence level. According to (4.129), the values t_r, t_a may include the interval 3σ, 2σ or, for example, 0.5σ. In the latter case, the probability distribution of the construction parameters, because of selection, is a section of a normal distribution and may be approximated by a uniform or trapezoid distribution, Figure 4.19 shows how the interval $\pm h_{\mathrm{lim}\alpha}$ narrows as the confidence level decreases. In the example of the resistance thermometer discussed, the confidence interval for $\alpha = 0.95$ will be narrower by 24% than the one shown in Fig. 4.18 for $\alpha = 0.99$. The random error distribution (characterized by a standard deviation $g(x)$) is calculated similarly.

---

[13] The dispersion of the parameter a is the effect of wire quality and usually characterizes the lots, not single resistors.

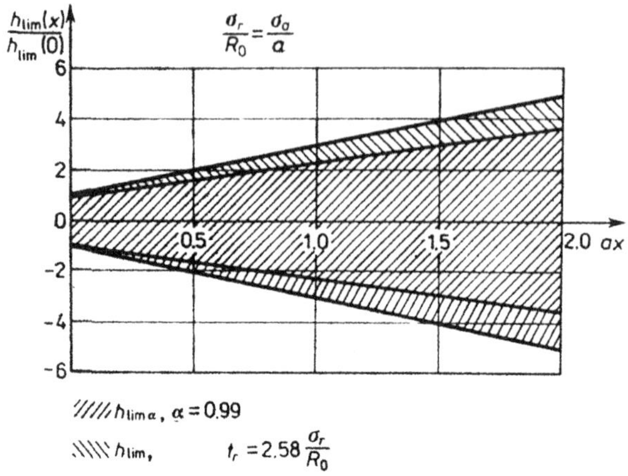

Fig. 4.18. The limiting distribution of a systematic error of a resistance thermometer.

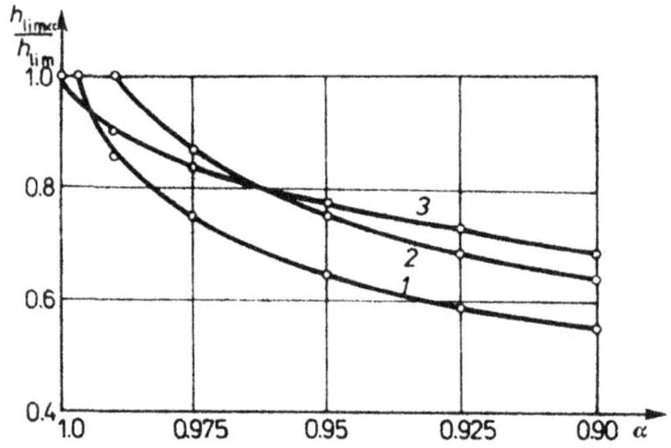

Fig. 4.19. The influence of the probability distribution of the construction parameters on the values of h_{lim} at different confidence levels: *1*—normal distribution $h_{lim} = 3\sigma$, *2*—normal distribution $h_{lim} = 1.96\sigma$, *3*—uniform distribution.

These calculations may be performed knowing the limiting values of the error sources (i.e. the compound errors $t_j g_j$):

$$(tg)_{lim} = \sum_{j=0}^{k} (t_j g_j). \tag{4.132a}$$

This is called the maximum error calculation principle, as the absolute values of the compound errors are added. An exact method of calculation consists of calculating the probability distribution of the total error and then of the limiting value at a given confidence level. In some cases a sufficient accuracy may be obtained more simply by calculating the variance of the total error as the sum of variances of component errors

$$\sigma^2 = \sum_{j=1}^{k} \sigma_j^2 . \tag{4.132b}$$

However, for nonlinear models and a different probability distribution of the error sources, such approximation is not accurate. Several examples of different distributions of errors are included in the works by Novitskii [170, 171].

4.5.3 Experimental Determination of the Limiting Distributions of Errors

The experimental determination of the distribution of systematic or random errors has a simple interpretation. N instruments should be investigated and separate error distributions ($h_i(x), g_i(x)$ at the points $x = x_j$, $j = 1, 2, \ldots, k$) should be calculated. Then the probability distributions $p[h(x_j)]$, $p[g(x_j)]$ should be determined, together with the limiting values $h_{\text{lim}}(x_j), g_{\text{lim}}(x_j)$ at a given confidence level. Next, the results are approximated by the functions $\hat{h}_{\text{lim}}(x), \hat{g}_{\text{lim}}(x)$. The limiting value is the calculated estimator, increased by half of the confidence interval of the estimation error. This procedure implies the following requirements for the sample sizes:

— number of instruments N as for the case of determination of the probability distribution, $N > 100$ (50);

— number of points k of the error-finding scale as for the estimation of a characteristic, i.e. $k \geqslant 5$ for each instrument;

— number of measurements n at a given point of the scale such that the systematic and random errors can be determined and the probability distribution of the random error can be derived; $n > 50$.

The sample size influences the estimation error which can be assessed using mathematical statistics. To obtain a reliable estimate, it is necessary to determine the probability distributions. Figure 4.11 shows that the distributions may differ. The methods with decreased numbers of measurements may also be applied here, but it is necessary to know whether the number of measurements may be decreased in a specific case. The determinant in this case is a piece of a priori information about the individual properties of an MS, i.e., about:

(1) the distribution of the variance of random errors $g^2(x)$,

(2) the model of the systematic error distribution,

(3) the ratio of systematic to random errors, expressed as the quotient of the variances (3.257) or the cross-correlation function of systematic error and total error [94].

For a measuring system, it is particularly important whether:

(a) a variance is constant,

(b) the variance changes according to a model (known at the level of coefficients); this case may be reduced to case (a) by the proper function transformation (3.246),

(c) the variance distribution is unknown—the number of measurements may not be reduced in this case.

In the case of constant variance (information probability $p = 1$), the effect of decreasing the number of measurements may be achieved by calculating the estimator of variance at different points x_j, $j = 1, \ldots, k$, with simultaneous estimation of $h_i(x)$. The probability distribution of random error may be obtained, either at one or two points of the measurement range, in order to verify the assumption $g(x) = \text{constant}$. It is the information on the model of systematic error distribution that decides whether a second-kind regression method may be used to identify the values of coefficients, as was discussed in Subsection 3.5.3. A rigid requirement for the model structure to conform to the real structure of an identified object is present here [68]. This condition allows avoiding the error of measurement method, while the structures not meeting the condition cause the model approximation error to be included in the model. The above requirement is essential for the methodology and should be satisfied in practice at the stage of deriving a characteristic $y = f_0(x)$ of a measurement system. When the errors are determined and the accuracy of the error value may be 5–10%, the requirement of the conforming structures is of minor importance. The systematic errors may be modelled with the simplest models, e.g. polynomials.

When the variance is assumed constant and the distribution of the systematic error is described by a qth order polynomial, formulae (3.259), (3.264) and (3.265) may be used in calculations. The model is not very complex and usually a polynomial of the order $q = 1, 2, 3$ is sufficient; the case when $q = 4$ is rather exceptional. An example of a result of estimation of $h(x)$ is shown in Fig. 3.40.

The result $\hat{h}_i(x)$ determined should be verified: the calculated coefficients should be significant. The F (Snedecor's) test (3.257) is applied for this purpose. Since the expected limiting distributions of systematic and random errors are known, the cross-correlation coefficient for the systematic and total error is also known

$$R^2_{\hat{h}\Delta} = \frac{\sum_{j=1}^{k} [\hat{h}(x_j) - \bar{h}](\Delta_j - \bar{\Delta})}{\sqrt{\sum_{j=1}^{k} [\hat{h}(x_j) - \bar{h}]^2 \sum_{j=1}^{k} (\Delta_j - \bar{\Delta})^2}} \tag{4.133}$$

The coefficients of the estimator of systematic error $\hat{h}(x)$ depend on the set of errors $\{\Delta_j\}$ (3.259). Thus, the numerator of expression (4.133) may be replaced by the sum od squares of the estimator deviations. Hence, the resulting expression [68]

$$R^2_{\hat{h}\Delta} = \frac{\sum_{j=1}^{k} [\hat{h}(x_j) - \bar{h}]^2}{\sum_{j=1}^{k} (\Delta_j - \bar{\Delta})^2} = \frac{s^2_{\hat{h}}}{s^2_{\Delta}} = R^2 \tag{4.134}$$

is more convenient for calculations, as the numerator multiplied by $1/k$ describes the sample variance of systematic error and the denominator (also multiplied by $1/k$) expresses the variance of total error. The correlation coefficient is related to the variable F (3.257):

$$F = \frac{R^2}{1-R^2} \frac{k-q-1}{q}. \tag{4.135}$$

Therefore, the cross-correlation coefficient may be applied instead of the F test [94]. In order to ensure that the regression coefficients are significant at the confidence level α, the number of measurements k should be sufficiently high, depending on the value of R (see Fig. 4.20). If the variance of the real systematic

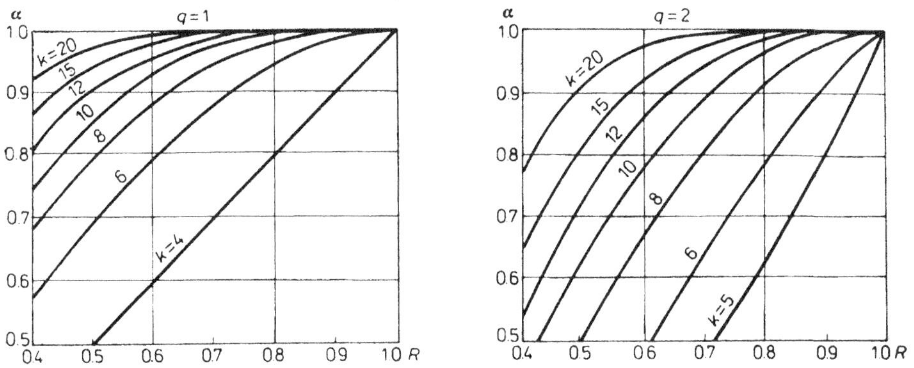

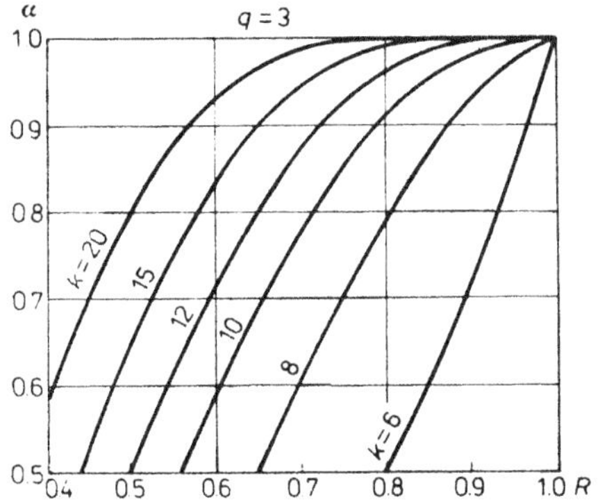

Fig. 4.20. The relationship of the significance level of regression coefficients and the cross-correlation coefficient of the systematic and total error for a given number of measurements.

error is smaller than the assumed value, the estimator coefficients are insignificant. The number of measurements should then be increased. Averaging of the results (elements of the set from an MS) is not a problem. The value of $\hat{h}_i(x_j)$ is known. Also the confidence interval $\Delta h_i(x_j)$ at a known confidence level (identical for many measurement systems) is known. Hence

$$\hat{h}_{av}(x_j) = \frac{1}{N}\sum_{i=1}^{N} \hat{h}_i(x_j), \tag{4.136a}$$

$$\hat{\sigma}_{av}(x_j) = \sqrt{\frac{1}{N-1}\sum_{i=1}^{N} [\hat{h}_i(x_j) - \hat{h}_{av}(x_j)]^2}, \tag{4.136b}$$

and thence, after introducing a confidence interval at a given α,

$$\hat{h}_{lim}(x_j) = \hat{h}_{av}(x_j) \pm t\hat{\sigma}_{av}(x_j). \tag{4.137a}$$

Since both components of formula (4.137a) are random variables, the variance of $\hat{h}_{lim}(x_j)$ is the sum of the component variables

$$\mathrm{var}\,\hat{h}_{lim} = \mathrm{var}\,\hat{h}_{av} + \mathrm{var}(t\hat{\sigma}_{av}) = \frac{1}{N}\hat{\sigma}^2_{1\Delta h} + t^2\hat{\sigma}^2_{\sigma\,av}. \tag{4.137b}$$

To assess these data, it is convenient to apply the sample range interpreted by order statistics (see Subsection 4.4.5). For a normal distribution, the relationship of the range L_k of a k-element sample and of a standard deviation is expressed by the formula [9]

$$\hat{\sigma} = \frac{L_k}{d_k}, \tag{4.138a}$$

while the variance of the estimated standard deviation is [6]

$$\mathrm{var}\,\hat{\sigma} = \hat{\sigma}^2\frac{D_k}{d_k}. \tag{4.138b}$$

The values of $d_k = E(L_k)$ and of D_k are given in Table 4.5 [6]. Here k is the number of elements in the sample, and in the problem discussed $k = N$, $\hat{\sigma} = \hat{\sigma}_{av}$, $\mathrm{var}\,\hat{\sigma} = \mathrm{var}\,\hat{\sigma}_{av}$. Thus, instead of (4.137b),

$$\mathrm{var}(\hat{h}_{lim}) = \frac{1}{N}\hat{\sigma}^2_{1\Delta h} + t^2\hat{\sigma}^2_{av}\frac{D_N}{d_N} = \frac{1}{N}\hat{\sigma}^2_{1\Delta h} + t^2\hat{\sigma}^3_{av}\frac{D_N}{L_N}. \tag{4.139}$$

The component $\hat{\sigma}_{1\Delta h}$ should be calculated from formula (3.265). The effectiveness of the estimation of standard deviation based on the range (4.138) is slightly lower than the sample estimation (4.134) (as shown in Table 4.5) and

$$e_k = \frac{\mathrm{var}\,S_k}{\mathrm{var}(L_k/d_k)}. \tag{4.140}$$

TABLE 4.5
The relationship of the expected value and variance of the sample range vs a number of indication checking points [6].

k	d_k	D_k	e_k
4	2.059	0.774	
5	2.326	0.747	0.96
6	2.534	0.719	
7	2.704	0.694	
8	2.847	0.672	
9	2.970	0.653	
10	3.078	0.635	0.86
11	3.173	0.620	
12	3.258	0.606	
13	3.336	0.594	
14	3.407	0.582	
15	3.472	0.572	0.77
16	3.532	0.562	

Next, the estimator $\hat{h}_{\text{lim}}(x)$ should be calculated from the results of (4.137a). The random component of the model (3.231) is the error of estimation of $\hat{h}_{\text{lim}}(x_j)$, whose variance is expressed by formula (4.139). When the variance is known, model (3.237) or (3.247) should be applied, depending on the distribution of variance. The diagram of the estimation scheme of a limiting distribution of systematic error is shown in Fig. 4.21.

From $\hat{h}_{\text{lim}}(x)$ and from the estimation error we obtain

$$h_{\text{lim}\alpha} = \hat{h}_{\text{lim}}(x) \pm t\hat{\sigma}_{h\text{lim}}. \tag{4.141}$$

Usually the function is smooth and may be a final result. However, if it is not smooth enough, proper smoothing should be applied. The smoothing principle should be a mean confidence level over the whole measurement range and a uniform distribution $p(x)$. The distribution of $h_{\text{lim}\alpha}$ is calculated from the formula

$$\alpha = \frac{1}{x_u - x_l} \int_{x_l}^{x_u} \alpha(x)\mathrm{d}x = \frac{1}{x_u - x_l} \int_{x_l}^{x_u} \int_{-h_{\text{lim}\alpha}}^{h_{\text{lim}\alpha}} p[h_{\text{lim}}(x)]\mathrm{d}x\mathrm{d}h. \tag{4.142}$$

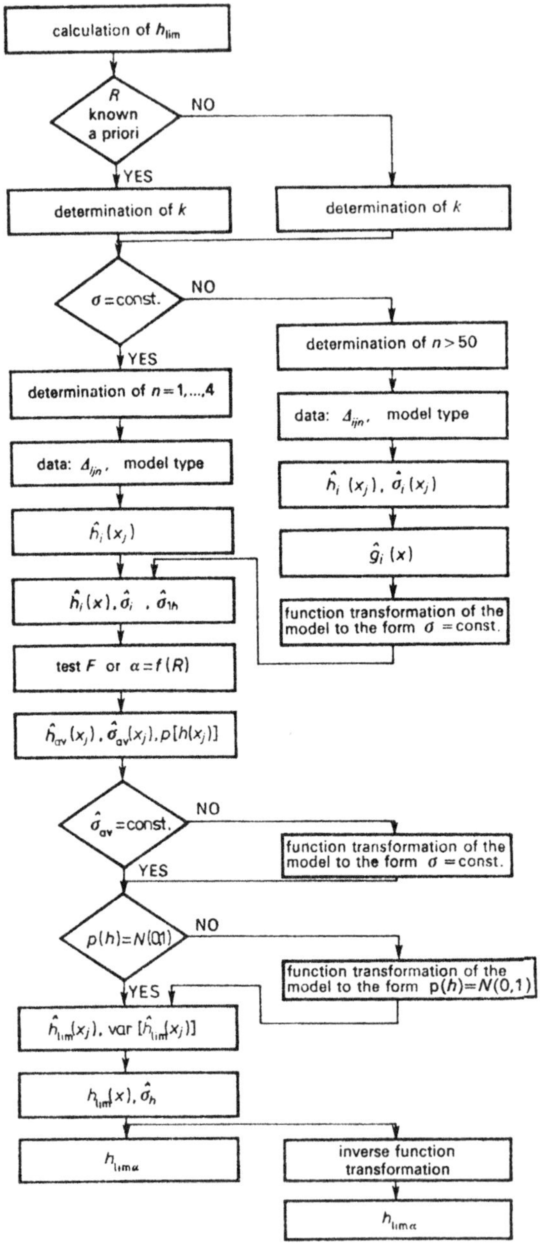

Fig. 4.21. Flowchart for calculating the estimator of a limiting value of systematic error of an MS type, based on experimental data.

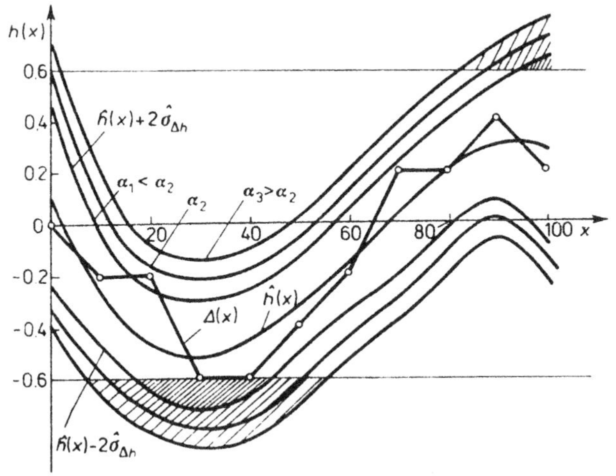

Fig. 4.22. The smoothing principle for the limiting distribution of systematic error.

In well-conditioned cases, weighted averaging may be introduced with weight coefficient $p(x) \neq$ constant. The principle is illustrated in Fig. 4.22. Then

$$\alpha = \int_{x_l}^{x_u} p(x) \int_{-h_{\mathrm{lim}\alpha}}^{h_{\mathrm{lim}\alpha}} p_h[h_{\mathrm{lim}}(x)]\,\mathrm{d}x\,\mathrm{d}h. \tag{4.143}$$

It must be stressed that the meaning of the limiting error distribution is a confidence interval at a given confidence level.

4.5.4 Interpretation of the Concept of Accuracy Class

As stated by the definition of accuracy class, the fundamental error of an MS does not exceed the limits $\pm\Delta_{\mathrm{all}}$ over the whole measurement range [40, 91]. Expressing metrological properties by the accuracy class is like applying a Lebesgue measure of a set B_α (1.66) with the value b^* equal to the reading. The difference is that the definition of accuracy class does not include the establishing of a confidence level α. We intend to improve this fault, based on the methodology of reading checking, by proper interpretation of the checking results.

The checking of readings is performed in reference conditions, at those points of a scale which are marked with numbers [91]. The points are usually denoted as x_j, $j = 1, \ldots, k$. The result of the checking is a set of errors $\{\Delta_j\}$. An MS conforms to the requirements of an accuracy class if

$$|\Delta_j| \leqslant \Delta_{\mathrm{all}} \quad \text{for} \quad j = 1, \ldots, k. \tag{4.144}$$

I. The following error properties are assumed:

(1) subsequent realizations of the results of reading checking are independent;

(2) the error Δ_j will have probability θ of exceeding an allowable value;

(3) at any of the checking points, an error may not exceed the range $2\varepsilon = 2\Delta_{\text{all}}$;

(4) only the $(k+l)$-th $(l \geqslant 1)$ checking result may exceed the range of allowable error and this point may not be realized in the checking process.

The checking results may then be described by a geometric distribution [92] and their probability density by the formula:

$$p(U = k+1) = (1-\theta)^k \theta. \tag{4.145}$$

The mean value of the random variable U is

$$E(U) = \frac{1}{\theta}, \tag{4.146}$$

and the variance is

$$\mathrm{var}(U) = \frac{1-\theta}{\theta^2}. \tag{4.147}$$

The most effective and unbiased estimator is

$$\hat{\theta} = \frac{1}{k+1}, \tag{4.148}$$

and when $l > 1$, $\theta < 1/(k+1)$ and

$$1-\hat{\theta} \geqslant \frac{k}{k+1}. \tag{4.149}$$

Some checked points (readings) of a measurement instrument whose accuracy class is verified are known a priori. Usually the number of such points is from 6 to 16, so the values of θ are included in the range $\langle 0.059, 0.200 \rangle$. On the other hand, when the value of θ is calculated from formula (4.148), the variance of the estimated parameter may be written as

$$\mathrm{var}(\hat{\theta}) = \frac{1}{(k+1)\sum\limits_u \left\{\left[\frac{\partial \log p(U,\theta)}{\partial \theta}\right]^2 p(U,\theta)\right\}} = \frac{\theta^2(1-\theta)}{k+1}, \tag{4.150}$$

according to the Cramer–Rao theorem of a lower limit of variance. In the formula, $p(U,\theta) = (1-\theta)^{U-1}\theta$. Note that in the expression

$$\frac{1}{\theta^2}\mathrm{var}(\hat{\theta}) = \frac{1-\theta}{k+1}, \tag{4.151}$$

which describes the relative error of the estimator, the dependence on θ, especially for small values, is weak. The properties of the estimator of parameter θ, expressed

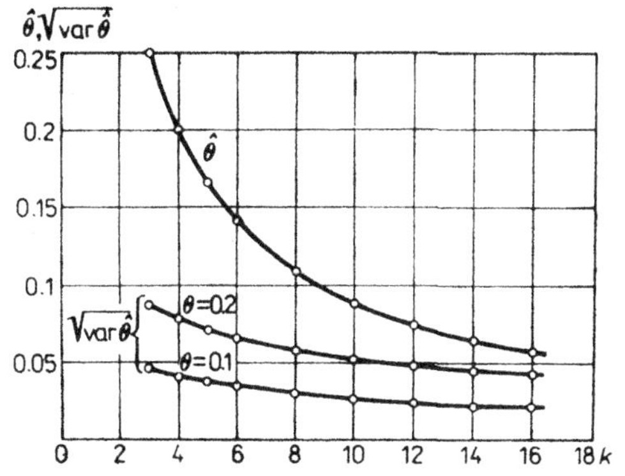

Fig. 4.23. The dependence of an estimator $\hat{\theta}$ and the standard deviation of $\hat{\theta}$ on the number of checking points.

by formulae (4.148) and (4.150), are shown in Fig. 4.23. Therefore the construction qualities of an instrument and the methodology of reading checking make it possible to calculate the probability that at a particular confidence level $1-\theta$ the allowable value of error will not be exceeded at k points (readings). The data are presented in Fig. 4.24.

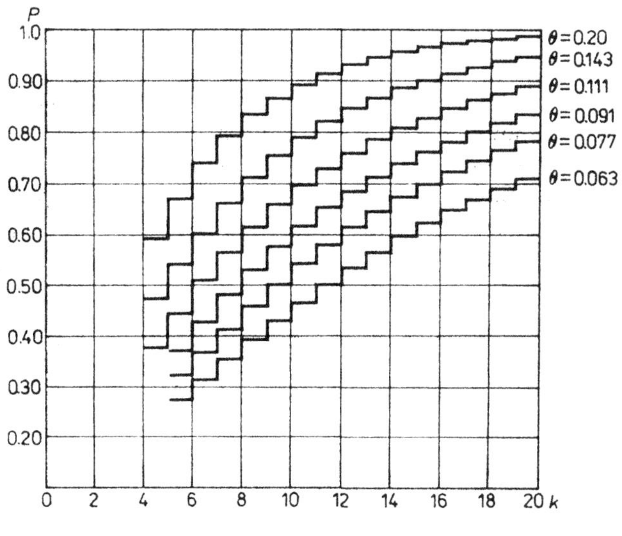

Fig. 4.24. The distribution function of a geometric distribution.

II. Given that the above assumptions valid, the results of reading checking may be interpreted as a realization of an *order statistic* [93]. An order statistic is a random variable $\tau_i^{(k)}$ set at the ith position in an ordered set of k random variables

$$\tau_1^{(k)}, \tau_2^{(k)}, \ldots, \tau_i^{(k)}, \ldots, \tau_k^{(k)}. \tag{4.152}$$

The set is ordered by the values of the realized elements of separate random variables. In the problem discussed, the realizations of random variables are the values of errors calculated in the process of reading checking for an instrument

$$\Delta_1^{(k)}, \Delta_2^{(k)}, \ldots, \Delta_i^{(k)}, \ldots, \Delta_k^{(k)}, \tag{4.153}$$

where $\Delta_i^{(k)} \leqslant \Delta_{i+1}^{(k)}$.

The set $\{\tau_i^{(k)}\}$ is a sufficient statistic [12]. The random variables $\tau_1^{(k)}$ and $\tau_k^{(k)}$ are called the external order statistics of a k-element sample [22]. The sample range

$$L = \tau_k^{(k)} - \tau_1^{(k)} > 0 \tag{4.154}$$

is also a random variable. A distribution of each order statistic and the joint distribution of both statistics may be determined. If the values of the statistics are $y_1, y_2, \ldots, y_k$, the joint distribution of the extermal statistics will be

$$g(y_1, y_k) = \begin{cases} k(k-1)\left[\int\limits_{y_1}^{y_k} p(x)\mathrm{d}x\right]^{k-2} p(y_1)p(y_k) & \text{for} \quad y_1 < y_k, \\ 0 & \text{for} \quad y_1 \geqslant y_k. \end{cases} \tag{4.155}$$

Using the above notation we interpret the independent random variables $X_1, X_2, \ldots, X_k$ with a common continuous probability density $p(x)$.[14] The values $p(y_1)$, $p(y_k)$ are the probability densities of respective variables X assuming a value y_i. The distribution of the random variable L (4.154), further denoted as $h(w)$, may be calculated from formula (4.155). When a common continuous distribution function $P(X)$ exists for the variables $X_1, \ldots, X_k$, then the random variable $Y = P(X)$ has a uniform distribution over the interval $\langle 0, 1 \rangle$ [22]. For the case when the values $y_1, \ldots, y_n$ are the realizations of random variables Y, formula (4.155) takes the simpler form

$$h(w) = \begin{cases} k(k-1) \int\limits_0^{1-w} w^{k-2}\mathrm{d}y_1 = k(k-1)w^{k-2}(1-w) & \text{for} \quad 0 \leqslant w \leqslant 1, \\ 0 & \text{for} \quad w < 0,\ w > 1. \end{cases} \tag{4.156}$$

[14] This is equivalent to the assumption that the random variable X is stationary in a narrow sense.

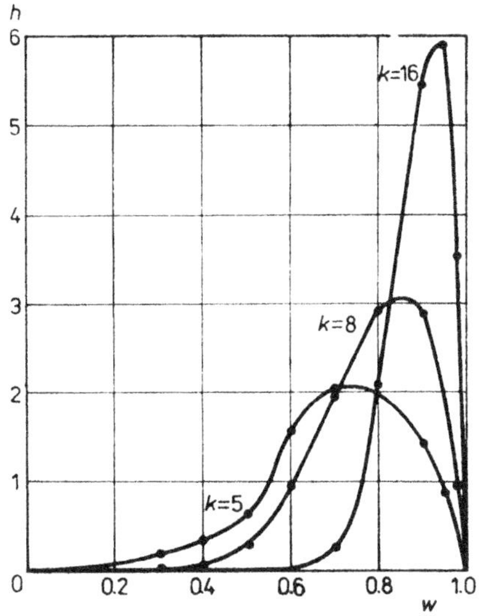

Fig. 4.25. The probability distribution of the element fraction within the limits of a sample range.

This distribution is shown in Fig. 4.25. The result of the reading checking, expressed in the form of order statistics (4.151) falls within tolerance limits which are extreme order statistics [22].

$$L_1 = \tau_1^{(k)}, \quad L_2 = \tau_k^{(k)}. \tag{4.157}$$

Some fraction $0 \leqslant W \leqslant 1$ of possible values of the reading error falls within the tolerance limits, i.e. within the sample range. This part is called the element fraction of the parent population of errors $\{\Delta\}$ (from which the k-element sample has been extracted). The fraction is [22]

$$W = \int_{L_1}^{L_2} \mathrm{d}P(x). \tag{4.158}$$

By denoting $x = \Delta$, we obtain

$$W = \int_{L_1}^{L_2} \mathrm{d}P(\Delta) = P(\Delta_k^{(k)}) - P(\Delta_1^{(k)}). \tag{4.159}$$

The properties of the random variable W are described by formula (4.156). Thus the probability that the fraction W has at least a value w is expressed by the formula [22]

$$P(W \geqslant w) = k(k-1)\int_{w}^{1} w^{k-2}(1-w)\mathrm{d}w \geqslant \alpha, \tag{4.160}$$

or, after the integral is calculated,

$$P(W \geqslant w) = 1 - kw^{k-1} + (k-1)w^k \geqslant \alpha. \tag{4.161}$$

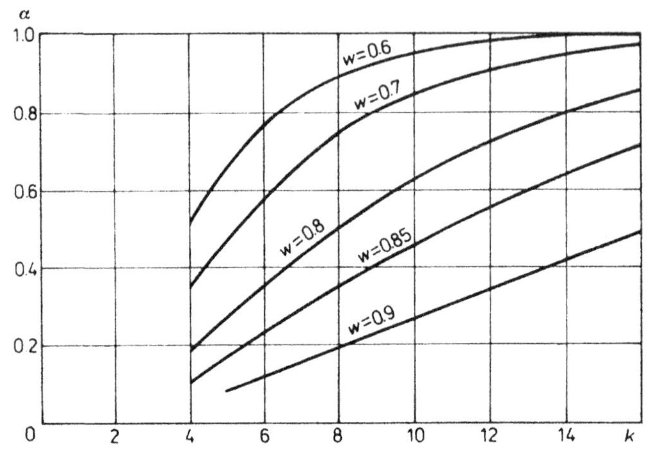

Fig. 4.26. The dependence of the probability $P(W \geqslant w) \geqslant \alpha$ on the number of checking points.

Formula (4.161) determines the number k of the checking points for which the probability $P(W \geqslant w) \geqslant \alpha$, for a given fraction W. The numerical data are presented in Fig. 4.26. The expected value of the random variable W (understood as a parameter describing the properties of measuring instruments) is

$$E(W) = \frac{k-1}{k+1} \tag{4.162}$$

and the variance is

$$\operatorname{var}(W) = \frac{k-1}{k+1} \frac{2}{(k+1)(k+2)}. \tag{4.163}$$

As the number of checking points increases, the standard deviation decreases relatively quickly

$$\frac{\sqrt{\operatorname{var}(W)}}{E(W)} = \sqrt{2}\frac{1}{\sqrt{(k+2)(k-1)}} \approx \frac{\sqrt{2}}{k}. \tag{4.164}$$

The unbiased estimator of the fraction W is the expression (4.162) [22]

$$\hat{W} = \frac{k-1}{k+1}. \tag{4.165}$$

The interpretation of parameter $\hat{W}$ is identical to that of parameter $1-\hat{\theta}$ when the class is interpreted by the geometric distribution model [92]. The numerical values are slightly different, because (4.149): $1-\hat{\theta} = k/(k+1)$. Thus, the value of the estimator $\hat{W}$ is determined only from the number of checking points. The accuracy class is then interpreted as follows: the range determined by the extreme order statistics should be accepted as the interval of allowable error, since for a given type of instruments, no error has exceeded this interval during the checking at k points. The fraction W expressed by the estimator (4.165) should be considered a characteristic of this type of instrument. The fraction determines a percentage $(W \times 100)\%$ of error realizations which do not exceed the allowable error value. According to formula (4.161), the certainty that $\hat{W} \geqslant w$ is equal to α. For $w = \hat{W}$ this certainty is relatively small, about 55%. To increase it, the number of checking points should be increased to $2k$ or $3k$. Then double checking increases the certainty to 90% and triple checking even to 98% [93].

The interpretation of accuracy class presented allows a convergent evaluation of the instrument's properties. In both cases it may be peinted out that the assumption that the subsequent checking results are independent is not fulfilled, since the systematic component of error causes correlation of the results. Actually, the subsequent results are correlated. The correlation is expressed by the correlation (3.164). When the subsequent checking points are spaced at the intervals $\xi = x_2 - x_1$, the cause of the correlation is a systematic error component $K_h(\xi)$ as well as a random error component $\mu_{2g}(\xi)\varrho(\tau)$, depending on the distance ξ and the time interval τ. The distribution of the systematic error is not known, and so this component is considered to be random. A similar situation occurs with the other component; the value of $\mu_{2g}(\xi)$ is usually constant and the factor $\varrho(\tau)$ is not determined and is not considered at the reading checking. These are practical arguments.

From a theoretical point of view, the above interpretation may be better grounded. In the case of a geometric distribution, the assumption at point 2 has been made that the frequency of the event A

$$A:\ (\Delta_j < -\Delta_{\mathrm{all}}) \vee (\Delta_j > \Delta_{\mathrm{all}}) \tag{4.166}$$

is θ. This assumption may be expressed differently. θ is the frequency of the event B

$$B:\ [\Delta_{rj} < -(\Delta_{\mathrm{all}} + a)] \vee [\Delta_{rj} > \Delta_{\mathrm{all}} - a], \tag{4.167}$$

where the index r denotes a random error and the value $|a| < \Delta_{\mathrm{all}}$. The subsequent realizations of the event B are not correlated when $\mu_{2g}(\xi)\varrho(\tau) = 0$, according to formula (3.164).

According to the former assumptions, the probability distribution of the event B is geometric, with the probability density (4.145). The value of a may be a function of x and need not be known; it is only the fact of occurrence of the event B that should be measurable. Assuming

$$a = \Delta_{sj},$$

while the systematic error is denoted as Δ_s, and adding the value $\Delta_{sj} = a$ to both sides of inequality (4.167), permits us to obtain inequalities (4.166) for $\Delta_j = \Delta_{sj} + \Delta_{rj}$. This means the equivalence of events A and B. Therefore, the frequencies of events A and B are described by the same probability distribution.

In the case of order statistics, the assumption was made that the variable X is stationary in a narrow sense. The random variable X is stationary if $h(x)$ and $g(x)$ are constant. In other cases the assumption is not fulfilled.

The interpretation of accuracy class presented should be understood as a characteristic of a specific random variable, which is a function (the sum) of two random variables: a random error and a systematic error described by a probability distribution according to geometric interpretation. The condition for such an interpretation is that the systematic errors are realized by uniform sampling of the domain of probability distribution or of the parameter x (which

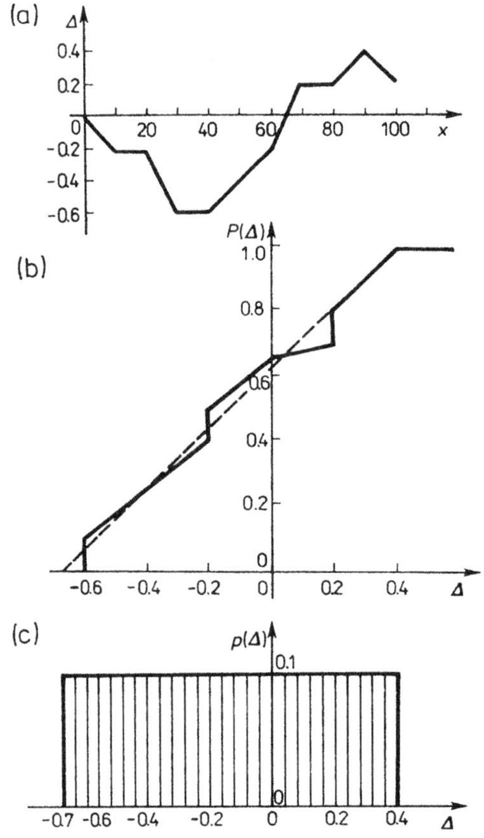

Fig. 4.27. The probability distribution of a systematic error: (a) distribution over the scale, (b) distribution function, (c) density function.

may be written as $p(h) = p[h(x])$ in the case of a uniform distribution. Figure 4.27 shows the probability distribution of systematic error for the case when the distribution of $h(x)$ is as in Fig. 3.40. The case $h(x) = ax+b$, where a and b are constants, is described by a uniform distribution $p(h)$. If the reading checking is performed at the points labelled with numbers, the requirement of uniform sampling is fulfilled.

The notion of accuracy class is applied in Class 5 of metrological characteristics. If it may be assumed that the decision about the way of expressing metrological characteristics is rational, then supposedly the dominating error is random. In the cases when a systematic error dominates, Class 3 of the characteristics should be applied.

4.6 The Basic Concepts of Inference in the Theory of Error

The formulated images of reality are referred to selected states of the selected qualities, i.e., to the reference states. The selected phenomena which reproduce these states are considered invariable. The above declarations concern the materials building the physical world, as well as the amounts of matter and energy existing in different forms.

The number and kind of basic reference states depend on the complexity of the reality being described, with respect to the actual state of knowledge of man as a subject of cognition. To facilitate practical activity, the appropriate derivative reference states are created and the appropriate phenomena are applied to reproduce these states. The derivative states and phenomena are selected by rational inference concerning rationalization of a practical human activity; also the possibilities of realization of these phenomena must be considered. Therefore, there are many standards of reference states, each of them having an approved reference to basic and primary standards. This reference is expressed by a nominal measure of the standard and by ambiguity of the reproduced reference state, as well as by the scheme by which a measure is transferred from a primary onto a secondary standard. These problems were presented in detail in Chapter 2. The problem of interpretation of measurement errors will be considered later.

The basic concept of the theory of error is to find ways of performing the measurements and of calculating the errors, while the physical limitations make it impossible to determine simultaneously the error and perform a measurement with an MS. These actions are thus separated in time, and result in random variations of the properties of an MS, transient effects, etc. Therefore, the staff carrying out the measurements must be aware of and forecast the temporal changes of errors and the variations due to changing conditions. An obvious way of making this task easier is to separate out error components which do not change or whose changes are easy to predict (the principle of the component changes is known). The remaining part of the error is a random component. This concept is the basis for dividing the errors into systematic and random categories. The value of a systematic error may be either constant or may change according to a known principle, and is known at the time of measurement. Efforts are being made to ensure that the greatest possible fraction of error has constant and reproducible properties. This leads to division into calibration error, supplementary error, dynamic errors, etc. The principles of variability of these errors rely on different factors, and thus various kinds of errors are determined and calculated differently.

On the other hand, a part of the errors (random in origin[15]) depend on the way in which the measurements are performed.

One of the concepts formulated by Gauss takes advantage of a property of random errors that positive and negative values of error are equally probable. Thus, addition of the errors reduces their total. A consequence of this fact is that the random error may be reduced by performing a series of measurements and by accepting the average value of these results as the measurement result. If the errors in separate measurements are independent and their variances are equal (i.e. the errors are stationary in a broad sense), the variance of the average value as a sum of random variables

$$\Delta_{av} = \sum_{i=1}^{n} \frac{1}{n} \Delta_i \tag{4.168}$$

is n times smaller, i.e.

$$\sigma_n^2 = \sum_{i=1}^{n} \frac{1}{n^2} \sigma_i^2 = \frac{1}{n} \sigma^2. \tag{4.169}$$

An extension of this concept is the application of different filters, especially in cases when the measured quantity is variable.

Another concept which helps to limit random errors is the frequent adjustment of an MS.[16] Since the effect of adjustment is not very obvious, the way in which it is performed will be discussed in more detail.

---

[15] The errors of a systematic nature become random when the factors which influence the error remain unknown, although they may be recognized. For this reason, the difference between the theoretical capabilities and the procedures applied must be stressed. This author does not accept the CIPM Recommendation (Cl–1981) [178], where the concept of systematic error is considered useless.

[16] We may also say, frequent calibration. The procedures of calibration and adjustment are discussed in Section 5.3. *Adjustments* are the adjusting actions in an MS that make the error assume a predetermined value, usually zero.

Let

$$\{\Delta_0(t)\} = g_0(t)\{\varepsilon(t)\}+h_0(t)+r(t) \tag{4.170}$$

be the error of an MS in reference conditions, for a given state of a measured quantity $x = x_a =$ constant. The components g_0, h_0, ε describe the fundamental error, according to model (3.136), and

$$r(t) = \sum_{i=1}^{n} r_i \mathbf{1}(t-t_i) \tag{4.171}$$

is the effect of adjustment of the MS at the instants t_i.

Case 1. Adjustment by an instantaneous value of error

At the instant $t = t_i$, the readings of the MS are corrected so that the value of error after the correction is zero. This means

$$r_i = -g_0(t_i)\,\varepsilon(t_i)-h_0(t_i), \tag{4.172}$$

and for $t > t_i$

$$\{\Delta_0(t)\} = g_0(t)\{\varepsilon(t)\}+h_0(t)-g_0(t_i)\varepsilon(t_i)-h_0(t_i). \tag{4.173}$$

Assuming $g_0(t) = g_0(t_i)$ and $h_0(t) = h_0(t_i)$, the variance of the error is

$$\begin{aligned}\sigma_\Delta^2(t) &= E\{\Delta_0(t)-E[\Delta_0(t)]\}^2\\ &= g_0^2\sigma_\varepsilon^2(t)+g_0^2\sigma_\varepsilon^2(t_i)-2g_0^2\varrho_\varepsilon(t-t_i) = 2g_0^2[1-\varrho(t-t_i)].\end{aligned} \tag{4.174}$$

Formula (4.174) means that directly after the adjustment was performed, when $\varrho(t-t_i) \approx 1$, the random error of an MS is zero. As time passes and the correlation coefficient of the process $\{\varepsilon\}$ decreases, the random error increases and for $t-t_i > t_{\text{lim}}$ ($\varrho = 0$), a constant value of variance $2g_0^2$ is reached. This result, shown in Fig. 4.28, may seem strange. The adjustment caused the component h_0

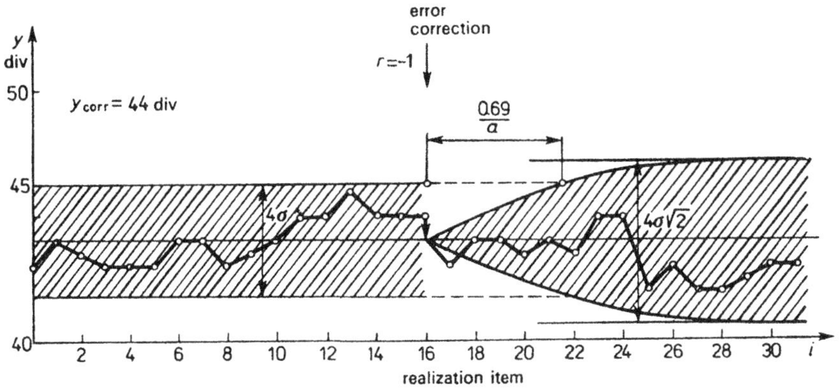

Fig. 4.28. The confidence interval of a random error after an adjustment by an instantaneous value of error.

of error to be fully corrected, while the value of the random component has been shifted by a random value $g_0\varepsilon(t_i)$. Thus, the variance of the random error is doubled. The adjustment by an instantaneous value of instability error causes the random error to decrease for the time for which $\varrho(t-t_i) > 0.5$. When $\varrho(\tau) = e^{-a\tau}$, this time is

$$T < \frac{1}{a}\ln 2 = \frac{0.693}{a}. \tag{4.175}$$

For the oxygen analyser with correlation coefficient (4.68), where $a = 0.23$, the interval is $T < 3$ hours.

Case 2. Adjustment by the average value of error

Before the adjustment is performed, the values of the fundamental error at instants t_i, $i = 1, 2, \ldots, N$, are determined. Next, the mean value is calculated, and the adjustment is performed at the instant t_j. The value for which the adjustment is performed is

$$r_j = -\frac{1}{N}\sum_{i=1}^{N}[g_0(t_i)\,\varepsilon(t_i)-h_0(t_i)]. \tag{4.176}$$

Assuming $g_0(t_i) = g_0$, $h_0(t_i) = h_0$ and $\varrho(t_{i+1}-t_i) = 0$ (which means that the subsequent results of the checking are not correlated), we get

$$\sigma_\Delta^2(t) = g_0^2+\frac{1}{N}g_0^2 = \frac{N+1}{N}g_0^2. \tag{4.177}$$

Even for small values of N, the variance of random error is slightly greater than the variance of the stochastic component. For example, for $N = 4$, $\sigma_\Delta^2 = 1.25g_0^2$ ($\sigma_\Delta = 1.12g_0$). The result (4.177) means that the adjustment should be performed after several checkings at time intervals $T_{\text{lim}} > 3/a$. In the example of the oxygen analyser, the time interval should be greater than about 12 hours.

Case 3. Adjustment with error prediction

If the purpose of an adjustment is to correct a determined component of error $h_0(T)$ and it is to be performed at time intervals much greater than T_{lim}, the value of r_j should be determined from the mean value of error. The assumption that $h_0(T) =$ constant is not allowed in this case. Based on the error measurements at time $T < T_0$, the value of $\hat{h}_0(T)_{T_0}$ may be determined for the instant T_0. Next, for $T > T_0$, the prediction $\check{h}_0(T)$ and the confidence intervals for the prediction may be found by extrapolation. The value of correction $r_0(T_0)$ may be determined from the condition of $\max_{r_0} T_E$ for $T \in T_E$, when

$$|\check{h}_0(T)_{T_0} \pm \check{\Delta}h_0(T)_{T_0}+r_0(T_0)| \leqslant \Delta_{\text{all}}. \tag{4.178}$$

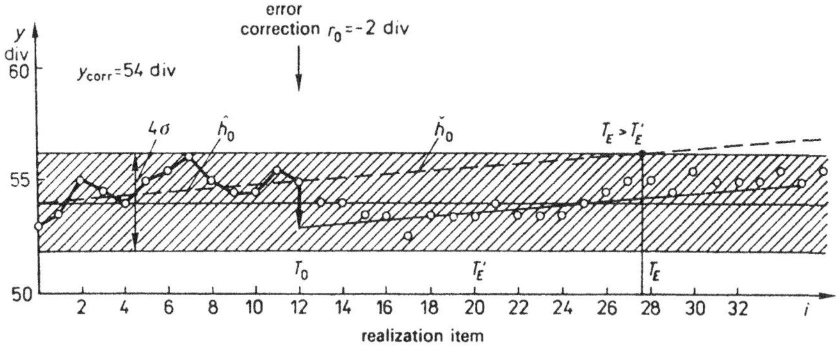

Fig. 4.29. Error changes after adjustment, with the prediction of a systematic error taken into account. T_E'—time after which the value of error exceeds the allowable value, with the prediction $\check{h}_0$ not taken into account.

Figure 4.29 demonstrates the manner of selecting $r_0(T_0)$ with the above condition satisfied. The problem of calculating T_E is presented in Chapter 5.

An interesting way of determining the correct value of a series of measurement results was proposed by Elyasberg [150]. In the classical approach, the result of a measurement series is a mean value of N single measurement results. Elyasberg's algorithm is based on the estimation of extreme results of individual measurements in a series. The result of each measurement is an interval $\langle x_{li}, x_{ui}\rangle$, where

$$x_{li} = x_i^* - \Delta_i, \quad x_{ui} = x_i^* + \Delta_i, \tag{4.179}$$

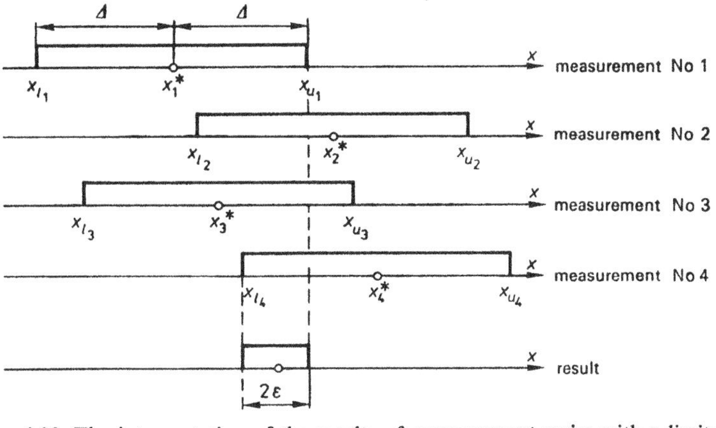

Fig. 4.30. The interpretation of the results of measurement series with a limited set of errors of a measurement system.

and x^* is the corrected result of the ith measurement, $i = 1, 2, \ldots, N$. The results for $N = 4$ are shown in Fig. 4.30. The result of a measurement series is

$$x^* \in \langle x_{l\,max}, x_{u\,min}\rangle, \tag{4.180}$$

where

$$x_{l\,max} = \max_i x_{li}, \qquad x_{u\,min} = \max_i x_{ui}.$$

The result may also be written as

$$x^* = \tfrac{1}{2}(x_{l\,max} + x_{u\,min}) \pm \varepsilon, \tag{4.181a}$$

$$\varepsilon = \tfrac{1}{2}(x_{u\,min} + x_{l\,max}) > 0. \tag{4.181b}$$

As may be seen from Fig. 4.30, $x_{lmax} = x_{l4}$, $x_{umin} = x_{u1}$. Since the result of a single measurement is a set X_i^* with a limited distribution [116], we see that in the method described the result of a measurement series is a product of the sets

$$X^* = X_1^* \cap X_2^* \cap \ldots \cap X_N^*. \tag{4.182}$$

This method is generally more sensitive to modelling errors; the empty result $X^* = \emptyset$ means that the modelling was erroneous [116]. According to Elyasberg, the method is of litle practical significance, as the barely interpretable result $\varepsilon < 0$ is obtained too often. This fact is the result of excessive measurement errors, occurring quite often. According to Elyasberg's assessment [150], the percentage of measurement results burdened with excessive errors in exploratory studies, and using new instruments, is 10–30%.

Some modifications of the method described have been presented by Skubis [183]. Apart from x_1, the measured quantities were $x_2 \approx x_1$ and the difference $d = x_1 - x_2$. The latter measurement is performed with practically no error. In this case, assuming that the distributions of the sets of measurement errors of the quantites x_1, x_2 are limited (Δ_1, Δ_2, respectively), the primary results are corrected and the uncertainty interval is diminished [183]. The measurement

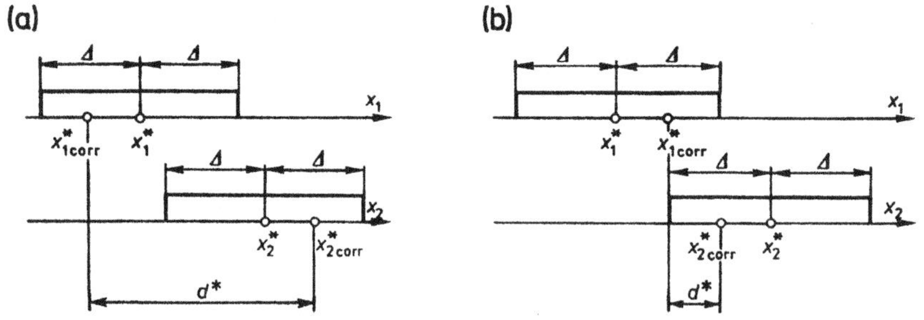

Fig. 4.31. The correction of results in the case of additional difference measurement: (a) $|x_1^* - x_2^*| < |d^*|$, (b) $|x_1^* - x_2^*| > |d^*|$.

results are shown in Fig. 4.31. The values x_1^*, x_2^* are the results of direct measurements of the quantities x_1 and x_2. With the result d^*, the results x_1^*, x_2^* should be corrected so as to obtain

$$x_{1\,corr}^* - x_{2\,corr}^* = d^*. \tag{4.183}$$

Therefore

$$x^*_{1\,\mathrm{corr}} = x^*_1 - \tfrac{1}{2}(x^*_1 - x^*_2 - d^*), \tag{4.184a}$$

$$x^*_{2\,\mathrm{corr}} = x^*_2 + \tfrac{1}{2}(x^*_1 - x^*_2 - d^*). \tag{4.184b}$$

Correction of the results also causes one of the bounds of the set of errors to be shifted over the same value, as shown in Fig. 4.31. In this case, it may be shown that

$$\mathrm{var}(x^*_{1\,\mathrm{corr}}) = \mathrm{var}(x^*_{2\,\mathrm{corr}}) = \tfrac{1}{4}(\sigma_1^2+\sigma_2^2)+\tfrac{1}{4}\sigma_d^2 \approx \tfrac{1}{2}\sigma^2_{1,2}. \tag{4.185}$$

The above assessment remains in agreement with the general rules, since x_1 is measured twice and thus the variance is decreased by half. Double measurement of x_1 also gives a result different from x^*_1, i.e., $(x^*_{11}+x^*_{12})/2$. The effectiveness of the method is hard to estimate unambiguously.

Systematic errors are a significant limitation to the accuracy of measurement. One of the concepts of eliminating them is the so-called *iterative measurement method* [141, 174]. The name 'iterative measurement' means that the measurement result approaches the correct value in succesive steps. The principle of the method is shown in Fig. 4.32.

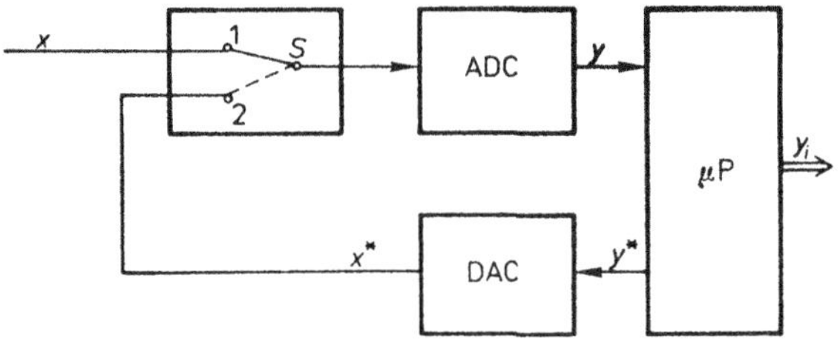

Fig. 4.32. An iterative measurement system.

A measurement converter (e.g. an analog-to-digital converter or ADC) is coupled to a storing/processing module μP (e.g. microcomputer). The system also includes an inverse conversion unit (e.g. a DAC) and a switch S. The conversion $x \to y$ takes place in the direct (main) path. The inverse conversion path is auxiliary, and its conversion errors are assumed to be negligibly small. In the case of DACs and ADCs, the assumption is usually fulfilled. At first, the measured quantity is x and the result is

$$y_0 = x+h+g\varepsilon_0. \tag{4.186}$$

The quantities h and $g\varepsilon_0$ denote the systematic and random errors, respectively; ε_0 is a realization of random process $\{\varepsilon\}$ in the zero iteration.

Next, the circuit input is switched into position 2 and the following cycle of iterations is performed.

I. $\quad y^*_1 = y_0; \quad y^*_1 \to x^*_1 \to y_1 = y_0+h+g\varepsilon_1. \qquad (4.187)$

The result of the first iteration is

$$y_{\text{I}} = y_0 + \Delta_1, \quad \Delta_1 = y_0 - y_1,$$
$$y_{\text{I}} = x + g(\varepsilon_0 - \varepsilon_1). \tag{4.188}$$

II. $y_2^* = y_{\text{I}}; \quad y_2^* \to x_2^* \to y_2 = y_2^* + h + g\varepsilon_2.$ (4.189)

The form of the result of the second iteration is

$$y_{\text{II}} = y_0 + \Delta_2, \quad \Delta_2 = y_2 - y_1,$$
$$y_{\text{II}} = x + g(\varepsilon_0 + \varepsilon_1) + g(\varepsilon_2 - \varepsilon_1). \tag{4.190}$$

III. $y_3^* = y_{\text{II}}; \quad y_3^* \to x_3^* \to y_3 = y_3^* + h + g\varepsilon_3.$ (4.191)

The result of the third iteration is

$$y_{\text{III}} = y_0 + \Delta_3, \quad \Delta_3 = y_3 - y_1,$$
$$y_{\text{III}} = x + g(\varepsilon_0 - \varepsilon_1) + g(\varepsilon_2 - \varepsilon_1) + g(\varepsilon_3 - \varepsilon_1), \tag{4.192}$$

and so on.

As we can see from the results, the first iteration will eliminate the systematic error if the value of the error is constant and independent of x. The random error increases with the number of iterations. Since the realizations of the process $\{\varepsilon\}$ may or may not be correlated, the influence of the random error may appear in different ways; the correlated errors are eliminated, while the uncorrelated errors grow with the number of iterations. This fact is a limitation of the method [147].

The errors of a measurement transducer may be interpreted differently. Assuming that the characteristic of the transducer is linear and a sensitivity error γ and a zero error v are introduced, we obtain [146]

$$y_0 = x(1+\gamma) + v. \tag{4.193}$$

The nth iteration result is

$$y_n = x[1 + (-1)^n \gamma^{n+1}] + (-1)^n v^n. \tag{4.194}$$

Altough the accepted assumptions make only two iterations sufficient to calculate x, γ, v, the structure of the formulae is a source of significant errors. Thus the above formula (4.194) should be considered an illustration of the convergence of the method when $\gamma < 1$.

The author of the iterative method of measurement presents a number of possible applications of it [200]. These are the measurements of the mean value, rms value, and higher-order moments of random variables, median measurements and measurements of random signal quantities [142]. One of the known applications of the iterative method is the rms value measurement; in this case, the application of one iteration in an AC/DC converter allowed a reduction of the error from 0.3% to 0.021% [156].

In nonlinear measurement transducers, the characteristic of the converter P_0 should be inverse to the original one [142]. The iterative measurements may be used for linearization of the characteristics, and the result is usually obtained in a few or a couple of dozen iterations, depending on the degree of nonlinearity of the primary characteristics [191]. Applications of the iterative method of measurement are limited to the domain of measurements of electrical quantities, and the accuracy of the method depends on the proportion of random errors in the total error.

In the theory of estimation, the effects disturbing the measurement results are still more often an object of interest. Many authors provide information on the abnormal part of the data. For example, Tukey [186] states that about 10% of engineering measurements are burdened with excessive errors. According to Daniel [212], the quantity of excessive errors in industrial data is 1–20%. Freedman [153] notes 5–7% excessive errors in seismographic data. In clinical studies, 8–12% of the results are abnormal.

Elyasberg [150] estimates abnormal results in exploratory space investigations at 10–30%. The classical statistical methods demand the rejection of results burdened with gross errors.

A new concept was introduced by Huber [158]. The idea is that the set of results has two sources. One is the measured object, and the probability distribution for this source is $p(x)$, and the other is an unknown object which disturbs the measurement; the probability distribution of this second source is $q(x)$. It is assumed that the results from the two sources are generated so that the probability of the data coming from the measured object and from the alien object are $1-\varepsilon$ and ε, respectively. Thus, the distribution of the measurement result is

$$\Pi(x) = (1-\varepsilon)p(x) + \varepsilon q(x), \quad 0 \leqslant \varepsilon \leqslant 1. \tag{4.195}$$

Usually, the distribution $\Pi(x)$ is considered to be known. To estimate the parameter θ of the distribution $p(x)$, both distributions $p(x)$ and $q(x)$ as well as ε must be known. The functions of probability density $p(x)$ and $q(x)$ may possibly be found as the functions φ of the class C, which model $p(x)$, and ψ, which model $q(x)$. Huber [158] introduced the minmax condition for the variance of the set of results:

$$\operatorname{var}(\varphi, \psi_0) \leqslant \operatorname{var}(\varphi_0, \psi_0) = \min_{\psi} \max_{\varphi} \operatorname{var}(\varphi, \psi). \tag{4.196}$$

The condition makes calculating ψ_0 for $\varphi \cong \pi(x)$ possible. A number of examples of solutions may be found in [159].

Huber's method is often compared with the modified Laplace method (least absolute values) [166]. The modifications concern some of the elements of the LSM, which are introduced here.

Current assessments of measurement results aimed at eliminating abnormal results make use of other methods called *linear*, or *nonlinear*, *filtration* [152]. The application of the median for mean value assessment is worthy of attention. Also the Laplace least absolute values method gives less dependence on abnormal results than the LSM [166]. The subject of so-called robust algorithms of measurement result handling has been developed intensively in the past few years.

Mathematical statistics brought the valuable tool of *experimental design* into metrology. Its ideas, however, did not meet with approval, since mathematical statistics employs some rules with unrealistic assumptions. A considerable number of the assumptions are never fulfilled in real experimental conditions, and work spent on proving the adequacy of the assumptions decreases the effectiveness of the experimental design.

For example, in factorial design the additive character of a random component with constant variance is assumed, independently of the value of the measured quantity. In this case the effectiveness of the factorial design is easily proved by showing the decrease of the variance of the error of the result or a decrease of the demanded number of measurements. Nalimov [76] gives a model example of two ways of measuring the masses of three objects *A*, *B*, *C*. The traditional (unifactorial) design is shown in Table 4.6 and the multifactorial design in Table 4.7. The assumed properties of the MS are

$$\sigma_{y_0} = \sigma_{y_1} = \sigma_{y_2} = \sigma_{y_3} = \sigma_{y_4} = \sigma. \tag{4.197}$$

TABLE 4.6

A unifactorial design for a measurement of the masses *A*, *B*, *C* [76].

Measurement No.	*A*	*B*	*C*	Result
1	−1	−1	−1	y_0
2	+1	−1	−1	y_1
3	−1	+1	−1	y_2
4	−1	−1	+1	y_3

$A = y_1 - y_0$; $B = y_2 - y_0$; $C = y_3 - y_0$.

Symbols: +1 means that an object is measured;

−1 means that an object is not measured.

The conclusion is as follows. In the unifactorial experiment, after 4 measurements are performed (the results shown in Table 4.6 are y_0, y_1, y_2, y_3), the variance of the error is

$$\sigma_A^2 = \sigma_B^2 = \sigma_C^2 = 2\sigma^2. \tag{4.198}$$

TABLE 4.7

A multifactorial design for the measurement of the masses *A*, *B*, *C* [76].

Measurement No.	A	B	C	Result
1	-1	-1	$+1$	y_1
2	$+1$	-1	-1	y_2
3	-1	$+1$	-1	y_3
4	$+1$	$+1$	$+1$	y_4

$$A = \tfrac{1}{2}(-y_1+y_2-y_3+y_4)$$
$$B = \tfrac{1}{2}(-y_1-y_2+y_3+y_4)$$
$$C = \tfrac{1}{2}(+y_1-y_2-y_3+y_4)$$

In the multifactorial experiment with the design shown in Table 4.7, the variance is two times smaller, with the same number of measurements, since

$$\sigma_A^2 = \sigma_B^2 = \sigma_C^2 = \sigma^2. \tag{4.199}$$

This is the interpretation by Nalimov. However, there is another aspect of the problem.

In the simplest case, the measurement error may be divided into the sensitivity and zero errors [91]. In a mass measurement by the null method, the zero error corresponds to the dead band of the detector. The error of the standard (weights) may be either the zero or the sensitivity error. Usually, the sensitivity error is dominant [197]. Inequality of the arms of the scales leads to the sensitivity error.

The conditions may be considered more typical when the dead band of the detector is negligible and the dominating error is the sensitivity error. Then, instead of assumption (4.197), it should be accepted that

$$\sigma_{y_0}^2 = 0, \quad \sigma_{y_1}^2 = \sigma_{y_2}^2 = \sigma_{y_3}^2 = \sigma^2, \tag{4.200a}$$

$$\sigma_{y_4}^2 = 9\sigma^2, \quad \text{when} \quad A \approx B \approx C. \tag{4.200b}$$

The variance of the error of mass measurement in the traditional unifactorial experiment is

$$\sigma_A^2 = \sigma_B^2 = \sigma_C^2 = \sigma^2, \tag{4.201}$$

and in the multifactorial experiment it is

$$\sigma_A^2 = \sigma_B^2 = \sigma_C^2 = \tfrac{1}{4}(1+1+1+9)\sigma^2 = 3\sigma^2. \tag{4.202}$$

The effectiveness of the multifactorial experiment is thus extremely different. Moreover, the fact of not performing the measurement of y_0 (which is a fragmentary check of the technical condition of the measuring instrument) is a logical fault, unacceptable in metrology.

This simple example has been chosen on purpose to make the results differ extremely. The experimental design cannot be recommended as a panacea, since reality is much more complex. There are some cases when the experimental design method gives the effect of decreasing the variance of random errors or of decreasing the number of necessary measurements. Each design should be analysed in detail before it is put into practice, in order to determine the properties of errors and the agreement of the mathematical assumptions with the experimental conditions. The design may be formulated only after this analysis is done and only then may the expected results of the variant designs be discussed.

In applying experimental design, we should remember that mathematical statistics and the experimental design, as its part, are helpless when systematic errors are being analysed. However, some ill-founded opinions concerning that subject are sometimes met. For example (Nalimow [76], p. 77): "On the grounds of general methodology, it is most important that the influencing factors with definitely different natures may be considered random variables. The evaluating property is not the nature of the factor but the formulation of the task and the conditions under which the experiment is performed. Formerly the experimenters, especially in metrology, carefully tried to classify the errors as random and systematic, depending on their nature. Such classification, however, makes sense no longer. Everything depends on the way in which the task is formulated..." Nalimov finally asks the question ([76], p. 199): "Now, the question may be asked why the mathematical theory of experiment has not been generally accepted by the experimenters? Why are so few experimental works found which are written in the language of mathematical statistics?..." Unfortunately, the language is not universal.

An attractive idea which it was thought in the 1960s would-dominate metrology was the application of *information theory* [78, 55]. The simple and imaginative apparatus of information theory was very encouraging, but relatively soon it became evident that some of the obstacles—still unsurmounted today—restrict applications of the theory to metrology. First of all, the information characteristics of the signals and systems are derived from the physical description as secondary characteristics. Instead of applying the characteristics in the form of physical models together with the information characteristics, it is quite sufficient to confine oneself to the former. Secondly, to define an elementary notion of the amount of information, it is necessary to determine the value of a quantum of a given physical quantity. The amount of information is rational when the number of quanta is a natural number or real number greater than 1. For example, to express the amount of information contained in a signal varying from 0 to 5 V, the value of the quantum may be assumed to be $q_1 = 10$ mV and the number of states $N_1 = 500$, or it may be $q_2 = 100$ mV and thus $N_2 = 50$.

In a way similar to the case of mapping (1.17), the use of a logarithmic scale changes the relations among the states (the measure of a quotient is the difference in the amount of information).

Basic difficulties occur in error analysis. The value of a quantum should be established a priori, and the smallest component of error remains unknown until the analysis is completed. The quantum may not be total error, for the partial errors are relative fractions. These difficulties may be overcome by performing the analysis with physical models and then applying information models to the final results only. This way, however, is indirect.

Chapter 5

Measurement Procedures

5.1 Definition of Measurement

There are several definitions of measurement, each stressing some particular property. The aim of measurement is to produce an image of the investigated fragment of reality, and therefore the more complex the reality, the more complex the image and the way in which it is created.

An elementary *object of measurement* is a phenomenon expressed by a state of one quantity. The phenomenon may be characterized by the states of several quantities, and the states may be variable. The purpose of measurement may be finding the relationships of the quantities. These may be causal connections, functional relations or stochastic relationships. The fact that the relationships occur may be proved in different ways.

In Chapter 1 the principles of formulation of the images of reality were presented. Some particular states of matter are distinguished as references and the image must refer to these states.

The definition of measurement applied by this author [91] is the following: "measurement is the set of activities which allows us to state that the value of a measured quantity x was $a < x < b$ under given conditions, at the moment when the measurement was performed using specific means and specific actions." The definition concerns the measurement of a state of an elementary object, but it may easily be expanded to cover time-dependent quantities and states expressed by more than one quantity. The measurement in this case consists of elementary actions which, possibly, can be regarded independently or which are interrelated.

Although any action which allows us to know that a particular relationship exists between the quantities u and v may be called a measurement, this range of measurements is usually considered to be an experiment.

If the object of measurement is a sequence of states described by a time function $x(t)$, the measurement consists of a sequence of elementary measurements. Often such measurement is called *dynamic measurement.* The measurement of the state of several quantities is a sequence of parallel elementary measurements, separated in space; this case is sometimes called *multiparametric*, or *complex, measurement*.

The statistical parameters of the states which are described by random variables are a specific object of measurement. These are the measurements of stochastic processes (*stochastic measurements*).

In all the cases mentioned, the above definition of measurement may be applied. The values *a* or *b* which determine the *measurement result* are the ranges of the measured quantity, occupying a well determined position in the scale of the quantity, relative to the standards defining the unit of measure.

The concept of a measurement model has also been introduced [91].

Two sets of quantities are involved in a measurement process: the set X of a measured quantity x and the set W of a known quantity w. The elements of this second set are ordered by their value, and marked with the index i. The measured quantity forms a finite or an infinite set, bounded at both ends. The set W is assumed to be finite, i.e. the consecutive elements w_i, w_{i+1} of the set differ by the value $w_{i+1}-w_i = 2\varepsilon_i > 0$ for every i, from $i = 0$ to N. The set W is generated by the standard quantity reproduced by the measuring instrument in

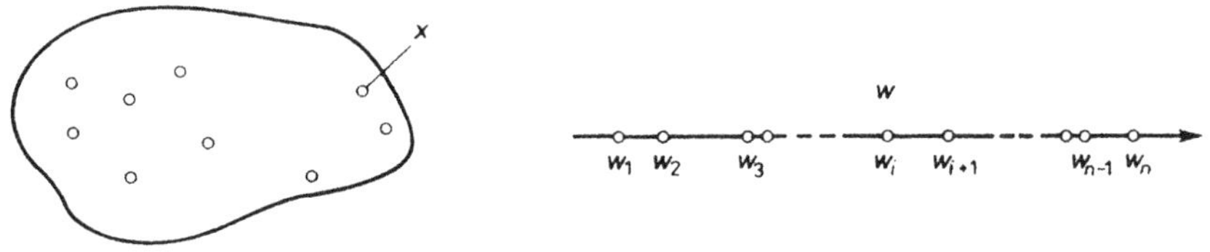

Fig. 5.1. Sets of quantities involved in the process of measurement.

the process of measurement. Both set are shown in Fig. 5.1. Within the measurement model, the measurement action is equivalent to the assignment of an element from the set W to an element x from the set X, with the same value. Since the set W is discrete, the assignment cannot be ambiguous. The result of the assignment is the inequality

$$w_i \leqslant x \leqslant w_{i+1}, \tag{5.1}$$

equivalent to the measurement result.

The model of measurement makes possible a formalization of the actions constituting the process of measurement. With respect to these actions, different *measurement methods* may be distinguished. It is only in the model of measurement that it may be seen that the individual measurement methods differ in the actions performed, in the structure of an MS and in the algorithm by which the measurement result is determined.

A characteristic feature of the *swing method* is that the element x points at the real axis of the set W, after the measured quantity is applied to the instrument. In the *difference method*, the element which points at the real axis is $x - x_c$.

In each of the *null methods*, the measured quantity is compared with the elements of set W, the difference between x and w is checked and then, by changing the elements w_i, such elements w are chosen as satisfy condition (5.1). In the *compensation method*, the comparison of x and w is direct and simultaneous. In the *comparison method*, an auxiliary numerical set K is used, which is realized by the ratio of two quantities. An MS may be balanced with the elements of either W or K. In the *substitution method*, the comparison of the elements x and w is not direct and simultaneous. Balancing of an MS is specific for the null methods; this means that a follow-up or programmed change of the elements of set W is carried out, until the measurement result is obtained.[1]

The structure of an MS for different measurement methods is shown in Fig. 5.2.

It is assumed that the reader has a basic knowledge of measurement methods and their properties [91].

When the measurement methods are presented in the form of functional diagrams, the following properties become essential.

1. Both a functional scheme of a method and a structural scheme modelling a given MS are models of the same system. A functional diagram determines the location and the function of a standard in a given MS.

2. The MS structure, describing the measurement method, is a minimum structure to which a mathematical model of the MS may be reduced. Removal of even a single unit from the functional scheme makes the measurement actions impossible to perform. This property is a consequence of the description of the measurement actions.

3. Application of the elements of the measurement method functional scheme at the stage of determining the measurement result allows unification of these actions. It is then easy to determine which of the properties of individual elements are really significant. This property considerably simplifies modelling of the MS.

The basis for the classification of measurement methods is the way in which the measured quantity x is compared with the standard quantity w. 'The way' means here the actions that must be performed in the process of measurement. A formalized description (verbally or mathematically) of these actions is called a *measurement procedure*. If the measurement consists of several stages, or if different variants are possible, there will be more procedures.

Until now, no rules of creation and unification of measurement procedures have been formulated. Also there are no principles of notation. An attempt to

[1] Automation of the measurement actions makes the external properties of the null method MS identical with those of the swing method.

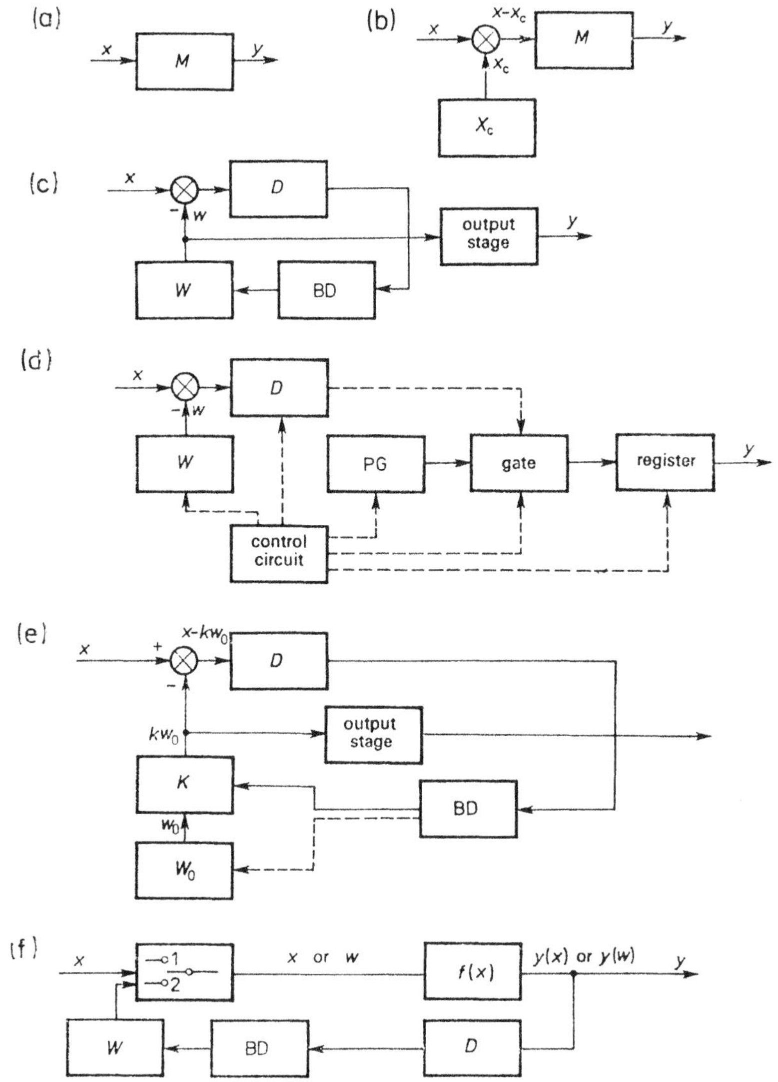

Fig. 5.2. The structure of an MS with different measurement methods applied: (a) swing method, (b) difference method, (c) null method with follow-up balancing, (d) null method with programmed balancing, (e) comparison method, (f) substitution method. W—source of the reference quantity, K—numerical set K, D—balance detector, X_c—source of the compared quantity, BD—balancing device, M—meter, PG—pulse generator.

determine the relation between an image and a distinguished state (standard) is in the work by Finkelstein [21], which deals with the problems discussed in Chapter 1.

An approach to the technique of formulation of the procedures is the scheme proposed by Hofmann [35], who also proposed a description of the so-called strategy of measurement. Applying the flowcharts used for description of computer programs, Hofmann presents the succesive steps and conditions by which the measurement results is obtained, starting from the design of a measurement program and ending with the processing of the result. The idea is a proposal but not worked out thoroughly enough to be applied to a specific task.

The problem of measurement procedures was considered by Jaworski. His work [41] includes a description of measurement methods and the elementary calculus of errors, and shows that there is still much to be done on this subject.

Rymaszewski [42] presents measurement procedures, calling them comparisons. The types of comparison he distinguishes are: a deterministic analog comparison—corresponding to the swing method, a deterministic digital comparison—corresponding to the compensation method, a deterministic dynamic comparison—used in A/D converters, and a stochastic comparison—not applied in practical instrument construction. A concise description of the comparison algorithms was presented using block diagrams.

The problem of measurement procedures—the name and its essence—was introduced by the present author [90]. Trivial procedures, concerning calibration and reading checking of an MS, are discussed on the examples presented in [91]. The problem is of great theoretical and practical importance. Theoretical, because it is a probe of the integrity of the whole theory of measurement and of the uniformity of the concepts of measurement science. All of the regulations and standards of specification concerning MS testing that appear in legal metrology, and are a practical activity, are fundamentally measurement procedures. The manner of formulating the measurement procedures is influenced by computer science—this is why the procedures were recorded in the form of flowcharts [35, 110]. This is not, however, the most important aspect.

The state of knowledge and science is strong evidence that measurement procedures exist, are known and are applied. The aim now is to unify and order them, possibly also simplify and formalize them. The following two procedures presented in Chapters 1 and 2 are essential for all the natural and technical sciences.

(1) Accepting the *metric system* as a basic scale for formulation of the images of reality; the scale determines the relationships among the values of a quantity and the relationships among the states of a quantity. The relationships between the values and the states are isomorphic.

(2) Establishing the standards of units of measure which reproduce the distin-

guished *reference states* of the images of reality. Each measurement result is related to these distinguished reference states through the scheme of verification of measuring instruments and through the hierarchy of standards.

5.2 Procedures Describing Measurement Methods

After the procedure for creating the images of reality is determined and the reference states are established, i.e., after the standards of units of measure are established, a further task is to determine a procedure of performing the measurements. The logical course of this process should be determined, together with the outcome considerations. Also the relations between the measurement result and a distinguished reference state (considered to be true) need to be determined. The measurement procedures are not the same for all measurement systems; for example, they are different for swing methods and for null methods. In all of the MS applying the null measurement method, the necessary part of the instrument is the reference quantity source. This fact is clearly seen from the equations describing the measurement [91]:[2]

$$|x-w| \leqslant \Delta x_D \qquad \text{compensation method,} \qquad (5.2a)$$

$$\begin{aligned} |x-kw| &\leqslant \Delta x_D \\ |x-kw_0| &\leqslant \Delta x_D \end{aligned} \qquad \text{comparison method,} \qquad (5.2b)$$

$$\begin{aligned} |y(x)-y(w)| &\leqslant \Delta y_D+\Delta y_n \\ |x-w| &\leqslant (\Delta y_D+\Delta y_n)|\varphi'(y)| \end{aligned} \qquad \text{substitution method,} \qquad (5.2c)$$

where x is the measured quantity, w is the reference quantity, k is an element from the numerical set K, Δx_D, Δy_D are the dead bands of the balance detector, $y(x)$ is the conversion equation applied in the substitution method, $\varphi(y)$ is the inverse function of $y(x)$, φ' is the derivative of φ and Δy_n is the change of the conversion equation in the time interval between the instant when $y(x)$ was substituted and the instant when $y(w)$ was substituted.

An approximate form of the measurement equation expressing a raw result is

$$x^* = w \qquad (5.3)$$

(or $x^* = kW$ in a comparison method).

The state of the measured quantity is directly equal to the state of the standard, within the limits of uncertainty of this standard and the uncertainty of comparison;

[2] These are the *measurement equations* which are the procedures of determination of a complete measurement result from the MS indications, called the raw measurement result.

the measurement equation clearly states this fact. The measurement equations corresponding to the listed measurement methods are presented in Table 5.1.

TABLE 5.1

Measurement equations for individual measurement methods

Item	Measurement method	Approximate measurement equation expressing raw result $x^* =$	Measurement equation $\{x^*\} =$	Corrected measurement result $x^*_{corr} =$	Random error components $\{\Delta_r\} =$
	Null methods				
1	Compensation	w	$w-\{\Delta_w\}-\{\Delta x_D\}$	$w-\Delta_{ws}$	$\{\Delta_{wr}\}+\{\Delta x_D\}$
2	Comparison	wk	$(w-\{\Delta_w\})(k-\{\Delta_k\})-$ $-\{\Delta x_D\}$	$wk-w\Delta_{ks}-$ $-k\Delta_{ws}$	$w\{\Delta_{kr}\}+k\{\Delta_{wr}\}+$ $+\{\Delta x_D\}$
3	Substitution	w	$w-\{\Delta_w\}-(\{\Delta y_D\}+$ $+\{\Delta y_n\})\|\varphi'\|$	$w-\Delta_{ds}$	$\{\Delta_{wr}\}+(\{\Delta y_D\}+$ $+\{\Delta y_n\})\|\varphi'\|$
	Swing methods				
4	Swing	y	$y-\{\Delta_M\}$	$y-\Delta_{Ms}$	$\{\Delta_{Mr}\}$
5	Difference	x_c+y	$x_c-\{\Delta_x\}+y-\{\Delta_M\}$	$x_r+y-\Delta_{xs}-$ $-\Delta_{Ms}$	$\{\Delta_{xr}\}+\{\Delta_{Mr}\}$

Notations: Δ_w—error of a standard quantity, Δ_k—error of realization of value k, Δ_M—error of a meter, Δ_x—error of a compared quantity, Δy_D, Δx_D—detector dead band, Δy_n—time change of the detector properties, x_c—compared quantity. *Indices*: s—systematic error, r—random error. It is assumed that the balance detector introduces no systematic error.

The equations in the forms given in the table take into account the errors introduced by the significant elements of an MS. A source of error lies in each element of an MS whose properties appear in the measurement equation. These are: the source of the reference quantity, the elements forming the numerical set K in a comparison method, the converter in a substitution method, a balance detector, a meter and a source of the compared quantity in the swing method. The elements necessary in the specific method are shown in Fig. 5.2.

A *corrected result* is obtained by introducing into the *raw result* the corrections eliminating the *systematic error*. The corrected result may be obtained from the measurement equation by averaging, which eliminates the random errors, i.e.,

$$x^*_{corr} = E\{x^*\}, \tag{5.4}$$

and the random error, according to the definition, is the difference

$$\{\Delta_r\} = \{x^*\} - E(x^*\}. \qquad (5.5)$$

The components of random error are listed in Table 5.1.

A *complete measurement result* is obtained after the confidence interval for the random error is calculated at a given confidence level α:

$$a < x^* < b, \qquad (5.6)$$

where

$$a = x^*_{\text{corr}} - \varepsilon, \quad b = x^*_{\text{corr}} + \varepsilon.$$

The following notation may be applied

$$x \rightarrow \{x^*\} = x^*_{\text{corr}} + \{\Delta_r\}, \qquad (5.7)$$

where the set $\{\Delta_r\}$ is characterized by the measure 2ε. In null methods, the reference state for the image of a measured quantity is a standard, which is a part of the MS. The procedure of referring to this standard is simple, and needs no additional interpretation. In the meters applying a swing method, however, the situation is different. A swing method meter does not include a standard. Although this is an unquestionable fact, some unjustified different interpretations may be found in the bibliography on this subject.

A *swing method instrument* is a device providing direct conversion of an input quantity x into an output quantity y with a determined accuracy. The conversion equation

$$y = f(x) \qquad (5.8)$$

describes the correspondence of the states y and x. The accuracy is expressed by the confidence interval (5.6). To make the instrument simpler to use, the readings $y_0(x)$ are given the values of the measured quantity x. However, the fact remains unchanged that the physical output quantity is y and the values of x have only been assigned to it. This fact may be treated as if the quantity y was expressed in the units of the measurand. The process of assigning the measure of x to the quantity y is called the *calibration of an MS*. Applying a quantity x to the input of the calibrated instrument and reading of the indication x^* permits writing the conversion equation in the form

$$x^* = \varphi(x) = \varphi[f_0(x)] = x, \qquad (5.9)$$

because usually $\varphi = f_0^{-1}$ within the limits of conversion error, i.e.,

$$P[|x^* - x| \leqslant \varepsilon(x)] \geqslant \alpha. \qquad (5.10)$$

In the *direct measurement*, the conversion equation of the measuring instrument has form (5.9). When the instrument used has a conversion equation in form

(5.8), the measurement is *indirect*, since the measurement result must be calculated by applying the function $\varphi(y) = f_0^{-1}$. This is the way in which the measurement system was presented in Fig. 3.1.

The procedure of measurement using a swing method meter determining the location of the standard may be understood to mean that the substitution method is performed using a swing method meter.

A classical description of the substitution method is as follows [91]. At instant T_1, the measurement quantity is provided to the MS and reading y_1 is obtained. Then, at the instant T_2, a standard quantity is provided at the input. The value of the standard quantity is regulated so that the reading obtained is $y(w) = y_1$. Thus

$$|y(x)-y(w)| \leqslant \Delta y_D+\Delta y_n. \tag{5.11}$$

The component Δy_D determines the limiting uncertainty of the comparison of $y(w)$ and y_1. The component Δy_n expresses the fact that the comparison of $y(x)$ and $y(w)$ is not simultaneous, but occurs after a time interval T_2-T_1, during which the conversion equation might have changed. Thus

$$\Delta y_n = |y(x, T_1)-y(x, T_2)|, \tag{5.12}$$

while

$$P(|\Delta y_n| \leqslant \varepsilon_n) \geqslant \alpha, \tag{5.13}$$

since it is impossible both to perform the measurement and to investigate the properties of the instrument at the same time, in order to determine Δy_n from definition (5.12).

The difference between the proposed description and the classical one is as follows. At the instant T_1, the standard quantity is applied to the swing method meter and the respective readings $y = f_0(w)$ are determined. Then, the identification of the model f_0 is made and the model is fixed by assuming the values $x^* = w$ to the readings y. The conversion equation takes the form of (5.9). The set of readings $Y = x^*$ is acknowledged and the instrument can reproduce the set W: this is the calibration procedure. At the instant T_2, after the measured quantity is provided, the value of x is compared with $x^* = w$, according to the measurement equation

$$|y(x, T_2)-y(w, T_1)| \leqslant \Delta y_D+\Delta y_n, \tag{5.14}$$

that is

$$|x-w| \leqslant (\Delta y_D+\Delta y_n)|\varphi'(y)|. \tag{5.15}$$

The instrument does not need another calibration and may be used as long as the value of error Δy_n does not exceed ε_n, according to requirement (5.13).

The instrument is applied as a swing method meter and the substitution procedure need not be repeated each time. The treatment of the swing method as a separate measurement method does not have a theoretical basis; the result of measurement performed with a swing instrument may be referred to the standards through the measurement equation of the substitution method. The distinction of the substitution method has practical reasons: the measurement actions are in this case different from the null method. This fact is clearly expressed by the measurement equation. After the measured quantity is applied to the instrument, the element x is recorded on the real axis of the set W—this is the way in which a mercury column in a glass thermometer, an ammeter pointer, etc. operates. In the process of measurement with a measuring instrument, the set W is reproduced in such a sense that the stability of the properties of the instrument is ensured by the memory of the instrument in a given period of time. In swing method meters, unlike the null method, no balancing is performed.

Calibration of an MS is a metrological procedure of providing a transducer of a quantity x into a quality y with the properties of a measuring instrument and it is applied in swing method meters only. The procedure is auxiliary, but must be performed before a measuring instruments is ready for use. There are three procedures resembling each other and having similar activity ranges: calibration, reading checking and adjustment of an MS. To avoid any misunderstanding, let us set the scene for all three.

Calibration is the determination and fixing of the characteristic of an MS. According to the previous definitions (Sections 4.1 and 4.2), the mathematical model $y_0 = f_0(x)$ is derived and the *characteristic* is the inverse function $x^* = \varphi(y_0) = f_0^{-1}(y_0)$. Another definition states that calibration is the process of giving the readings of an MS the measure of a measured quantity in strictly determined conditions (called reference conditions). The measure is transferred with finite accuracy, expressed by the limiting error, which is a metrological property of the calibrated system.

Reading checking is an action performed to prove that the readings of an MS agree with the fixed characteristic within the limits of an allowable error of the MS. If the readings do not conform to the characteristic, the MS must be eliminated, as it may no longer be used for measurements.

Adjustment of an MS is a correction of the construction parameters of an MS, performed in order to obtain a desired mathematical model. After the adjustment, the characteristic satisfies the equality $\varphi = f_0^{-1}$ and the readings of the MS conform to the fixed characteristic.

Calibration and adjustment result in a change of the properties of an MS. Reading checking is a qualifying procedure, deciding if the MS is correct or not. Thus every calibration (at the end of the manufacturing process or after repair) and every adjustment should be followed by readings checking to prove that the calibration or adjustment was performed correctly. On the other hand, a negative result of readings checking may lead to disposal, repair or adjustment of an MS. The relations of the above procedures are shown in Fig. 5.3.

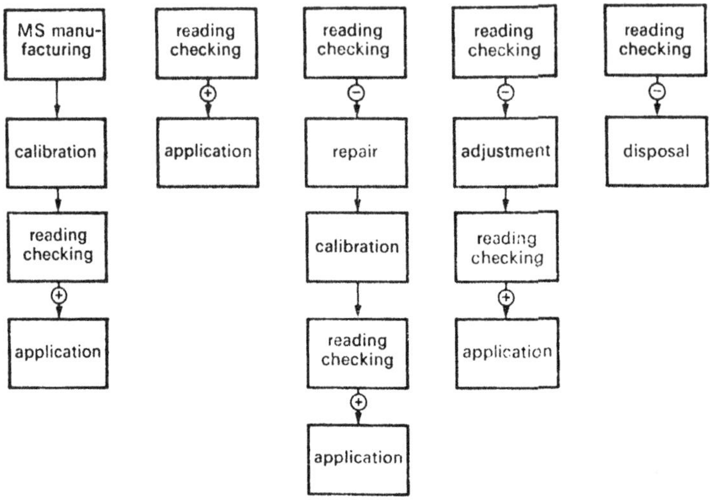

Fig. 5.3. The sequence of calibration, reading checking and adjustment procedures.

5.3 Calibration Procedures

The purpose of calibration is to determine and to fix the characteristic of an MS with a predetermined uncertainty. Before calibration, the MS characteristic should be known at the level of coefficients. The inaccuracy is expressed as one of the classes of metrological characteristics, usually as a limiting error distribution over the measurement range, i.e., the confidence intervals, including all sources of errors in reference conditions.

The course of calibration depends on the way in which the characteristic is fixed.

Method I (*proposed name—algorithmic*). The mathematical model of an MS is determined in an ordinary (natural) way and the characteristic $x^* = \varphi(y)$ is determined on that basis. Then, the characteristic is fixed by any method, for example by drawing a scale, producing a diagram, writing a processing program (in computerized systems), etc.

Method II (*proposed name—constructional*). According to the mathematical model known a priori at the level of numerical values of coefficients, the characteristic is fixed with any method before the calibration is performed. In the calibration process, the construction parameters are corrected so that the model obtained conforms to the assumed one. Since only 2 (maximum 3) parameters may be adjusted independently, this method is usually applied in MSs with a linear characteristic. In the example of the converter shown in Fig. 3.3, the number of degrees of freedom is n. The characteristic is fixed by selecting the values of the resistors.

Method III (*mixed*). The method is usually applied in cases of nonlinear characteristics. Because of the established constraints, such as the scale length, stan-

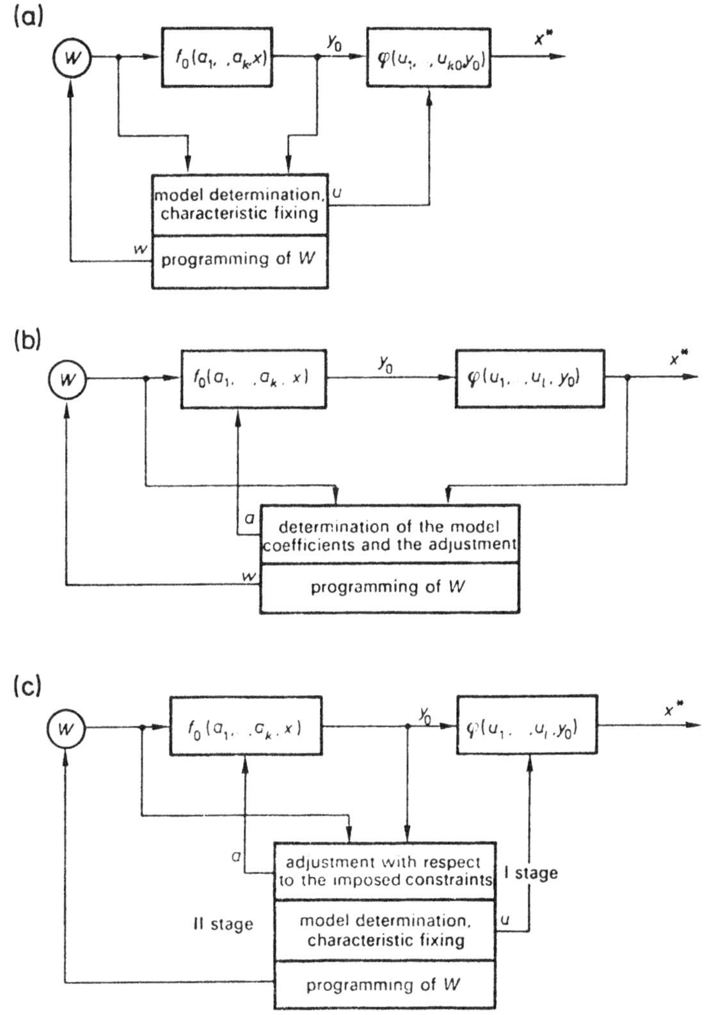

Fig. 5.4. Methods of fixing the characteristics of an MS: (a) algorithmic, (b) constructional, (c) mixed.

dard value of the output signal, etc., two points of the measurement range (the start and end points) should be fixed by realizing a proper construction. The other part of the scale is fixed naturally, using the algorithmic method. This method of calibration is carried out in two stages.

The diagrams showing the methods of fixing the calibration result are shown in Fig. 5.4.

Calibration may be performed by the indirect or direct method. The choice of the method influences the means, technique and course of calibration. Indirect calibration may be applied in some of the MSs with a relatively simple model, expressed by parameters which are easily and very accurately measurable.

The *calibration is direct* when the standard quantity applied to the calibrated MS is the same kind as the measured quantity. The characteristic is determined from the readings. The case of direct calibration is presented in Fig. 5.5. The

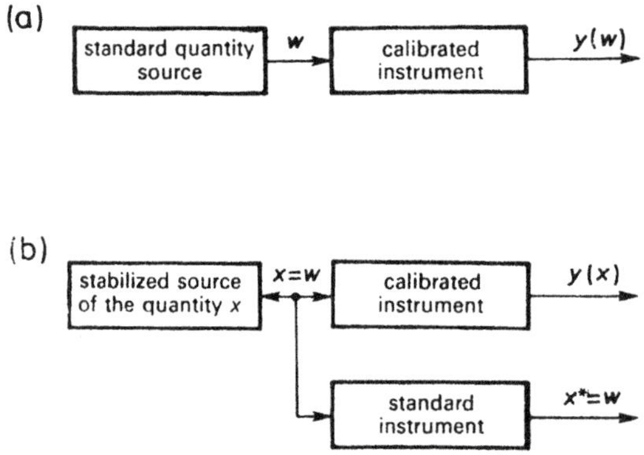

Fig. 5.5. Direct calibration: (a) with a standard applied, (b) with a standard instrument applied.

source of the standard quantity is either a standard or a source of a measured quantity with a constant value. In this last case, the value is measured with a reference meter, of adequate accuracy.

The *calibration is indirect* when the characteristic is derived from the mathematical model. The model must be identified with the accuracy level of the numerical values of coefficients, with adequate certainty.

Direct calibration is commonly used and the calibration procedure is known; a detailed example may be found in [91]. The problems of the number of calibration points and their selection as well as of the accuracy of the standard are presented in Chapter 6.

The direct-method calibration procedure with algorithmic fixing of the characteristic:

1. Establish the reference conditions for the MS.

2. Apply a standard quantity with a value suitable for the point x_1.
3. Note the reading of the standard MS and of the calibrated MS.
4. Apply a value x_2 of the standard quantity.
5. Note the readings as in step 3.
6. Perform steps 2 and 3 the necessary number of times at all the points $x_1, \ldots, x_N$.
7. From the results obtained, determine the mathematical model $y_0 = f(x_0)$ of the calibrated MS.
8. From the model obtained in step 7, determine the characteristic $x^* = \varphi(y_0)$.
9. Fix the determined characteristic.
10. Perform the reading checking of the MS.

The direct-method calibration procedure with constructional fixing of the characteristic:

1. Establish the reference conditions for the calibrated MS.
2. Apply the value x_1 of the standard quantity of the MS.
3. Adjust the parameter r_1 so that the reading of the MS is x_1.
4. Perform steps 2 and 3 for the values $x_2, \ldots, x_N$. The adjusted parameters are $r_2, \ldots, r_N$, respectively.
5. If the adjusted parameters interact, steps 2, 3 and 4 should be repeated such number of times that in the last repetition no parameter needs to be adjusted.
6. If the adjusted parameters do not interact, repeat steps 2, 3 and 4 as reading checking without adjustment.

The indirect method is generally applied to orifice plates [103] and ultrasonic flowmeters [135], relatively rarely to electroconductometric transducers [79] and to unique instruments when there are no standards with an accuracy suitable for direct calibration [96]. In some cases the sets of instruments in indirect measurements may be considered as an indirectly calibrated set for direct measurements.[3]

The indirect calibration procedure:

1. Formulate a mathematical model of an MS at the level of coefficients.
2. Prove that the accuracy of the formulated model is proper.
3. Measure the values of the construction parameters and estimate their uncertainties.
4. Calculate the numerical values of the coefficients of the model.
5. Determine and fix the characteristic of the MS.
6. Determine the metrological properties of the MS in reference conditions as well as the operating conditions of the MS.

---

[3] When the definition of indirect measurement is $x = f(A, B, C)$ and the values A and B are constants, then $x = f(C)$ may be a direct measurement after the set of instruments is calibrated. An example may be the direct method of air humidity measurement.

The *formulation of the model* should be adequate to the purpose of the measurement with the calibrated MS.

The *purpose of measurement* is usually stated in claims as follows.

1. A physical quantity $x = x_m$ is to be measured; the assumed characteristics of the quantity are the probability density $p(x)$ and the spectral density of x, $S_x(\omega)$. The $p(x)$ is usually assumed uniform, log-uniform, hyperbolic or different [78], and the spectral density is usually considered in the frequency interval $(0, \omega_{\text{lim}})$ or (ω_1, ω_2).

2. The associated conditions of the measured quantity: physical environment, associated phenomena, energy of the source of the measured quantity, etc.

3. Measurement conditions—known or assumed sets of influencing quantities $x_1, \ldots, x_{m-1}$ and the characteristics of these sets as in claim 1 above.

4. Links, connecting the MS with the rest of the cybernetic system; the links determine the distribution of the MS (location of the source of the measured quantity and of the information receiver), the form of the output information, matching the cybernetic system.

5. Measurement system environment—the constraints resulting from the national system of measures (legal metrology), state of technology, operating conditions and maintenance of the MS, repairs, etc.

6. Demanded accuracy of the MS with the conditions as in claims 2, 3, 4 above; the following values should be determined: $\|\varrho\|$—the metric of the distance of the sets $\{x\}$, $\{x^*\}$, ϱ_{all}—the allowable distance of the above sets, where $\{x^*\}$ is the set of measurement results and $\{x\}$ is the set of correct measurement results, as in (4.10). The set $\{x\}$ is assumed in claim 1 of the purpose of measurement. The accuracy may be expressed by the value of allowable error Δ_{all} at a given confidence level and with determined error metric.

The purpose of the system may be stated more widely. In this case it may be decomposed into its component aims, which are considered separately [86].

The first stage is formulation of the mathematical model of an ideal MS. A specific system or testing post is investigated, applying the idea arising from a simple concept and developed through bibliographic studies. Usually the model of an ideal measurement system may be found in bibliography, but its form is rarely complete and ordered.

The model of an ideal MS is formulated on the basis of its structural diagram, by writing the conversion equations for causal relations between the output quantity of the MS and the measured quantity. The relations are the consequence of the application of a concrete measurement principle. The influence of the physical environment is neglected, since the reference conditions and the disturbing effects are regarded and treated separately. After the synthesis of the structural diagram and of the conversion equations is performed, a model in form (3.10) is obtained.

The second stage is the formulation of a model of the real MS. The structural diagram is extended by considering the phenomena described in the measurement purpose claims 2 and 3. Different phenomena that might occur are considered, such as temperature changes, power supply changes, humidity, radiation and parasitic effects (induced voltages, chemical reactions, etc.). These are qualified either as causes or as effects and are introduced in the previously established structural units or as additional units of the structural diagram. After the conversion equations of each unit are written, they should be converted to a form explicit for the dependent variables of the units $j = 1, \ldots, n$:

$$\begin{aligned} y_1 &= f_1(x_1, \ldots, x_m, y_2, y_3, \ldots, y_n), \\ &\cdots\cdots\cdots\cdots\cdots\cdots\cdots\cdots\cdots \\ y_j &= f_j(x_1, \ldots, x_m, y_1, \ldots, y_{j-1}, y_{j+1}, \ldots, y_n), \\ &\cdots\cdots\cdots\cdots\cdots\cdots\cdots\cdots\cdots \\ y_n &= f_n(x_1, \ldots, x_m, y_1, \ldots, y_{n-1}). \end{aligned} \tag{5.16}$$

If it is impossible to write the equations in an explicit form, the equations should be left in an implicit form

$$\zeta_j(x_1, \ldots, x_m, y_1, \ldots, y_n) = 0. \tag{5.17}$$

Synthesis of equations (5.16), which, in fact, is the elimination of the dependent variables $y_1, \ldots, y_{n-1}$, leads to the model (3.8) for $y = y_n$. In the case when the synthesis yields mathematical problems, the model is left in form (5.17). The formulated mathematical model (3.9) describes the external properties of the measurement system. The notation of the model is incomplete, as the construction parameters $a_1, \ldots, a_j, \ldots, a_k$ appear as the coefficients in equations (3.9) or (5.17). The complete notation of the model is (3.8).

Formulation of a real model of an MS is performed step by step. The created models $F_1, F_2, \ldots$ differ in the number of disturbances accounted for and in the minuteness of detail.

In the subsequent, third stage, a model conforming to the requirements should be selected from the sequence of formulated models $F_1, F_2, \ldots, F_r, \ldots$ ordered so that the models F_r and F_{r+1} differ by one source of error. To select a proper model of an MS, the error sequence

$$\Delta_r = F_r - F_{r-1} = y_r - y_{r-1} \tag{5.18}$$

should be ordered for $r = 1, 2, \ldots$ and for identical properties of the measured quantity x (according to the claim 1 of the purpose of measurement). The model F_0 is ideal (3.10). In the notation as (5.18), Δ_r is a random variable resulting from the random changes of $x_1, \ldots, x_{m-1}$ or $a_1, \ldots, a_k$.

To select a model, it is sufficient to investigate the limiting values of errors $\Delta_{r\,\mathrm{lim}}$, using the differential form of equation (3.8)

$$\mathrm{d}y = \sum_{i=1}^{m-1} \frac{\partial F}{\partial x_i}\,\mathrm{d}x_i + \sum_{j=1}^{k} \frac{\partial F}{\partial a_j}\,\mathrm{d}a_j + \sum_{j=1}^{k} \frac{\partial F}{\partial a_j} \sum_{i=1}^{m-1} \frac{\partial a_j}{\partial x_i}\,\mathrm{d}x_i . \tag{5.19}$$

In the series of errors $\Delta_{r\,\mathrm{lim}}$, $r = 1, \ldots, r$, the error sources whose values are much smaller than the value of allowable error Δ_{all} determined as in claim 6 of the purpose of measurement are neglected. This condition may be formulated more precisely by applying the *small error criterion*: the errors $\Delta_{r\,\mathrm{lim}}$ 5% or 10% of total error are neglected. The error $\Delta_{r\,\mathrm{lim}}$ is negligible when:

$$\Delta_{r\,\mathrm{lim}} \leqslant 0.05\Delta_{\mathrm{all}} \quad \text{for the linear metric,} \tag{5.20a}$$

$$\Delta_{r\,\mathrm{lim}} \leqslant 0.2\Delta_{\mathrm{all}} \quad \text{for the square metric.} \tag{5.20b}$$

The value in the small error criterion depends on the number of components of error—the more components the total error has, the smaller the neglected error should be.

As a result of neglecting some of the sources of errors, a new sequence of models F_q and a new sequence of errors Δ_q are obtained, $q < r$. If the condition $\Delta_{\mathrm{lim}} < \Delta_{\mathrm{all}}$ is met, the proper model is F_q. The value of the limiting error Δ_{lim} is determined from the total error

$$\Delta = \sum_{q=1}^{q} \Delta_q \tag{5.21}$$

which is the sum of the random variables Δ_q. Most of the physical phenomena which are the sources of errors formally come from the mathematical model in form (3.8).

Potentially, each of the influencing quantities listed in claim 3 of the measurement purposes is a source of error. Also the phenomena listed in claim 2 of the purposes are error sources and appear in equation (3.8), together with the construction parameters a_j.

Every parameter a_j, $j = 1, 2, \ldots, k$, is potentially burdened with some error. The components of the error are modelled by equation (3.112).

To be distinguished from the external sources of errors caused by the changes of $x_1, \ldots, x_{m-1}$, the internal sources of errors Δa_{j0} and Δa_{jn} are marked in the structural diagram of an MS by circling a symbol of a parameter [91].

Synthesis of the errors of an MS is not an easy task. When the contribution of each error to the total error was determined, each source of error could be regarded separately, assuming that there were no other sources. Now, all the sources must be regarded simultaneously. Most often, the values of all the coefficients in equation (5.19):

$$k_i = \frac{\partial F}{\partial x_i}, \quad k_j = \frac{\partial F}{\partial a_j} \tag{5.22}$$

are constant. Calculating the total error is then reduced to adding the random variables dx_i, da_j with known or assumed probability distributions, the variables being multiplied by constant coefficients k_i or k_j, namely

$$\Delta y = \sum_{q=1}^{q} \Delta y_q = \sum_{i=1}^{m-1} k_i \Delta x_i + \sum_{j=1}^{k} k_j \Delta a_{jo} + \sum_{j=1}^{k} k_j \Delta a_{jn}. \tag{5.23}$$

The singularities of errors should be considered when the errors are added. The mean values of the variables $\Delta x_i = x_i - x_{io}$ for $i = 1, \ldots, m-1$ may equal 0, and then they are the error sources. When $E(\Delta x_i) \neq 0$, a supplementary systematic error appears as well.

The causes of the sources of errors Δa_{jo} are beyond the analysed MS. They arise from the properties of the measurement systems measuring the values of the parameters a_j. In the analysed MS only one value of Δa_{jo} $(j = 1, \ldots, k)$ is randomly realized. The value is constant, yet unknown.

When the mean values of the random variables Δa_{jo} equal 0 and the variances are σ_{jo}^2, then

$$E\left\{\sum_{j=1}^{k} k_j \Delta a_{jo}\right\} = 0, \tag{5.24}$$

which means that the expected value of error caused by the sources Δa_{jo} is equal to zero. The interval of this error may be estimated after the variance has been calculated. Let us reduce all the error sources to one common set with variance σ_0^2, by applying weights so that

$$\sigma_0^2 = w_j^2 k_j^2 \sigma_{jo}^2 \quad \text{for} \quad j = 1, \ldots, k \tag{5.25}$$

and e.g. $w_1 = 1$. Then, when

$$\operatorname{cov}\{\Delta a_{jo}, \Delta a_{lo}\} = 0 \quad \text{for} \quad j, l = 1, \ldots, k, \quad j \neq l, \tag{5.26}$$

and for the fulfilled condition (5.24), we obtain

$$\operatorname{var}\left\{\sum_{j=1}^{k} k_j \Delta a_{jo}\right\} = \operatorname{var}\left\{\sum_{j=1}^{k} \frac{w_j k_j \Delta a_{jo}}{w_j}\right\} = \sigma_0^2 \sum_{j=1}^{k} \frac{1}{w_j^2}. \tag{5.27}$$

With probability α, the systematic error caused by the error sources Δa_{j0} does not exceed the interval

$$\pm t\sigma_0 \sqrt{\sum_{j=1}^{k} \frac{1}{w_j^2}} \,. \tag{5.28}$$

The error sources $\Delta a_{j\,\mathrm{in}}$ (3.112) have the most complex characteristics.

Because of ageing and wearing effects in the elements of an MS, the instability error $k_j \Delta a_{j\,\mathrm{in}}$ is a non-stationary random variable. Thus the analysis of this error should be limited to a time interval T_1, during which the sources of the instability error are stationary in origin and the mean value is zero, i.e.,

$$E\{k_j \Delta a_{j\,\mathrm{in}}\}_{T_1} = 0, \tag{5.29a}$$

$$\operatorname{var}\{k_j \Delta a_{j\,\mathrm{in}}\}_{T_1} = \sum_{j=1}^{k} k_j^2 \sigma_{j\,\mathrm{in}}^2 = \text{const}. \tag{5.29b}$$

However, the evaluation of the metrological properties of the MS may not be limited to the interval T_1 (when conditions (5.29) are fulfilled). The value of the variance (5.29b) in a different time interval T_2 may be different, as compared with the value at the time interval T_1, and condition (5.29a) may not be satisfied. The mode of modelling an instability error is described by formulae (4.84), (4.85).

Among three kinds of errors (5.23):

— supplementary errors arising from changes of x_i (within the tolerance limits of reference conditions);

— the errors of the rated values of construction parameters (Δa_{j0});

— the instability error ($\Delta a_{j\,\mathrm{in}}$),

only the second source produces a systematic error in a given specimen of an MS; the other two may include both systematic and random components.

The following example explains the indirect calibration method.

An ultrasonic flowmeter. Indirect calibration method

The principle of operation of an ultrasonic flowmeter is shown in Fig. 5.6. The flowmeter consists of two probes, mounted in a pipeline with flowing water, and a converter. The converter includes electronic circuits which emit and receive an ultrasonic beam and circuits for measuring the beam transient time. Each probe contains a piezometric transducer, acting alternately as an emitter and a receiver of the ultrasonic beam.

The transient time of the ultrasonic beam from probe 1 to probe 2 is

$$t_{12} = \frac{D}{\cos\alpha}\,\frac{1}{c + w\cos\alpha}, \tag{5.30a}$$

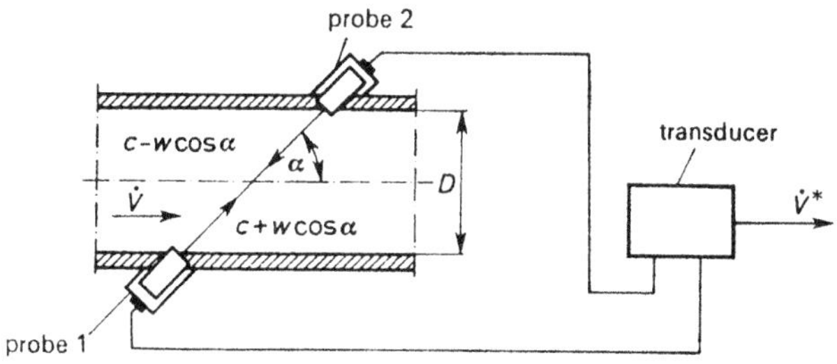

Fig. 5.6. An ultrasonic flowmeter—a simplified diagram.

and from probe 2 to probe 1

$$t_{21} = \frac{D}{\cos\alpha}\frac{1}{c-w\cos\alpha}, \tag{5.30b}$$

where D is the pipeline diameter, α is the angle between the ultrasonic beam direction and the pipeline axis, w is the water velocity (mean value along the path of the ultrasonic beam) and c is the ultrasound velocity in the liquid. Thus, the difference of the times is proportional to the velocity of the liquid:

$$\Delta t = t_{21}-t_{12} = \frac{D}{\cos\alpha}\frac{2w\cos\alpha}{c^2-w^2\cos^2\alpha} \approx \frac{2Dw}{c^2}, \tag{5.31}$$

for $w \ll c$. Also, denoting

$$t = t_{21}+t_{12} = \frac{D}{\cos\alpha}\frac{2c}{c^2-w^2\cos^2\alpha} \approx \frac{2D}{c\cos\alpha}, \tag{5.32}$$

we have

$$\frac{\Delta t}{t^2} \approx \frac{2Dw}{c^2}\frac{c^2\cos^2\alpha}{4D^2} = \frac{w\cos^2\alpha}{2D}, \tag{5.33}$$

independent of the ultrasonic beam velocity.

The details of construction of the electronic system will not be discussed here. The output signal y, either analog or digital, is proportional to $\Delta t/t^2$ and thus the reading is proportional to w. Since the measured quantity is usually the flow rate $\dot{V}$ of liquid in the pipeline, the mathematical model of the flowmeter is more complex. The conversion stages are shown in Fig. 5.7.

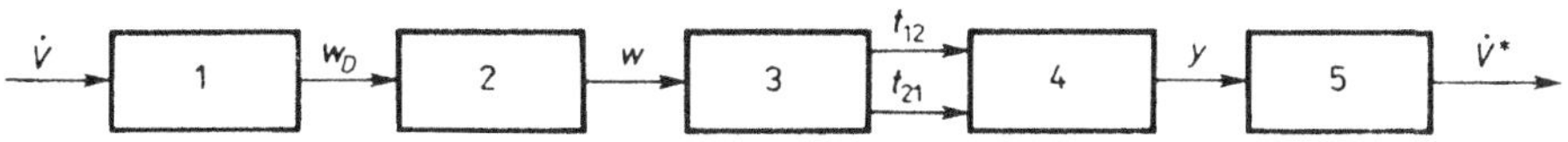

Fig. 5.7. The conversion diagram of an ultrasonic flowmeter.

The measured value $\dot{V}$ is converted into the value of the velocity of the liquid (precisely, into its mean value in the cross-section of the pipeline)

$$w_D = \frac{4}{\pi} \frac{\dot{V}}{D^2}. \tag{5.34}$$

The relationship of the mean velocity along the path of the ultrasonic beam and the mean velocity in the pipeline is expressed by the formula

$$w = \frac{1}{K} w_D, \tag{5.35}$$

where K depends on the flow nature, described by Reynolds' number, and on the pipeline roughness. The procedure for calculating K is shown in Fig. 5.8, where

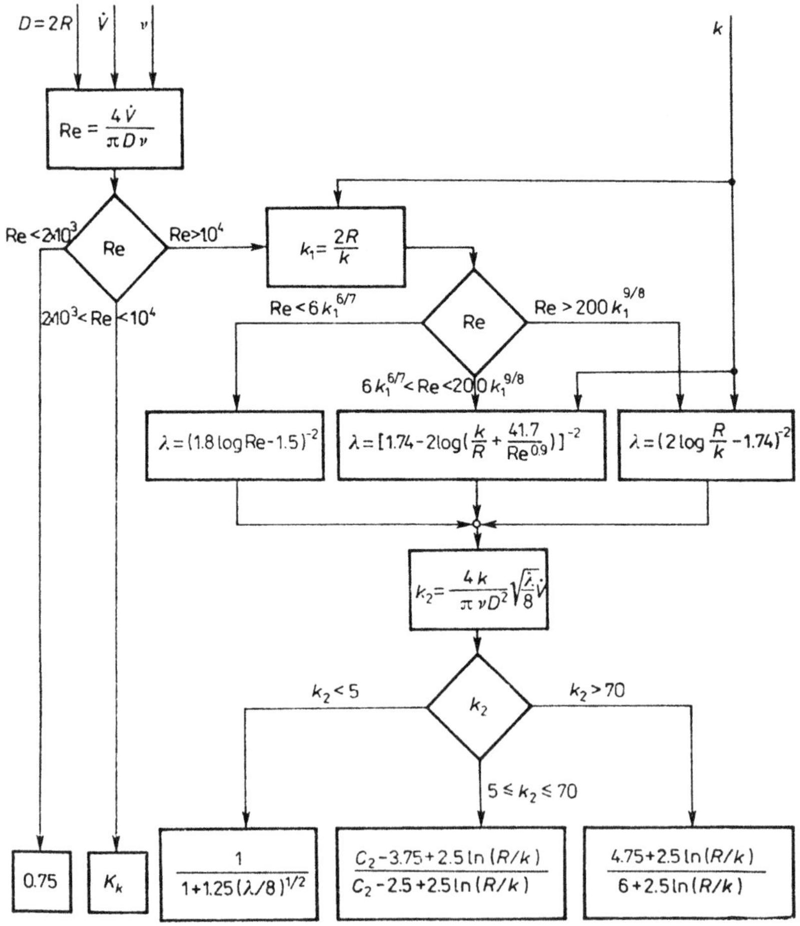

Fig. 5.8. A flowchart for calculation of the coefficient K characterizing the second conversion stage of an ultrasonic flowmeter [135].

$R = D/2$ is the pipeline radius, k is the roughness, ν is the kinetic viscosity, k_1, k_2 are coefficients separating the areas of turbulent flow, λ is the linear flow resistance coefficient, and C is a constant coefficient.

The equations for stage 3 were given above (5.30), (5.33). The equation for stage 4 is

$$y = k_4 \frac{\Delta t}{t^2}. \tag{5.36}$$

Thus, the form of the mathematical model is

$$y = \frac{2}{\pi} \frac{k_4 \cos^2\alpha}{D^3 K} \dot{V}. \tag{5.37}$$

To identify the model, i.e. to calibrate the flowmeter, the values of D, α, and k should be measured and the value of K calculated. The coefficient k_4 is set in a converter using a special control, and the value of k_4 may be read. Usually the value of D is also preset. The value of K depends on the Reynolds' number, as shown in Fig. 5.9. Therefore, the model is nonlinear and should be determined

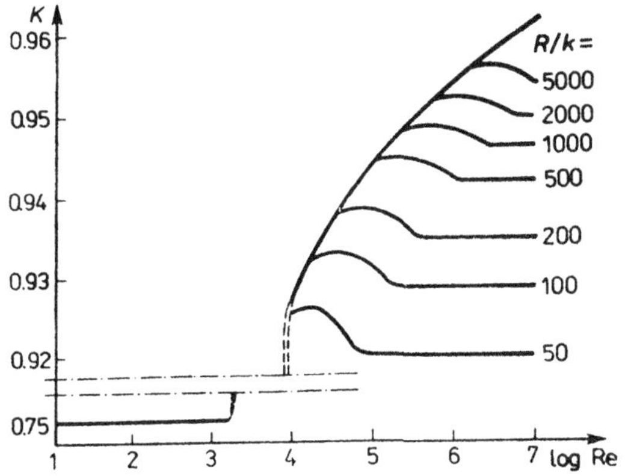

Fig. 5.9. The relationship K vs Reynolds number.

stepwise. The model is sufficiently accurate in so-called normal installation conditions of the flowmeter and, with proper mounting tolerances obeyed, allows neglect of the additional effects.[4]

The characteristic is usually approximated by a straight line

$$\dot{V}^* = a_0 + a_1 y = a_0 + \frac{\pi}{2} \frac{KD^3}{k_4 \cos^2\alpha} y. \tag{5.38}$$

[4] See Polish Standard PN/M-42370: Liquid Volume Flux Measurement with Ultrasonic Flowmeters. (Proposition of the standard to be applied experimentally from 1986.)

The coefficients a_0, a_1 are selected so that the approximation error

$$\Delta = V^* - \dot{V} \tag{5.39}$$

conforms to an optimality criterion in the whole measurement range, for example

$$\int_{\dot{V}_1}^{\dot{V}_2} \Delta^2 \mathrm{d}\dot{V} = \min \quad \text{or} \quad \min_{a_0, a_1}[\max \Delta(\dot{V})]\Big|_{\dot{V}_1}^{\dot{V}_2}. \tag{5.40}$$

In the cases when the characteristic is fixed in the construction, the values of a_0, a_1 are preset using zero and sensitivity controls. Experiments performed on 500 and 1800 mm diameter pipelines show that the indirect method of calibration of an ultrasonic flowmeter makes it possible to reach an accuracy level of about 1% [135]. However, the reader must realize that the conditions stated above are not sufficient to obtain such accuracy.

5.4 Reading Checking and Adjustment of Measurement Systems

Reading checking is necessary when there is doubt about the nominal metrological properties of an MS being preserved. Also, the checking is done after every calibration and adjustment, as well as periodically. Wider range reading checking is performed when the time stability of the metrological properties is investigated for forecasting purposes.

The nominal metrological properties are determined together with the reference conditions and fundamental and supplementary errors for every MS. The fundamental error is expressed by the confidence intervals, which may be dependent on the value of the measured quantity. This fact is illustrated in Fig. 4.10.

According to the definition of the confidence interval, the probability of the error being not greater than the limiting value is below unity, for example 0.9 or 0.95. There are no universal rules for expressing the confidence interval and the confidence level. For example, a notion of accuracy class is used, which is equivalent to the confidence interval as in Fig. 4.10a, but the definition of accuracy class does not describe the confidence level explicitly. The concept has been interpreted in Section 4.5.4.

The reading checking procedure includes applying a standard quantity to the MS in reference conditions,[5] a readout, determination of an error and checking whether the error exceeds the allowable error, i.e. the confidence intervals. The re-

---

[5] Reading checking under conditions different from the reference conditions is allowed, but the limits of allowable error must be determined. The certainty of such checking is obviously smaller.

sult of the checking is positive if the error does not exceed the confidence interval at any checking point.

There are no accepted rules to say at how many points of the scale the error should be investigated, how the points should be distributed over the scale and what the number of repetitions should be. Usually it is recommended to check the readings once or twice at all the points of the scale which are labelled with numbers. This recommendation is, of course, valid for analog instruments. In digital equipment, the number of points depends on the principle of measurement. For example, a dual-slope digital voltmeter should be checked at two points, while a successive approximation voltmeter should be checked at all the points which are the weights of the output signal code.

Reading checking includes:

(1) providing the reference conditions for the MS being checked;

(2) applying the value x_1 of the measured quantity;

(3) readout of the result indicated by the MS (if necessary, also by a standard MS);

(4) performing steps (2) and (3) at points $x_2, \ldots, x_n$, once or twice, as necessary;

(5) determination of the errors of readings of the MS at all checkpoints;

(6) evaluation of the results and qualification of the MS.

A positive result of reading checking means that for some time after this procedure, the metrological properties of the MS will conform to the nominal data. The length of this time interval is not known precisely. In some of the MSs it may be a few hours, and in others several years. The time interval may be estimated experimentally; it is called the correct work period of an MS or up-time, in the sense of reliability. As a parameter characterizing an MS, the time is a random variable. To determine rationally the time for the next reading checking, more extensive investigations of the MS are necessary. The design of such experiments has been presented in Subsection 4.5.1.

The service time with no reading checking may be derived from models (4.82)–(4.87) and (3.147), (3.148). The time should be found for which the allowable error will be exceeded over the whole measurement range, at which the problem is serious. However, proper adjustment allows extending the useful time interval.

The time changes of error depend on the value of x and on the parameters $r_1, \ldots, r_l$, controlled in the adjustment process. When the predictions of the confidence interval of the error are known with a given probability P_z, the predicted confidence interval will exceed the known limiting value of the allowable error Δ_{all} after the time T, dependent on

$$T = f(P_z, \Delta_{all}, x, r_1, \ldots, r_l), \tag{5.41}$$

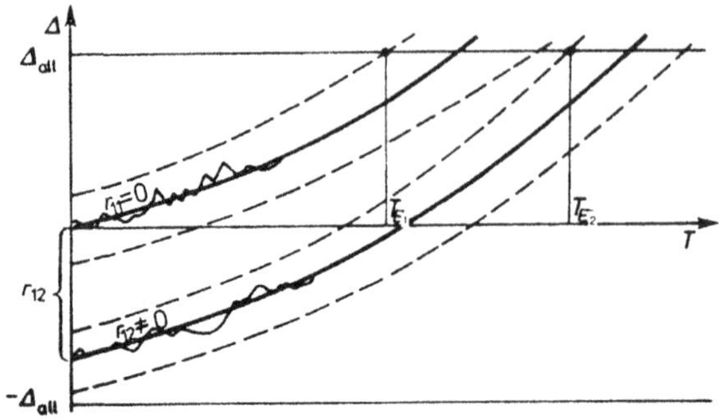

Fig. 5.10. The influence of the value of the adjusted parameter r_1 on the service time. The width of the confidence interval depends on P_z. The thin line marks the error changes, the thick line the estimator and the prediction. The confidence intervals are marked with dotted lines.

which is illustrated in Fig. 5.10. The service time to the next checking is understood as a time T_E, fulfilling the condition

$$T_E = \max_{r_1 \dots r_l} \min_{x} T. \tag{5.42}$$

Olszewska [80] has determined the service time fulfilling condition (5.42), by applying models (4.86) and (3.148).

The component correlation function Z_t' of each of the processes $\{\tilde{k}\}$ and $\{\tilde{x}_0\}$, as well as the cross-correlation function, was modelled by the function

$$R(T) = \sum_{j=1}^{N} A_j e^{-\alpha_j T} \sin(\omega_j T + \varphi_j). \tag{5.43}$$

The coefficients of the model (3.147) were determined by the regression method. Then the confidence intervals of the model and their predictions were calculated [80]. The time T_E was determined by the trial-and-error method, by dividing the allowable error Δ_{all} into two components

$$\Delta_{all} = x\Delta_{k\,all} + \Delta_{0\,all}. \tag{5.44}$$

From the prediction of the process $\{\tilde{x}_0\}$ and from the second component of condition (5.44), the value of the time T_{01} was calculated. The time T_{k1} was calculated from the prediction of $\{\tilde{x}_0\}$ and the first component of the allowable error, while the values of the times T_{01}, T_{k1} were calculated by changing the settings of the correcting elements r_0, r_k with the trial method, so that the maximum values of T_{01}, T_{k1} were obtained (Fig. 5.11).

Next, in the second stage, the ratio of both components of condition (5.44) was changed and the values of T_{02}, T_{k2}, T_{03}, T_{k3}, etc. were calculated repeatedly, in the search for the state $T_0/T_k = 1$. This action is shown in Fig. 5.12. The value of T_E obtained is underestimated because condition (5.44) was considered a sum

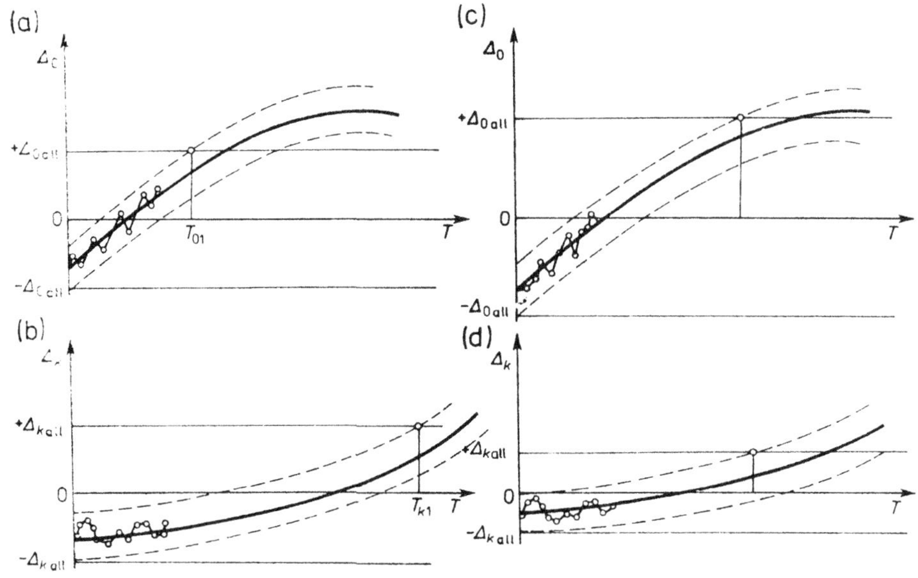

Fig. 5.11. The influence of the ratio of zero and sensitivity error on the values of T_0, T_k: (a), (b) improper ratio, (c), (d) the ratio chosen properly.

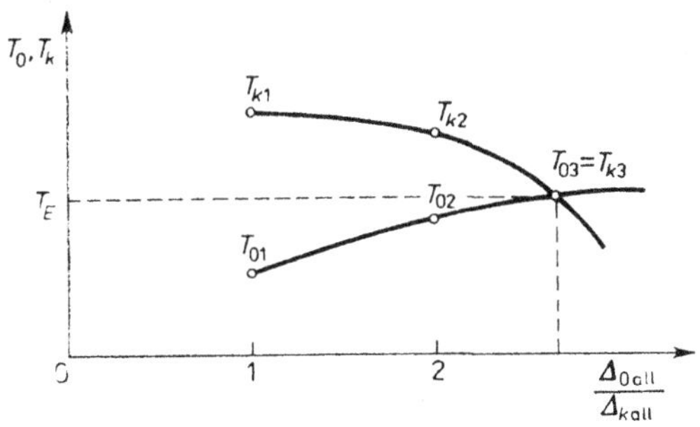

Fig. 5.12. Subsequent trials of finding the condition $T_0 = T_k$.

of absolute values. In some cases, if the signs of zero and sensitivity errors are opposite, the time T_E is greater. However, the estimation may be corrected [80].

Theoretical grounds have been given for the method of determining the time T_E and the settings of the elements in the process of adjustment. A simulation performed has shown that [80]:

(1) the estimates of T_E were correct,

(2) by a proper selection of the settings r_0, r_k, the MS service time without readings checking can be extended.

The method of estimation of the service time presented demands much experimental work and calculations. It is recommended to calculate the service time for Classes 1 and 2 of the metrological characteristics only. An approximate estimation is applied for the other classes, based on the statistical data of the parametric reliability of an MS of a given type.

The adjustment procedure is applied in two cases:

(1) as a preliminary step before the measurement system is put into operation;

(2) to correct the properties of an MS.

The first case includes the steps listed in the service manual of an MS. These are:

— switching on;

— setting the mechanical zero of the analog meters;

— horizontal or vertical positioning;

— adjustment of the supply currents and voltages;

— adjustment of the correction elements appropriate to the operating conditions, such as ambient temperature, atmospheric pressure, etc;

— setting of the measurement range, measurement time, etc;

— setting of the selected parameters, such as power supply frequency, filter parameters, level or code of the output signal, etc;

— zero and sensitivity adjustment appropriate to the transducers used, etc.

The properties of an MS are corrected when the result of the reading checking is unsatisfactory and the adjusting elements are provided in the MS, as shown in Fig. 5.3. Usually the corrected parameters are the zero point and the slope (sensitivity) and, rarely, the nonlinear components of the characteristic. The range of the correction depends on the kind of MS and its construction, and is usually a few to one hundred percent.

Since the result of the reading checking is known before the adjustment is performed, and so also are the required properties of the MS, the values of correction parameters may be calculated. For example, if the readings at points x_1 and x_2 are x_1^* and x_2^* respectively, the values of the correction elements settings are calculated from the equations

$$x_1^* = (k_0 - r_k)x_1 - r_0, \tag{5.45a}$$

$$x_2^* = (k_0 - r_k)x_2 - r_0, \tag{5.45b}$$

where r_0 is the value of the setting of the zero correcting element and r_k is the respective value for the slope. Hence, the values are

$$r_k = k_0 - \frac{x_2^* - x_1^*}{x_2 - x_1}, \tag{5.46a}$$

$$r_0 = x_1^* - \frac{x_2^* - x_1^*}{x_2 - x_1} x_1 = x_2^* - \frac{x_2^* - x_1^*}{x_2 - x_1} x_2. \tag{5.46b}$$

If none of the points x_1, x_2 has a value equal to zero, the correction by the value calculated from (5.46) is performed efficiently in one step, whereas experimental correction by the trial method at points x_1, x_2 involves interaction and the correction requires several steps.

In the case when a random error occurs in an MS, the problem of correction is more complex than was presented above. The method of selecting the settings during adjustment was presented in Section 4.6 and in the discussion of the service time calculation algorithm given above.

5.5 Measurement Procedures for Constant Quantities

A measurement procedure should include a description of all the important actions that provide a measurement result with the qualities of a true image of the reality investigated. A measurement procedure should be preceded by a measurement design procedure, presented in Chapter 6. At the design stage, the purpose of the measurement and the use of the result obtained should be stated. This means that by using the said procedure, its aims will be attained. This also implies the use of an appropriate measuring instrument, that is, one which does not introduce an error of method, and the provision of proper operating conditions for the instrument.

This author stresses the opinion that a measurement should not be performed if a positive conclusion of the design stage has not been previously achieved. Otherwise the action is purposeless, and both the means and the effort are wasted. Measurement is performed only if there is no other simple and inexpensive way to reaching the said goal.

Before starting the measurement, the metrological properties of the instrument should be known. When the measured quantities are constant, it is necessary to know the static characteristics of an MS according to one of the classes presented in Section 4.2.

The number of sources of errors may vary considerably, depending on the measurement method. The sources have been listed in Table 5.1; the notations of the functional units in different measurement methods are as in Fig. 5.2.

The conditions under which a measurement is conducted are, of course, known at the time and the limiting errors under these conditions are taken into account. In the calculus of error the rule should be obeyed that errors, systematic in origin but having an unknown value, are considered random.

5.5.1 *Single Measurement*

The procedure of a single measurement may be as follows:

(1) check whether the measurement conditions conform to the design assumptions;

(2) if a reading checking or an adjustment needs to be performed, it should be carried out;

(3) apply the measured quantity to the MS;

(4) wait until the transient effects elapse (swing method) or balance the MS (null method);[6]

(5) read the raw measurement result (or record the result);

(6) determine the measurement result, and then

(a) determine the corrected result by adding the corrections to the raw result (the correction is equal to the systematic error but taken with the opposite sign),

(b) determine the result by stating the confidence interval (the limits of allowable error).

The measurement result may be determined differently, depending on the data expressing the metrological characteristics of the MS, which are the only basis for estimation of an error. When the corrected result is being determined, the corrections introduced should eliminate both calibration error and supplementary errors. Depending on the class of characteristics applied, the corrections include the following.

Class 1. The systematic error is described by formula (4.28) or its components (4.32) and (4.41). In addition, the error may be calculated as the sum of the estimator of the error at the moment of calibration and of the prediction of systematic error

$$h(T) = \hat{h}(T_0)+\check{h}(T). \tag{5.47}$$

The known values here, allowing calculation of the correction, are: a raw measurement result, the values of the influencing quantities and the parameters of the construction properties of the MS (4.72), (4.73).

Class 2. The estimator of the calibration error (4.69) and the supplementary errors (4.72), (4.73) are known. The prediction of error is not known in this case.

Class 3. The correction eliminates the calibration error and the supplementary errors.

---

[6] Balancing is performed using the balancing algorithms, adequate for a given kind of MS. The algorithms may te called balancing procedures.

Classes 4 *and* 5. No corrections should be introduced.

A complete measurement result is obtained after the confidence interval for the random errors is determined.

Class 1. The components of a random error are:

— an instability error described by the function

$$g(T) = \hat{g}(T_0)+\check{g}(T) \tag{5.48}$$

(which is the so-called estimator of standard deviation at the instant of calibration T_0), by the prediction of the deviation changes in the time interval T, and by the probability distribution $p_\varepsilon(\Delta)$ under measurement conditions $x_1, \ldots, x_{m-1}, x$;

— the confidence interval of the estimator $\hat{h}(x_1, \ldots, x_{m-1}, x, T_0)$ and of the prediction $\check{h}(x_1, \ldots, x_{m-1}, x, T)$ with a known probability distribution $p_h(\Delta)$;

— the instability of the measurement conditions or an error of measurement of these conditions.

The probability distributions of the influencing quantities $p_i(x_i)$ should be known. Therefore, from the formulae given in Table 3.1, the probability distribution $p_{\text{sup}}(\Delta)$ may be calculated. The random error distribution is then

$$p_r(\Delta) = p_\varepsilon(\Delta) * p_h(\Delta) * p_{\text{sup}}(\Delta), \tag{5.49}$$

whence the confidence interval of random error is calculated at a given confidence level.

Class 2. The confidence interval is calculated as for Class 1, but the prediction error is not taken into account, as the prediction is unknown.

Class 3. The instability error and supplementary errors are considered in the error calculus. The instability error should include the uncertainty of the calibration error.

Class 4. Since only the limiting values of the systematic errors are known, the systematic errors should be regarded as random. The properties of the components of error are:

— the random error in reference conditions, with an unknown probability distribution $p_g(\Delta)$ but with a known limiting value $g_{\text{lim}}(x)$;

— the systematic error with an unknown value but with a known limiting value $h_{\text{lim}}(x)$;

— a supplementary error with the known limiting values of the construction parameters of error, e.g. $k_{i\,\text{lim}}$ as in (4.41), and with a known probability distribution of the influencing quantities $p_i(x_i)$, $i = 1, \ldots, m-1$.

The calculus of errors is ambiguous, for the data are not sufficient for determining the confidence interval exactly.

Method 1. Calculation of the limiting value of the total error

$$\Delta_{\text{lim}} = \sqrt{h_{\text{lim}}^2+g_{\text{lim}}^2+\sum_{i=1}^{m-1} (k_{i\,\text{lim}}\Delta x_{i\,\text{lim}})^2}, \tag{5.50}$$

where g_{lim} is the confidence interval of the random error and is not a standard deviation, as in the model (4.27). When the limiting values are established as confidence intervals with identical confidence levels, the confidence of the calculated interval $2\Delta_{lim}$ is the same. The relationships, however, are approximate, since the probability distributions of the components of error are not the same.

Method 2. The calculations may be exact, but a simplifying assumption should be made concerning the probability distributions of the component sources of errors. For example, the distribution of the random error is normal for an analog meter, uniform for most digital meters if the quantization error prevails, triangular (Simpson's) when the counting error prevails, etc. Most often, the probability distribution remains unchanged for different values of x.

The distribution of the systematic error changes along the scale. It may be assumed that every value of error may be realized with the same probability and $p = 1/2h_{lim}$, i.e., the distribution is uniform.

The properties of the supplementary error are different. When applying the parameters A_i (4.73) and δ_i (4.72), it should be realized that:

— the dependence of the zero value A_i arises from the small precision of assembling; within the limits $\pm A_{i\,lim}$, the dispersion is considerable;

— the influence of δ_i is a typical property and the deviations from specification data do not exceed 10%.

The parameters $A_{i\,lim}$ and δ_i should be applied to calculate the error. The distributions of $p_i(x_i)$ are known, and thus the formulae from Table 3.1 may be applied to determine $p_{sup}(\Delta)$.

Method 3. The limiting values of the systematic and random errors (h_{lim} and g_{lim}, respectively) are added. This classical approach has neither theoretical nor practical bases. The effect of the addition of the maximum errors is described by formula (5.57), manifesting itself in the significant increase of confidence, so that the calculated limiting value of error will not be exceeded. The present author does not recommend this method.

Class 5. The characteristic gives a complete answer, since it includes a confidence interval of the total error in reference conditions. The confidence interval should be extended by the value of the supplementary error, calculated as in the Method 2 of Class 4.

When the metrological characteristic is expressed as an accuracy class, the interpretation given in Subsection 4.5.4 should be applied.

5.5.2 Multiple Measurement—The Measurement Series

Here the single measurement procedure is repeated completely N times, so that the physical indepedence of subsequent measurements is ascertained.

The aim of multiple measurement has to be stated directly at the design stage, and may be one of following.

A. Increasing the accuracy of measurement

The starting measurement conditions were assumed with the probability $P = 1$:

$$x = \text{const}, \tag{5.51a}$$

$$|\Delta| \leqslant \Delta_{\text{lim}}. \tag{5.51b}$$

The raw result of a measurement series is the mean value of the raw results of the single measurements

$$x_N^* = \frac{1}{N}\sum_{i=1}^{N} x_i^*. \tag{5.52}$$

Each of the single measurements is burdened with a systematic error $\Delta_{si} = \Delta_{s1}$ and with a random error Δ_{ri}, which is a realization of a random process with a mean value $E(\Delta_{ri}) = 0$ and with variance $\text{var}(\Delta_{ri}) = \sigma_1^2$. Since

$$x_i^* = x_{\text{corr}}^* + \Delta_{s1} + \Delta_{ri}, \tag{5.53}$$

$$x_N^* = \frac{1}{N}\sum_{i=1}^{N}(x_{\text{corr}}^* + \Delta_{s1} + \Delta_{ri}) = x_{\text{corr}}^* + \Delta_{s1} + \frac{1}{N}\sum_{i=1}^{N}\Delta_{ri}. \tag{5.54}$$

The raw result of a measurement series is a random variable with a mean value

$$E(X_N^*) = x_{\text{corr}}^* + \Delta_{s1} \tag{5.55}$$

and variance

$$\text{var}(X_N^*) = \text{var}\sum_{i=1}^{N}\left(\frac{1}{N}\Delta_{ri}\right) = \frac{1}{N}\,\text{var}\,\Delta_{ri}, \tag{5.56a}$$

that is

$$\sigma_N^2 = \frac{1}{N}\sigma_1^2. \tag{5.56b}$$

The above formula is not sufficient for determining a measurement result, because in the general case, the probability distributions of single measurement errors and of the errors of a measurement series are not the same.

The error of a single measurement may have any distribution. The kind of distribution of an error of a measurement series depends on the number of measurements in the series, according to the central limit theorem and the law of large numbers. At the limit, for $N \to \infty$, the distribution approaches the normal distribution. In Fig. 5.13 an example is presented showing how the distribution of

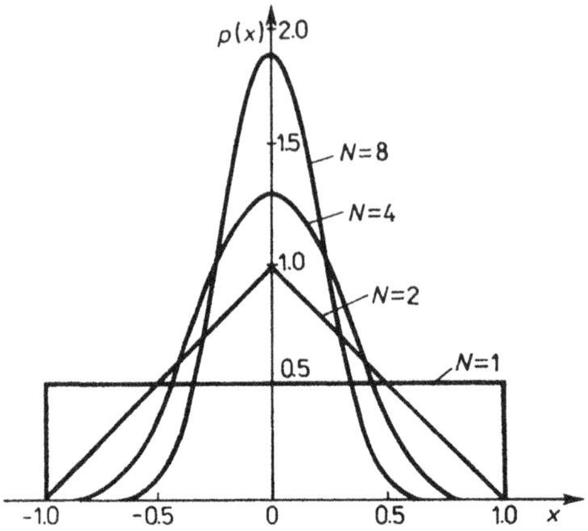

Fig. 5.13. Changes of probability distribution of a measurement series result in the case when the distribution of a single result is uniform.

error of a measurement series changes, when the distribution of a single measurement is uniform. In the example, the ratio of the confidence interval to the standard deviation changes as shonw in Table 5.2, and depends on the confidence level.

TABLE 5.2

Values of m for normal and uniform distributions

N	Values of m		
	Uniform distribution		Normal distribution m independent of α
	$\alpha = 0.90$	$\alpha = 0.95$	
2	1.32	1.21	1.41
4	1.81	1.69	2.00
8	2.64	2.36	2.83

There are, however, two exceptions. When the distribution of error of a single measurement is normal or of Poisson type, the distribution of the error in a series of measurements is the same. For $N > 30$, the Poisson distribution may be approximated by a normal distribution with approximation error below 0.1%.

A common interpretation is that the purpose of a multiple measurement is to reduce the random errors. When the mean value of the single measurements in the series is taken as a result of the measurement series, the variance of the error of the result is N times smaller than the variance in a single measurement.

For a given confidence level, the limiting random error of the result of a series is m times smaller than the limiting random error of a single measurement. For normal and Poisson distributions, $m = \sqrt{N}$; the values of m for a single measurement error are given in Table 5.2. For other distributions, the values of m should be calculated.

The effect of a multiple measurement, i.e. reduction of random error, depends on the proportion of the random error in the total error of an MS. For Classes 1, 2 and 3 of metrological characteristics, the exact value of the error of a single measurement result may be calculated: the systematic error is the same as in a single measurement, and the random error is diminished m times. For Class 4 the effect is similar, the difference is that the value of m is not known precisely because the probability distribution of random error is not known. In the case of Class 5 of metrological data, if the ratio of the random error to the systematic error is unknown, it is impossible to say a priori whether a multiple measurement will reduce the error, as compared with a single measurement. We may say so only a posteriori, after a measurement series is performed. With the assumption $P(x = \text{constant}) = 1$, a multiple measurement makes the determination of a random error of an MS possible. Depending on the number of measurements, we may determine either the limiting value of the random error of a single measurement, or its probability distribution and then the uncertainty interval. The error of the result of multiple measurement is then determined identically as for Classes 1, 2, 3 of metrological characteristics. It is obvious that a small dispersion of subsequent measurement results implies a small part of the random error and then increasing of the length of the series brings no expected effect.

The result of a multiple measurement may also be interpreted differently. A multiple measurement increases the confidence that the result of a measurement series remains within the limits of allowable error of an MS, that is, within the interval of a single measurement error. When the result of a series of N measurements ($N = 1, 2, \ldots, N$) has a distribution $p_N(\theta_1, \theta_2, \ldots)$, where θ_j is a parameter of the distribution, the confidence level is

$$\alpha_N = \int_{-\Delta_{all}}^{\Delta_{all}} p_N(\Delta, \theta_1, \theta_2, \ldots)\mathrm{d}\Delta = \psi(N, \Delta_s, \sigma_1). \tag{5.57}$$

The parameters of the distribution are

$$\theta_1 = \frac{1}{N}\sum_{i=1}^{N} \Delta_i = \Delta_s = \text{constant}, \quad \theta_2 = \sigma_N^2 = \frac{1}{N}\sigma_1^2, \tag{5.58}$$

where σ_1^2 is a variance of a random error in a single measurement.

When the distribution of a random error is normal, the dependence of the confidence level on the number of measurements is described by the data presented

TABLE 5.3

The dependence of the confidence level α on the number of measurements N for different values of the systematic error. The distribution of the random error is normal. The limiting error is $\Delta_{lim} = \pm 2\sigma_1$

N	The values of α for				
	$\Delta_s = 0$	$\Delta_s = 0.5\sigma_1$	$\Delta_s = 1.0\sigma_1$	$\Delta_s = 1.5\sigma_1$	$\Delta_s = 2.0\sigma_1$
1	0.9445	0.927	0.840	0.691	0.5
2	0.9953	0.983	0.921	0.761	0.5
3	0.9995	0.995	0.958	0.808	0.5
4	0.9999	0.9987	0.977	0.841	0.5
5		0.9996	0.987	0.879	0.5
6			0.993	0.889	0.5
8			0.998	0.922	0.5
10			0.999	0.943	0.5

in Table 5.3. If the probability distribution of random errors and the limiting value of systematic error are known a priori, the relationship $\alpha_N = \psi(N)$ may be calculated. The data are included in Classes 1 and 2 of metrological characteristics. For Class 4 the estimation may be approximate, as the distribution of random errors is not known. The dependence of confidence on the number of measurements and on the value of systematic error is shown in Figure 5.14. Figure 5.14a corresponds to the data from Table 5.3. Fig. 5.14b shows how a known value of systematic error causes the error distribution to be asymmetric with respect to the confidence interval. In Fig. 5.14d the influence of the number of measurements on the confidence is shown for the case when the distribution of a systematic error as in Fig. 5.14c is independent of N. The distribution of random error is as shown in Fig. 5.14a. The interpretation of the distribution of systematic error is shown in Fig. 4.29. Data like this should be included in Class 4 of the metrological characteristics.

For Class 5 the estimations cannot be performed a priori. It is only possible to analyse a posteriori the results of measurement series. Reduction of metrological characteristics does not give grounds a priori for the effectiveness of a measurement series.

The manner of mathematically processing the results of measurement series is widely presented in the bibliography [8, 126, 22] and will be not discussed here.

B. Investigation of the stability of a measured quantity x

Before a measurement is performed, it is known that

(a) $|\Delta| \leqslant \Delta_{lim}$ with the probability $P = 1$,

(b) $x =$ constant with the probability $P < 1$.

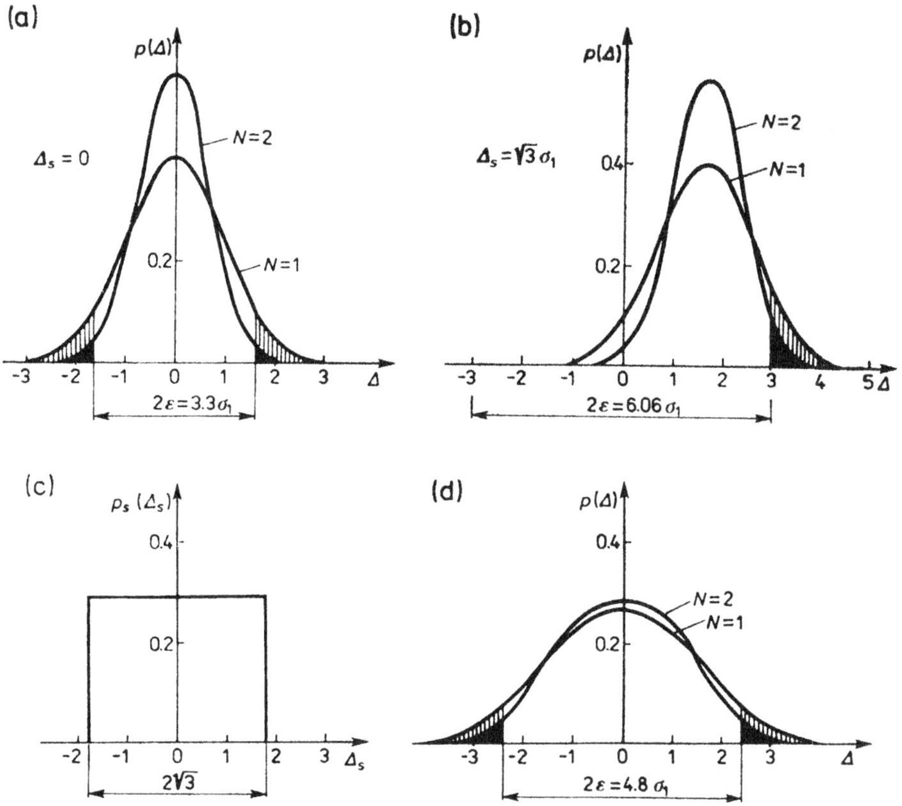

Fig. 5.14. The influence of a systematic error on the confidence interval (at the confidence level $\alpha = 0.9$ in a single measurement) for $N = 2$: (a) systematic error equals zero, (b) the value of the systematic error falls within the interval $\langle -\sqrt{3}\sigma_1, \sqrt{3}\sigma_1 \rangle$ and is added algebraically to the random error, (c) systematic error distribution, (d) statistical adding of random and systematic errors from (a) and (c).

In this case the aim is to measure the value of x and to verify the hypothesis that $x =$ constant. The set of single measurement results $\{y\}$ is the realization of the sum of the random process $\{x\}$ and of the random process of the random error $\{\Delta_r\}$ of an MS. The process $\{\Delta_r\}$ is known from the metrological characteristics of the MS at the level of

(a) the probability distribution of random error—in Classes 1, 2, 3;

(b) the limiting value of random error (confidence interval)—in Class 4;

(c) the allowable value of error—in Class 5.

The process $\{y\}$ is the result of measurement, and thus

$$\{x\} = \{y\} - \{\Delta_r\}, \tag{5.59a}$$

$$\operatorname{var}\{x\} = \operatorname{var}\{y\} - \operatorname{var}\{\Delta_r\}. \tag{5.59b}$$

For the case when the value of x varies considerably, it is easy to answer the question whether x = constant, based on variance calculations. When the changes of x are small, the hypothesis may be formulated classically, by checking whether $\{y\}$ and $\{\Delta_r\}$ are realizations of the same population [9, 48]. In the case when the distributions of both sets Y, Δ_r are normal, the F statistic (3.257) may be applied.

The method of verification of the hypothesis should be determined at the stage of measurement design, because the possible answers must be known in advance in order to select the method, the number of measurements in the series, etc.

It should be stressed that the purpose of measurement formulated in this way imposes no limitation on the effect of increasing the accuracy by multiple measurement or on the effect of increasing the confidence of the result. The result may be processed as in case A only after the hypothesis that x = constant is verified.

C. Investigations of errors

The third case concerns the reading checking procedure. We have (a) x = constant with the probability $P = 1$, (b) $P(|\Delta| \leqslant \Delta_{\text{lim}}) < 1$. In this case

$$\{\Delta\} = \{y\} - x, \tag{5.60a}$$

$$E\{\Delta\} = \Delta_s, \tag{5.60b}$$

$$\operatorname{var}\{\Delta\} = \operatorname{var}\{y\}. \tag{5.60c}$$

The interpretation is similar to that of case B. The tests are performed to prove that $\operatorname{var}\{\Delta\} > \operatorname{var}\{\Delta_r\}$ or that the value of the limiting error is small enough, i.e., that Δ_{lim} of the sample is greater than Δ_{all}. For this purpose, proper statistical tests are applied.

These measurements may be performed as partial reading checking, for example when $\Delta_s = 0$ or when the metrological properties may be described by Classes 2 or 3 of the characteristics. Then, if the allowable value of random error is exceeded, the nominal values are also exceeded. This means that the properties of the MS have changed and probably the systematic error has changed as well. The important condition is $P(x = \text{constant}) = 1$, with the accuracy being adequately greater than the errors of the MS.

5.6 Measurement Procedures for Time-Dependent Quantities

The aim of measurement of a time-dependent quantity may be

(1) the determination of an analytical form of the measured waveform, or

(2) the determination of a characteristic parameter of the waveform. In this case, the assumption about the form of the waveform should be verifiable.

An example of the above goals may be the measurement of the dynamic properties of different objects or the measurement of characteristic parameters, for example as in chromatography. It is assumed that the measured waveform is either continuous or discrete in time. In comparison with the measurements of constant quantities, two more sources of error appear here:

— dynamic error,

— recording and readout error (readout is understood here as the replaying of an implicit record, e.g. a magnetic tape).

The measurement procedures are presented assuming that the measuring instruments have been selected properly at the measurement design stage, which means that their accuracy is sufficient. The algorithms for processing the result should be determined at the design stage.

5.6.1 Single Measurement

The measurement procedure is as follows:

(1) carry out any adjustments of the instruments,
(2) apply the measured quantity to the MS,
(3) record the investigated waveform,
(4) process the measurement result:
  (a) read the raw result,
  (b) introduce the corrections to eliminate systematic errors,
  (c) determine the values of the coefficients of the model,
  (d) test the significance of the calculated coefficients,
  (e) determine the uncertainties of the estimated coefficients.

The method of adjustment of a given MS is described in the technical specification of the instrument. The description includes preparing the recorder for work, setting up the adjustable parameters such as measuring range, speed of the information carrier or sampling interval (in the case of digital recording), zero and sensitivity adjustment, selection of an attenuation coefficient, selection of input filters and their parameters, testing and switching on the time marker, etc.

The measurement is initiated with a start signal, setting off the recording and the measured signal. In the case of digital recording, a raw measurement result is printed. In the case of analog recording, analog processing may be applied, or the results may either be processed graphically (on the information carrier, e.g. recorder tape) or numerically, by reading the results in numerical form at a given sampling interval. The time scale is produced by subsequent intervals of the time marker. When the adjustment of an MS is correct, the systematic errors—except for the dynamic error—are eliminated. The mode of processing the results differs, depending on the purpose of measurement, the form of the measurement results and on the dynamic properties of the MS.

A. Continuous measurement result, with a negligibly small dynamic error
Usually, the purpose of measurement is to determine such characteristic parameters of the measured time-dependent quantity as its maximum value, graph coordinates ratio, etc., or to determine the parameters of the function describing the quantity, such as time constant, amplitude and period of harmonic oscillations, etc. These parameters may be determined graphically, either directly from the graph, or after performing a function transformation. The methods are widely known [91, 28, 81] and need not be described here.

When the results are graphically interpreted, the errors are determined by definition as the sum of the MS error described by the metrological characteristics, the error of the carrier scale and the readout error. The error of the carrier scale is caused by changes in the dimensions of the carrier (due to photochemical processing, humidity, etc.) and by other factors. Since any parameter of the time function is most often an interval crossing the points 1 (t_1, x_1) and 2 (t_2, x_2), the errors of the coordinates are therefore determined as in the case of measurement of a constant quantity and the error of the parameter is the sum of errors at points 1 and 2.

When the identification of a waveform is performed, the estimators of the model coefficients are determined by special graphical procedures (for example, time constants may be determined from the response of the second-order inertial element driven by a unit step function), or by a trial-and-error method. After an estimator of the waveform is drawn on the carrier, it may easily be seen whether the estimator is determined correctly and whether the model is appropriate. From the graphical construction and from the errors at characteristic points (or the readout errors derived point by point from the metrological characteristic), the confidence intervals for the estimated function may be calculated. However, the intervals usually cannot be drawn on the carrier because of the fact that, due to the scale of the drawing, the effect may not be clear. This limitation makes the evaluation of errors difficult and is a weak point of the graphical methods of processing the measurement result.

Numerical processing of the results is based on the N-element sample (t_1, y_1), (t_2, y_2), ..., (t_N, y_N). The estimator $\hat{y}(t)$ of the assumed function $y(t)$ should be calculated. When the function $y(t)$ is linear with respect to the estimated coefficients, the regression method or the least squares method is applied. However, the linear form is hardly obtainable in the method of function transformation. In this case either use of an algorithm, analog processing or some other more complex method should be applied (such as the trial-and-error method). For example, for a model

$$y(t) = \sum_{q=1}^{q} A_q e^{-a_q t}, \tag{5.61}$$

where $a_1 < a_2 < a_3$, the function transformation

$$z = \ln y \tag{5.62}$$

may be applied and the values A_1, a_1 may be obtained from the linear part of $z(t)$. Next, the function $y(t) - A_1 e^{-a_1 t}$ may be transformed to the form

$$z_1(t) = \ln[y(t) - A_1 e^{-a_1 t}], \tag{5.63}$$

and the values of A_2, a_2 may be calculated as previously. Next, calculate $z_2(t)$ as in (5.63), etc. [51]. If, for example,

$$y(t) = \frac{k}{\sqrt{2\pi}\,\sigma} e^{-(t-t_0)^2/2\sigma^2}, \tag{5.64}$$

the coefficients may be calculated from the formulae:

$$\hat{k} = \frac{1}{N} \sum_{i=1}^{N} y_i(t_i), \tag{5.65a}$$

$$\hat{t}_0 = \frac{1}{Nk} \sum_{i=1}^{N} t_i y_i, \tag{5.65b}$$

$$\hat{\sigma} = \frac{1}{N-1} \cdot \frac{1}{k} \sum_{i=1}^{N} (t_i - \hat{t}_0)^2 y_i. \tag{5.65c}$$

When the coefficients are calculated numerically, estimation of error becomes a serious and complex problem. The corrected results are introduced, and the unknown sources of systematic errors are regarded as random errors.[7] The error of each element of the sample is considered known, and is determined as described in Section 5.5. Then the algorithm of function transformations, the calculation algorithms for the estimators and the determined model form a definition of an indirect measurement which allows calculating the errors of the model. In the case of the estimators (5.65) and model (5.64), the definition of indirect measurement may be written in an analytical form and the calculations may be performed, yet they are still complex. In a case such as (5.61) or in the trial-and-error method, the only way is to determine the error of the model by simulation, i.e., by repeated calculations of the estimators from the data $y_i + \Delta y_i$ ($i = 1, \ldots, N$), where Δy_i are the generated errors, which simulate the random error. To determine the

[7] The calculus of errors should be performed separately for the systematic errors regarded as random and for the random error, since the effects are different. The systematic error arises from the systematic error of the model.

probability distribution of the error, a considerable number of simulations (50–100) need to be performed, and the cost of computation increases significantly.

Selection of the sample is not easy. On the one hand, the limitation is a minimum number $N > q$, where q is the number of coefficients of the model. On the other hand, increasing of the sample size improves the accuracy of estimation, but increases the cost of computation. The influence of the sample size on the error is complex, since the subsequent realizations of errors, at small time and space intervals, are correlated; the effect may be seen in the form of correlation function (3.162). What is more, the nonlinearity of the models causes the sensitivity of the estimator to a measurement error to be different at different points. Not only the number of points but also their location influences the error. Usually, uniform sampling is applied in order to simplify the calculations.

B. The dynamic errors are not negligibly small

Processing of the measurement error in this case includes determining the dynamic error and, after the measurement is performed, introducing corrections. According to the models of an MS which were described in Chapter 3 with the Volterra operator, and considering the definition of dynamic error (4.59), the task of algorithmic correction of dynamic errors in linear systems is performed as shown in Fig. 4.5. The recorded waveform $y(t)$ is an image of the measured time-dependent quantity $x(t)$. The image is distorted, as the dynamic properties of a MS are not ideal (4.50). The properties of an MS are described by the kernel of the Volterra operator $k(t, \tau)$ (3.26) or by the frequency response $K(j\omega)$ (in the frequency domain). The procedure $h(t, \tau)$ of reproducing the measured quantity is an algorithm of dynamic error correction and is a conjugated kernel of an MS fulfilling condition (4.58). The error due to incomplete correction is described by formula (4.59). There are three classes of methods of reproduction of the state of a measured quantity [133]:

(1) approximation methods, in which the correct time function is calculated from formula (4.55), applying the integrand approximation;

(2) iterative methods; the first approximation is the raw measurement result;

(3) algebraic methods of solving equation (4.55).

1. Approximation of the raw measurement result and of the conjugated kernel may be performed either in the time or frequency domains. The approximating functions usually applied are rather simple [133]. For example, if

$$y(\omega) = e^{-\alpha|\omega|} \tag{5.66a}$$

and

$$H(\omega) = e^{+\beta|\omega|}, \tag{5.66b}$$

then

$$x^*(t) = \frac{1}{2\pi} \int_{-\infty}^{\infty} e^{-(\alpha-\beta)|\omega|} e^{j\omega t} d\omega = \frac{\alpha-\beta}{\pi[(\alpha-\beta)^2+t^2]} . \tag{5.67}$$

Also if

$$y(\omega) = e^{-\alpha^2\omega^2/4} \tag{5.68a}$$

and

$$H(\omega) = e^{+\beta^2\omega^2/4}, \tag{5.68b}$$

then

$$x^*(t) = \frac{1}{\alpha^2-\beta^2} \frac{1}{\sqrt{\pi}} e^{-t^2/(\alpha^2-\beta^2)}. \tag{5.69}$$

A property of the latter approximation is that the waveforms $x(t)$ and $y(t)$ are described by the Gauss function. In the time domain, the form of the conjugated kernel of the operator is (5.68b) and the time function of the raw result may be approximated with a polynomial, applying the Maclaurin series

$$y(t) = \sqrt{\pi} \sum_{n=0}^{\infty} a_n 2^n t^n \tag{5.70}$$

with the coefficients

$$a_n = \frac{y^{(n)}(0)}{\sqrt{\pi}\, 2^n n!} . \tag{5.71}$$

Then the form of the reproduced waveform is a sum of the Hermite series (see Table 3.2), namely [133]

$$x^*(t) = \sum_{n=0}^{\infty} a_n H_n(t). \tag{5.72}$$

A more general solution of the problem is obtained by applying the orthogonal series $\psi_i(t)$, which is an eigenfunction of the kernel $k(t-\tau)$ of the operator. For the series we have

$$x^*(t) = \lim_{N\to\infty} \sum_{i=0}^{N} x_i^* \psi_i(t) \tag{5.73}$$

and

$$\lambda_i \psi_i(t) = \int_{-T}^{T} k(t-\tau)\psi_i(\tau) d\tau. \tag{5.74}$$

The eigenvalues of the kernel are

$$\lambda_i = \int_{-T}^{T} \psi_i^2(\tau)\mathrm{d}\tau \tag{5.75}$$

since

$$\int_{-T}^{T} \psi_i(\tau)\psi_j(\tau)\mathrm{d}\tau = \lambda_i \delta_{ij}, \tag{5.76}$$

$$\delta_{ij} = 1 \quad \text{for} \quad i = j, \qquad \delta_{ij} = 0 \quad \text{for} \quad i \neq j, \qquad i,j = 0, 1, 2, \ldots, N.$$

The time function of the raw result is approximated by the series

$$y(t) = \sum_{i=0}^{N} y_i \psi_i(t), \tag{5.77}$$

where

$$y_i = \int_{-T}^{T} y(t)\psi_i(t)\mathrm{d}t. \tag{5.78}$$

Because of the properties (5.73), (5.74), the following relationship holds:

$$y(t) = \int_{-T}^{T} k(t-\tau)x(\tau)\mathrm{d}\tau = \int_{-T}^{T} k(t-\tau)\sum_{i=0}^{N} x_i\psi_i(\tau)\mathrm{d}\tau = \sum_{i=0}^{N} x_i\lambda_i\psi_i(t). \tag{5.79}$$

Therefore,

$$x_i^* = \frac{y_i}{\lambda_i} \tag{5.80}$$

results from the comparison of (5.79) and (5.77). The values (5.80), substituted in (5.73), give an approximation of the reproduced waveform $x^*(t)$ [133], for example

$$\psi_i(t) = \cos\left(\beta_i \frac{t}{T}\right), \quad i = 1, 3, 5, \ldots, \tag{5.81a}$$

$$\psi_i(t) = \sin\left(\beta_i \frac{t}{T}\right), \quad i = 2, 4, 6, \ldots, \tag{5.81b}$$

defined over the interval $\langle -T, T\rangle$.

The accuracy of reproduction of the measured waveform is strongly influenced by disturbances. If a raw result $y(t)$ is the sum of an ideal waveform $n(t)$ and the disturbances $v(t)$

$$y(t) = n(t)+v(t), \tag{5.82}$$

then, in the case of disturbing white noise with variance σ_v^2, not correlated with the waveform $n(t)$, the variance of the reproduction error will be [133]

$$\sigma^2 = \sigma_v^2 \sum_{i=1}^{N} \frac{1}{\lambda_i}. \tag{5.83}$$

As the eigenvalues decrease with increase of i, formula (5.83) shows that extending the approximating series improves accuracy only over a limited range.

2. The iterative methods of reproduction of a measured waveform may also be realized in the time or frequency domain. One of the methods is to expand the calibration procedure into the Neyman series

$$\varphi = I + \sum_{k=1}^{\infty} (I-F)^n = I + \sum_{n=1}^{\infty} \Phi_n(y), \tag{5.84}$$

where F is the modelling operator of the MS (3.1), φ is the calibration procedure (3.2) and $I = \varphi(F) \equiv 1$ as in (3.4). We have

$$x^* = \varphi(y), \tag{5.85}$$

and thus, after a substitution (5.84),

$$x^* = y + \sum_{n=1}^{\infty} \Phi_n(y), \tag{5.86}$$

which means that the reproduced waveform $x^*(t)$ is obtained from $y(t)$, considered a zero iteration. The next iterations are performed n times, with a component $\Phi = I - F$. However, the convergence of iterations in applying this series is not satisfactory [133].

Iterations in the frequency domain are easier to perform when the Fourier transform is applied to equation (4.55). Then

$$x^*(t) = \frac{1}{2\pi} \int_{-\infty}^{\infty} y(j\omega) H(j\omega) e^{j\omega t} d\omega. \tag{5.87}$$

Because

$$H(j\omega) = \frac{1}{K(j\omega)} = \left[\frac{1}{K(j\omega)} - 1\right] + 1, \tag{5.88}$$

the first component may be expanded into a power series

$$\frac{1}{K(j\omega)} - 1 = \sum_{n=1}^{N} a_n U_n(j\omega) = \sum_{n=1}^{N} a_n (j\omega)^n. \tag{5.89}$$

By applying the property

$$\frac{1}{2\pi}\int_{-\infty}^{\infty}(j\omega)^n y(j\omega)e^{j\omega t}d\omega = \frac{d^n y(t)}{dt^n} \tag{5.90}$$

and substituting (5.90) instead of (5.87) in (5.89) and (5.88), we obtain

$$x^*(t) = y(t)+\sum_{n=1}^{N} a_n \frac{d^n y(t)}{dt^n}. \tag{5.91}$$

In the case of a first-order inertial element

$$K(j\omega) = \frac{1}{1+j\omega T}, \tag{5.92}$$

only the coefficient $a_1 = T$ is different from zero, and so

$$x^*(t) = y(t)+T\frac{dy(t)}{dt}. \tag{5.93}$$

The above formula is the calibration procedure of type (4.21a). Expression (5.89) may be approximated by the series

$$\sum_{n=1}^{N} a_n \sin(\tfrac{1}{2}nc\omega). \tag{5.94}$$

This leads to the approximation formula [133]

$$x^*(t) = y(t)+\sum_{n=1}^{N}\frac{a_n}{(2j)^n}\Delta_c^n y(t), \tag{5.95}$$

where $\Delta_c^n y(t)$ is an nth order difference of the waveform $y(t)$ from the raw result, over the interval $(-c/2, c/2)$.

3. The algebraic methods of reproducing the measured quantity are derived from the matrix form of the conversion equation

$$Y = KX, \tag{5.96}$$

where Y is a known vector of the raw result coordinates, X is the required vector reproducing the waveform and K is the matrix of the coefficients of an MS response to a pulse excitation. Hence

$$X^* = K^{-1}Y \tag{5.97}$$

is a corrected result. If the number of data N is greater than the number of corrected results Q, the least squares method should be applied. The result is then

$$X^* = (K^TK)^{-1}K^TY. \tag{5.98}$$

The Gauss, Gauss–Seidel, Galerkin, saddle-point or other methods may be applied in the calculations. One of the methods is the application of the pseudo-inverse Moore matrix [133].

Reproduction of the waveforms with correction of the dynamic errors is difficult. It is necessary to prove that the applied method leads to a solution, and to check that the solution is the only stable and allowable one. In iterative calculations, the convergence rate is an important parameter. The Tikhonov regularization method [185] is particularly interesting.

5.6.2 Multiple Measurement

The cases A, B, C discussed in Section 5.5 are valid here as well. Because of the observation interval of the measured waveform, the random errors may reveal themselves independently in one realization of the waveform or in time-spaced realizations. The character of the random errors of an MS is determined by the correlation coefficient $\varrho_\varepsilon(\varepsilon)$ (3.138) and by the spectral density $S_\varepsilon(\omega)$ of the random process $\{\varepsilon\}$.

A multiple measurement of a waveform is performed relatively often, in order to investigate the dispersion of the waveform (case B). In case C, test signals the most often applied are: unit step function $\mathbf{1}(t)$, pulse excitation, sine excitation, and others, described by elementary functions.

Chapter 6

The Design of Measurements and Experiments

6.1 Selection of Standards for Calibration of Measuring Instruments

To determine the accuracy of the standards used for calibration of a measuring instrument, the relationships derived in Section 2.2 will be applied. The essence of that section was the process of transferring a measure of a primary standard onto a secondary standard, with the smallest error possible. The primary and the secondary standards were one-dimensional.

In a calibration process, the calibrated instrument reproduces a set of the measured quantity, expressed as a measurement range. The aim is to achieve a predetermined accuracy of the calibrated instrument, using the smallest possible number of standards or a primary standard with the least possible accuracy. The problem should not be considered individually for a specimen instrument but for a class of instruments. The case of direct calibration with algorithmic fixing of the result is taken into consideration. The following errors are neglected:

— calibration result fixing error,

— an error of applying a standard quantity to an instrument,

— the dynamic errors,

— the errors due to parasitic effects, such as galvanic or thermoelectric voltages, parasitic capacities and couplings, installation leakages, etc.

First, let us derive the required computational formulae, as an extension of those given in Section 2.2.

6.1.1 Transfer of a Measure of a Standard onto Measuring Instruments

When the measure of a standard is transferred onto a measuring instrument, some differences in error propagation occur with respect to the description in Section 2.2. A primary standard is either a standard or a standard (reference) measuring instrument. The nominal value of measure of the primary standard is w_{10} and the errors are described by the model (2.2), including a systematic error

h_1 with an unknown value and a random error $g_1\{\varepsilon_1\}$. The role of the secondary standard in this case is played by the calibrated instrument, whose model is

$$\{y\} = f(x)+g_2\{\varepsilon_2\}, \tag{6.1}$$

where y is an output quantity, x is a measured quantity and the component $g_2\{\varepsilon_2\}$ describes a random error of the instrument according to the model (2.8b), but $g_2 = g_2(x)$, and the error is expressed in the values of the measured quantity. The calibrated instrument is in reference conditions, and therefore the supplementary errors are eliminated.

After a state w_{1j} of a standard quantity is applied to the calibrated instrument, the readout is y_j. It is assumed that no readout and comparison errors affect the result of comparison, as the readout error is included in the component $g_2\{\varepsilon_2\}$ and the comparison error is assumed to be equal to zero. Therefore no comparator errors h_0, $g_0\{\varepsilon_0\}$ appear in the equation describing the comparison results.

The metrological characteristic of the instrument, i.e., $\hat{f}(x)$, $\hat{h}_2$ and $\hat{g}_2\{\varepsilon_2\}$, should be determined from the metrological characteristics of m primary standards and from the model $f(x)$ of the instrument, known at the level of coefficients.

Let us assume a two-stage estimation procedure, consisting of:

(i) Determination of the instrument characteristic which is currently fixed. The LSM should be used. The characteristic is

$$f_0(x). \tag{6.2}$$

(ii) Determination of the errors of the instrument:

(a) random error

$$g_2\{\varepsilon_2\}, \tag{6.3a}$$

(b) systematic error

$$h_2(x) = f(x)-f_0(x), \tag{6.3b}$$

while the results

$$\Delta_{2j} = y_j-f_0(x_j) \tag{6.4}$$

are known at the points $j = 1, \ldots, m$ and the model is known at the level of coefficients, for example

$$h_2(x) = W_q(x) = b_0+b_1x+b_2x^2+ \ldots +b_qx^q. \tag{6.5}$$

To determine the errors of the calibrated instrument, the regression method may be applied. The general model of the systematic error is

$$h_2(x) = W_q(x)+\eta, \tag{6.6}$$

where η is the random component of the model. Depending on the course of calibration, the properties of the realizations of η may differ significantly.

Let us determine the sources of errors of both of the components in equation (6.6). At the moment of calibration, the following relationship holds:

$$y = w_2 = f(w_1), \tag{6.7}$$

which means that no error occurs during the comparison of standards. With the errors considered, the form of equation (6.7) is

$$w_{20}+h_2+g_2\,\varepsilon_2 = f(w_{10}+h_1+g_1\,\varepsilon_1). \tag{6.8}$$

The nominal value of the reading of the instrument is, according to (6.2),

$$w_{20} = f_0(w_{10}). \tag{6.9}$$

By subtracting (6.9) from (6.8), we obtain

$$h_2 = f(w_{10}+h_1+g_1\,\varepsilon_1)-f_0(w_{10})-g_2\,\varepsilon_2. \tag{6.10}$$

The function f may be expanded into a Taylor series at the point w_{10}

$$f(w_{10}+h_1+g_1\,\varepsilon_1) = f(w_{10})+(h_1+g_1\,\varepsilon_1)f'(w_{10})+\tfrac{1}{2}(h_1+g_1\,\varepsilon_1)^2 f''(w_{10})+\ \dots \tag{6.11}$$

It is sufficient to take into consideration in further calculations only the first two terms of the series. Then, after substituting (6.11) in (6.10), the systematic error is described by the formula

$$h_2 = f(w_{10})-f_0(w_{10})+(h_1+g_1\,\varepsilon_1)f'(w_{10})-g_2\,\varepsilon_2. \tag{6.12}$$

After substitutions

$$W_q(x) = f(w_{10})-f_0(w_{10}) \tag{6.13}$$

and

$$\eta = (h_1+g_1\,\varepsilon_1)f'(w_{10})-g_2\,\varepsilon_2, \tag{6.14}$$

the assumed model (6.6) is obtained as

$$h_2 = W_q(x)+\eta.$$

The expected value and the variance of error at the calibration point w_{1j} may be easily calculated by substituting (6.13) and (6.14) in (6.6). We find

$$E(h_2) = \hat{W}_q(w_{10})+f'(w_{10})\overline{h}_1 = \hat{f}(w_{10})-\hat{f}_0(w_{10})+f'(w_{10})\overline{h}_1 \tag{6.15}$$

since $E(g_1\varepsilon_1) = E(g_2\varepsilon_2) = 0$. The symbol $\hat{W}_q(x)$ denotes the mean value of $W_q(x)$ and the symbol $\overline{h}_1$ the mean value of the systematic error of the primary standard, related to the set of standards of that type. The first component in (6.15) is the error of approximation of the characteristic of the instrument. The second component is an uncertainty of the measure of the primary standard transferred onto the calibrated instrument, considered as a secondary standard. The variance of error is

$$\mathrm{var}(h_2) = \mathrm{var}[\hat{W}_q(w_{10})] + [f'(w_{10})]^2 \mathrm{var}\, h_1 + g_1^2[f'(w_{10})]^2 + g_2^2, \tag{6.16}$$

or

$$\mathrm{var}(h_2) = \sigma^2_{\Delta h_1} + \mathrm{var}(h_1') + g_1'^2 + g_2^2 = \sigma^2_{\Delta h_1} + \sigma^2_{\eta 1}, \tag{6.17}$$

with the substitutions

$$\sigma^2_{\Delta h_1} = \mathrm{var}[\hat{W}_q(w_{10})], \tag{6.18a}$$

$$\mathrm{var}(h_1') = \mathrm{var}[f'(w_{10})h_1], \tag{6.18b}$$

$$g_1' = f'(w_{10})g_1. \tag{6.18c}$$

The above formulae concern the case of a single determination of a measure of the calibrated instrument, at one point of the measurement range. In the case of n-fold determination of measure with the same standard w applied independently, n equations (6.12) are obtained, $i = 1, \ldots, n$:

$$h_{2i} = f(w_{10}) - f_0(w_{10}) + (h_1 + g_1 \varepsilon_{1i}) f'(w_{10}) - g_2 \varepsilon_{2i}. \tag{6.19}$$

Thus, after adding the equations and dividing them by n, we have

$$h_{2\mathrm{av}} = \frac{1}{n}\sum_{i=1}^{n} h_{2i} = [f(w_{10}) - f_0(w_{10})] + h_1 f'(w_{10}) +$$

$$+ g_1 f'(w_{10}) \frac{1}{n}\sum_{i=1}^{n} \varepsilon_{1i} - g_2 \frac{1}{n}\sum_{i=1}^{n} \varepsilon_{2i}. \tag{6.20}$$

The expected value ofe rror $h_{2\mathrm{av}}$ is the same as (6.15) and the variance, instead of (6.17) is

$$\mathrm{var}(h_{2\mathrm{av}}) = \sigma^2_{\Delta h_n} + \mathrm{var}(h_1') + \frac{1}{n} g_1'^2 + \frac{1}{n} g_2^2 = \sigma^2_{\Delta h_n} + \sigma^2_{\eta_n}. \tag{6.21}$$

Apart from an n-fold decrease of random errors, described by the standard deviations g_1, g_2, the variance of the estimation error $\sigma^2_{\Delta h}$ decreases as well, since the variance of a random component is smaller for n measurements than for one. In formulae (6.17), (6.21), this fact is denoted by the indices 1 and n for improved clarity.

The calibration of an instrument is performed with the primary standards $W_1, \ldots, W_j, \ldots, W_m$ with different measures w_{10j} ($j = 1, \ldots, m$). The form of the equations describing the result, when each standard is realized once, is

$$w_{20j} + h_{2j} + g_{2j}\varepsilon_{2j} = f(w_{10j}) + (h_{1j} + g_{1j}\varepsilon_{1j}) f'(w_{10j}). \tag{6.22}$$

After subtracting a nominal characteristic (6.9), the equation is

$$w_{20j} = f_0(w_{10j}).$$

Then, after substitutions as in (6.11) and (6.18), m error equations are obtained, $j = 1, 2, \ldots, m$, with known values of the errors Δ_{2j} and w_{10j}

$$h_{2j} = [f(w_{10j}) - f_0(w_{10j})] + h'_{1j} + g'_{1j}\varepsilon_{1j} - g_{2j}\varepsilon_{2j}. \tag{6.23}$$

When the variance of the random component in equations (6.23) is constant

$$\sigma_\eta^2 = \mathrm{var}(h'_1) + g_1'^2 + g_2^2, \tag{6.24}$$

$$\sigma_{\eta_j}^2 = \mathrm{const}, \quad \text{for} \quad j = 1, \ldots, m, \tag{6.25}$$

the regression method may be applied to determine $h_2(x)$. Let us denote

$$X_0 = [1 \; x_0 \; x_0^2 \; \ldots \; x_0^q], \tag{6.26a}$$

where X_0 is a matrix describing the selected points of the measurement range of the instrument and x_0 is the current coordinate x with the value x_0. When there are more values, the number of rows of the matrix increases

$$X = \begin{bmatrix} 1 & x_1 & x_1^2 & \ldots & x_1^q \\ 1 & x_2 & x_2^2 & \ldots & x_2^q \\ \ldots & \ldots & \ldots & \ldots & \ldots \\ 1 & x_j & x_j^2 & \ldots & x_j^q \\ \ldots & \ldots & \ldots & \ldots & \ldots \\ 1 & x_k & x_k^2 & \ldots & x_k^q \end{bmatrix}. \tag{6.26b}$$

The matrix describes the model $W_q(x)$ and the distribution of measure of the primary standards over the scale of the instrument ($x_j = w_{10j}$). If n_j repetitions are performed at some of the points, then $k = \sum_{j=1}^{m} n_j \geqslant m$. For this reason, the second index k is introduced:

$$M = \frac{1}{k} X^T X, \tag{6.26c}$$

where M is the standardized *Fisher matrix*, independent of the number of calibration points k [68] but dependent on their distribution.[1] Generally, $k \geqslant m$ (this fact will be explained later). Denoting

$$\Delta_2^T = [\Delta_{21} \Delta_{22} \ldots \Delta_{2j} \ldots \Delta_{2k}], \tag{6.26d}$$

$$b^T = [b_0 b_1 \ldots b_q], \tag{6.26e}$$

[1] Sometimes a thesis may be found in the bibliography on this subject that a normalized variance of the regression function $d(x)$ is independent of the number of measurement points [68]. As shown in Fig. 6.3, the properties of $d(x)$ are more complex. The independence of $d(x)_{max}$ of the number of points may be assumed in cases of a G-optimum design.

the systematic error is calculated from the formula

$$\hat{h}_2(x) = \boldsymbol{X}_0 \hat{\boldsymbol{b}}, \tag{6.27}$$

while

$$\hat{\boldsymbol{b}} = \frac{1}{k} \boldsymbol{X}^T \boldsymbol{M}^{-1} \boldsymbol{\Delta}_2. \tag{6.28}$$

The confidence interval of the regression function is calculated from the well known formula (3.265) [68]

$$\sigma^2_{\Delta h_k} = \boldsymbol{X}_0^T (\boldsymbol{X}^T \boldsymbol{X})^{-1} \boldsymbol{X}_0 \sigma^2_\eta = \frac{1}{k} \boldsymbol{X}_0^T \boldsymbol{M}^{-1} \boldsymbol{X}_0 \sigma^2_\eta. \tag{6.29}$$

In specific cases, the variance σ^2_η is not known and therefore its estimator, calculated from the experimental data, is applied

$$\sigma^2_{\eta_k} = \frac{1}{k-q-1} \sum_{j=1}^{k} (\Delta_{2j} - \hat{h}_{2j})^2 = \frac{1}{k-q-1} \sum_{j=1}^{k} [\Delta_{2j} - \hat{W}_q(w_{10j})]^2. \tag{6.30}$$

The result is then

$$\hat{\sigma}^2_{\Delta h_k} = \frac{1}{k} \boldsymbol{X}_0^T \boldsymbol{M}^{-1} \boldsymbol{X}_0 \frac{1}{k-q-1} \sum_{j=1}^{k} (\Delta_{2j} - \hat{h}_{2j})^2. \tag{6.31}$$

Also the random error of the instrument should be determined from the calibration results. The error is not $\{\eta\}$ as in (6.14), because η is realized only in the calibration process. When applied, the instrument generates only a random error with variance g_2^2. After introducing the variance estimator (6.30) into formula (6.24), all the components in the formula thus obtained are known and the random error variance may be calculated

$$\hat{g}_2^2 = \hat{\sigma}^2_{\eta_k} - \mathrm{var}(h_1') - g_1'^2. \tag{6.32}$$

Thus the metrological properties of the calibrated instrument are determined by:

(1) the characteristic, whose model is

$$f_0(x) = a_0 + a_1 x + \ldots + a_r x^r \tag{6.33}$$

and whose coefficients are calculated by the LSM,

(2) the systematic error (6.27) with a known value,

(3) the systematic error with an unknown value and with variance expressed by formulae (6.29) and (6.31),

(4) the random error generated by the calibrated instrument, with variance g_2^2, according to (6.32).

Unlike the classical result (3.267), the random error of the calibrated instrument is only a part of the random component, as shown in Fig. 6.1. The ratio of the random errors appearing in the calibration process and during operation of the instrument may be obtained from formulae (6.24) and (6.32).

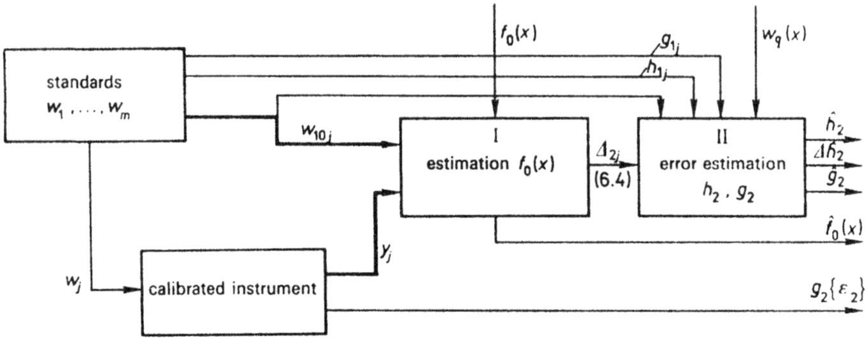

Fig. 6.1. A scheme of estimation of the error of a calibrated instrument.

6.1.2 Selection of the Calibration Points

From the formulae derived in Subsection 6.1.1, the influence of the number of calibration points and the number of repetitions on the error of the calibrated instrument may be evaluated. The remaining problem is the influence of the position of a measurement point on the calibration error, i.e., the selection of the value w_{10j} for the same number m ($j = 1, \ldots, m$). In this case there are no rules which could be written in an analytical form. The tools for solving this problem are the experimental design methods [68, 44].

Optimum designs are known in which the optimality criteria concern different properties of the result. The designs are [68]:[2]

type A—the minimized optimality index is the mean variance of the estimators of the model coefficient b (6.28),

type D—the volume of the ellipsoid of concentration of the estimators of the model coefficients is minimized,

type E—the major axis of the above ellipsoid is minimized,

type G—the minimized index is the maximum value of the variance of the regression function estimation error, i.e. $\sigma^2_{\Delta h}(x)$.

The best solution for the problem considered is a G-optimum design, since the variance of the error of estimation of the systematic error is to be minimized

[2] A systematic presentation of the method and its mathematical grounds are given in [44], [68].

in the calibration process. For the remaining designs there is either no physical interpretation or there are some disadvantages which make their application difficult [17]. The reference point is traditionally considered a uniform distribution of calibration points. In Fig. 6.2 the standardized variance

$$d(x) = X_0^T M^{-1} X_0 = k \frac{\sigma_{\Delta h}^2}{\sigma_\eta^2} \tag{6.34}$$

is shown, for $k = 7$ and $k = 11$, with $\sigma_\eta^2 = 1$ and $q = 3$, the taking points uniformly distributed over the interval $\langle 0, 1 \rangle$[3] [17]. A characteristic feature in this case is that the value of variance is greatest at the ends of the measurement range. The dependence of the maximum value of the standardized variance (6.34) on the number of calibration points is shown in Fig. 6.3.

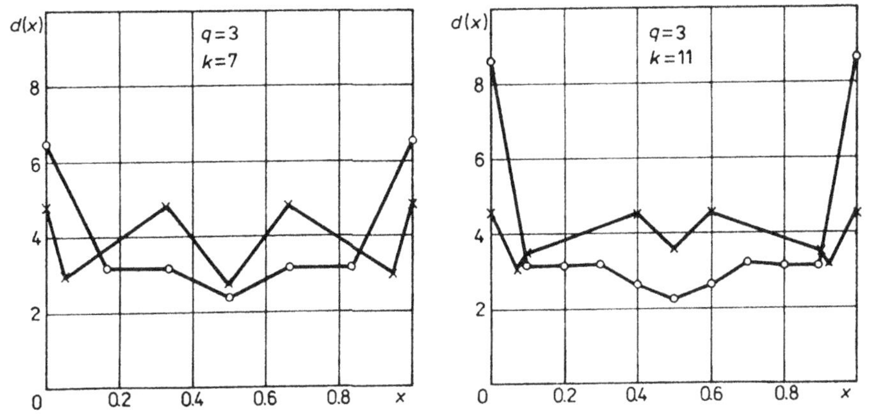

Fig. 6.2. The distribution of the standardized variance of a systematic error. Small circles mark the variance for a uniform distribution of points, crosses mark the case of a *G*-optimum design of points distribution.

The *G*-optimum design implies different distribution of the points [68, 44]. The calibration points are in this case selected so that

$$\min_X \sigma_{\Delta h\,\max}^2 = \min_X \max_x \sigma_{\Delta h}^2(x). \tag{6.35}$$

From (6.29), we may write

$$\min_X \max_x \sigma_{\Delta h}^2 = \min_X \max_x X_0^T M^{-1} X_0 \frac{1}{k} \sigma_\eta^2 \tag{6.36}$$

---

[3] In the *D*-optimum and *G*-optimum designs, the scale change does not influence the optimality of the design. Therefore, application of the normalized range $\langle 0, 1 \rangle$ gives universal results.

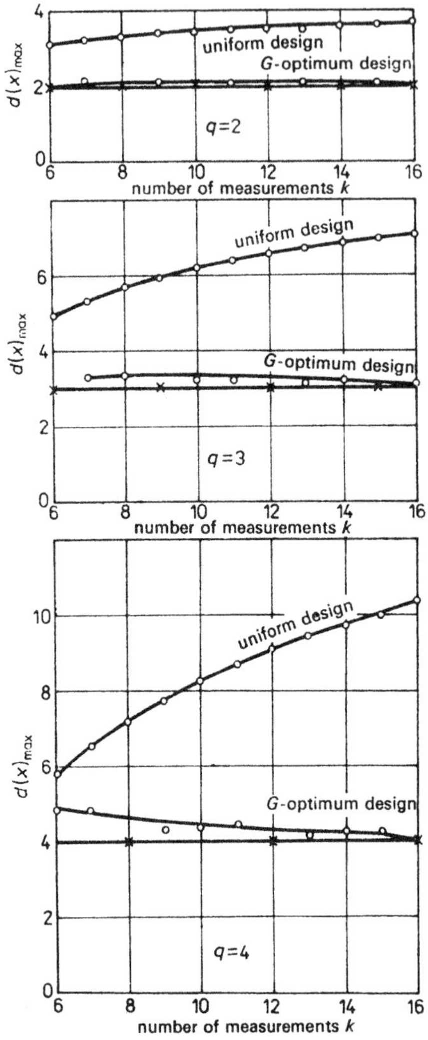

Fig. 6.3. The dependence of the maximum standardized variance of estimation error on the number of measurements for the cases of uniform distribution of points and G-optimum design. The crosses mark the designs with the number of points equal to the number of model coefficients.

instead of (6.35). For a given number of points k and a given variance σ_η^2 of the random component, the design should provide a distribution of points such that

$$\min_X \max_x d(x) = \min_X \max_x \sigma_{\Delta h}^2, \tag{6.37}$$

where $d(x)$ is the standardized variance of the estimation error (6.34).

The G-optimum designs fulfilling condition (6.37) for the qth order polynomials (6.5) are shown in Table 6.1 ($k = 6, \ldots, 16$). The property of the designs is that some of the points repeat and thus $m < k$ standards are applied in the calibration process. The numbers m are also presented in Table 6.1. Because of the small difference of the values of design points, the values of m given in parentheses may be applied in practice. This means that $m = 6$ to 8 standards are sufficient for the calibration to be performed.

There is also another advantageous property of the G-optimum designs. The distribution of the measurement points is independent of the values of the coefficients of the model; instead, it depends on the class of the model, i.e., in the case of a polynomial, on the order of the polynomial. Therefore the design may be determined a priori, and is universal for a given model. Thus the G-optimum designs may be tabulated (as shown, for example, in Table 6.1).

An example of the distribution of the variance of estimation error is shown in Fig. 6.2 (for $q = 3$, $k = 7$ and $q = 3$, $k = 11$). As compared with the traditional uniform distribution of points, due to the G-optimum design, the variance decreases at the ends of the measuring range and increases in the main part of the range. The maximum value of the variance is reached at several points.

The effect of the use of G-optimum designs is shown in Fig. 6.3, where the dependence of the maximum value of standardized variance on the number of points is shown. In G-optimum designs, $d(x)_{max}$ is almost independent of the number of points, while for a uniform distribution of points the variance increases.[4] The diagrams shown in Fig. 6.3 make a direct comparison of the variances of estimation error possible for different distributions of points when the number of points remains the same. The dependence is due to both factors in (6.29), i.e. $d(x) = X_0^T M^{-1} X_0$ and σ_η^2/k. The properties of the first factor are shown in Fig. 6.3. The second factor involves a relationship which is much more complex than the form of the formula. In G-optimum designs some of the points are repeated n_j times and

$$k = \sum_{j=1}^{m} n_j \geqslant m. \tag{6.38}$$

The numbers k and m are equal when every calibration point is realized once; the case $n_j = n$ for every $j = 1, \ldots, m$ is similar. In the case $k > m$, $k \neq nm$,

[4] It is sometimes claimed that the standardized variance of the regression function is independent of the number of measurement points [68]. The figure shows that the properties of $d(x)$ are in fact more complex. Independence of $d(x)_{max}$ from the number of points may be assumed for the case of G-optimum design.

TABLE 6.1

A G-optimum design of determination of the coefficients of a qth order polynomial

$q = 2$

k	Accumulation points over a normalized interval ⟨0, 1⟩																	m
6						0	0	0.500		0.500	1.0	1.0						3
7						0	0	0.256	0.500	0.744	1.0	1.0						5
8					0	0	0.135	0.477		0.523	0.865	1.0	1.0					6
9					0	0	0	0.500	0.500	0.500	1.0	1.0	1.0					3
10				0	0	0	0.264	0.440		0.560	0.736	1.0	1.0	1.0				6
11				0	0	0	0.135	0.471	0.500	0.529	0.865	1.0	1.0	1.0				7
12			0	0	0	0	0.500	0.500		0.500	0.500	1.0	1.0	1.0	1.0			3
13			0	0	0	0	0.257	0.470	0.500	0.530	0.743	1.0	1.0	1.0	1.0			7
14		0	0	0	0	0.135	0.469	0.490		0.510	0.531	0.865	1.0	1.0	1.0	1.0		8
15		0	0	0	0	0	0.500	0.500	0.500	0.500	0.500	1.0	1.0	1.0	1.0	1.0		3
16	0	0	0	0	0	0.257	0.473	0.483		0.517	0.527	0.743	1.0	1.0	1.0	1.0	1.0	8 (6)

$q = 3$

k	Accumulation points over a normalized interval ⟨0, 1⟩																	m
6						0	0.090	0.344		0.656	0.910	1.0						6
7						0	0.055	0.264	0.500	0.736	0.945	1.0						7
8					0	0	0.276	0.276		0.724	0.724	1.0	1.0					4
9					0	0	0.155	0.289	0.500	0.711	0.845	1.0	1.0					7
10				0	0	0.089	0.275	0.342		0.658	0.725	0.911	1.0	1.0				8
11				0	0	0.056	0.276	0.289	0.500	0.711	0.724	0.944	1.0	1.0				9 (7)
12			0	0	0	0.276	0.276	0.276		0.724	0.724	0.724	1.0	1.0	1.0			4
13			0	0	0	0.155	0.283	0.289	0.500	0.711	0.717	0.845	1.0	1.0	1.0			9 (7)
14		0	0	0	0.089	0.277	0.279	0.342		0.658	0.721	0.723	0.911	1.0	1.0	1.0		10 (8)
15		0	0	0	0.052	0.264	0.268	0.272	0.500	0.728	0.732	0.736	0.948	1.0	1.0	1.0		11 (7)
16	0	0	0	0	0.276	0.276	0.276	0.276		0.724	0.724	0.724	0.724	1.0	1.0	1.0	1.0	4

assumption (6.25) is not satisfied, as the variances $\sigma^2_{\eta_j}$ (6.21) are different for different n_j.

The assumption (6.25) arises from the classical approach to regression analysis: the first-kind regression is a conditional expected value at the point w_{0j} and the second-kind regression is solved with the LSM, with assumption (6.25) and with model (6.6) being known with the probability $P = 1$. Assumption (6.25) is also necessary in order to determine the variance σ^2_η = constant and then the confidence interval for the regression. This is why $m > q+1$. If $\sigma^2_{\eta_j} \neq$ constant, it would be necessary to determine $q+1$ coefficients and m values of variance; the number of measurements should in this case be sufficient. Thus the values

$$\hat{h}_{2\mathrm{av}\,j} = \frac{1}{n}\sum_{i=1}^{n_j} \Delta_{2ij} \quad \text{and} \quad \hat{\sigma}^2_{\eta_j} = \frac{1}{n_j-1}\sum_{i=1}^{n_j} (\Delta_{2ij}-\hat{h}_{\mathrm{av}\,j})^2 \tag{6.39}$$

are calculated at the points w_{0j}, and the confidence interval is also determined at these points. When $\sigma^2_{\eta_j}$ = constant, the coefficients of the regression function are determined with the LSM from w_{0j}, $y_{\mathrm{av}\,j}$. The constraints of the model and of the constant variance make it possible to determine the coeffcients and the variance from a smaller number of measurement results.

In the $k > m$ case discussed, the difference of variances $\sigma^2_{\eta_j}$ is no limitation to the regression method, because the variance ratio is known from formula (6.21) and the value of n_j is also known from

$$\sigma^2_{\eta_j} = \frac{1}{n_j}\sigma^2_\eta, \tag{6.40}$$

and thus only one value of σ^2_η is being determined. Therefore either

— the WLSM with m points, or

— the LSM with k points

may be used to estimate the model coefficients, as the weights in the WLSM are proportional to n_j (inversely proportional to the variance) and both methods give the same result. Thus, as may be supposed, formulae (6.26) and (6.28) with k repetitions may be found in the bibliography on the regression method. The method is adequate when the subsequent realizations of the random component η are independent.

In the case of calibration with repetitions of some of the points, relationship (6.40) is not valid because of the component h_1 in formula (6.21). As the G-optimum design concerns only m different points (m standards with systematic errors h_1 realized independently), definition (6.24) of the variance of the random component may be applied:

$$\sigma^2_{\eta\,mk} = \frac{1}{k}\sigma^2_\eta = \frac{1}{m}\mathrm{var}(h'_1)+\frac{1}{k}g'^2_1+\frac{1}{k}g^2_2. \tag{6.41}$$

By marking the variance with indices m and k we stress the analogy to expressions (6.21) and (6.29), and thus we show the number of realizations of the error sources. As the number of standards and the number of points increase, the variance of the estimation error decreases directly with the number of points. By combining formula (6.41), the relationship of $d(x)_{max}$ and the number of calibration points, the relationship shown in Fig. 6.4 is obtained for a uniform distribution of points

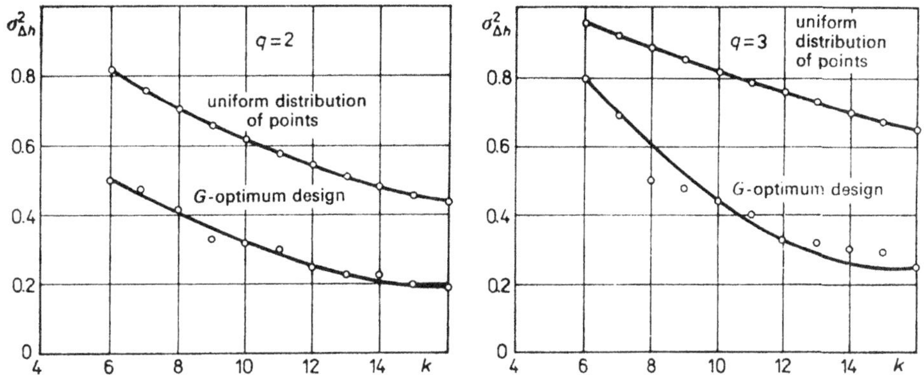

Fig. 6.4. The effect of decreasing the maximum variance of estimation error due to the application of a G-optimum design to the distribution of points (lower curve), compared with the case of uniform distribution of points (upper curve).

and for the case of G-optimum design. In the latter case, the variance minorant is the curve $(q+1)/k$. The assumption (6.25) is not fulfilled for these designs with repetitions. Therefore the LSM should be applied, but the relations of the variances at individual points should be known. Hence the determination of g_2 must be effected by successive approximations.

There can be no doubt that the application of G-optimum designs decreases considerably the number of calibration points, while providing at the same time the same value of the maximum variance of the error of estimation of the systematic error of the calibrated instrument. The theoretical basis of these designs indicates that more effective designs may possibly be reached.

6.1.3 The Boundary Problems of G-Optimum Designs

When the methods of *experimental design* are applied, the existing limitations must always be realized.

1. The experimental design is a derivative of mathematical statistics. Therefore the formulation of the purpose, the assumptions which are made and the indicated uses, are not in fact rooted in reality. To present results in the field of physics

or the technical sciences, and in particular in the field of metrology, profound knowledge of both the mathematical and the physical aspects of the tasks is necessary, as well as an understanding of the physical interpretation of the results.

2. The development of the experimental design methods is tending towards simulation experiments, optimization, etc., and not towards the needs of metrology. For example, one of the basic concepts is optimum use of the independent variables space. For this reason, only some of the methods are applicable in metrology. At present, there are few metrological applications of the experimental design methods.

3. The methods developed concern some particular tasks and thus are not universal and may not give a formula for obtaining good (optimum) results. This is a consequence of the variety of aims, tasks, conditions, etc. One can also come across heuristic methods with no mathematical basis.

Nevertheless, the influence of the experimental design on measurement science is positive in respect of some aspects. These include the ways of formulating the tasks, use of the information a priori, research programming and methods of processing the results, as well as taking full advantage of the information gathered in the course of the experiment and in the form of its results.

An important practical and theoretical problem within the experimental design is to reduce the number of measurement points. The problem was solved for D- and G-optimum designs [50, 19]; the designs show some similarity. The necessary condition for the optimality of the D-design is to fulfil the optimality requirements of the G-design. For the designs concentrated in a finite number of points, the information matrix $\boldsymbol{M}$ is equivalent to the analogous matrix of continuous design (with some assumptions fulfilled, concerning the model of the regression function). In this case, a discrete design may be considered continuous.

A characteristic of a continuous design is the distribution $p_j(x_j)$ of the experiment points, expressing the frequency of realization of a point x_j of the independent variable.

A property of a continuous G-optimum design with accumulation points $x_1, \ldots, x_j, \ldots, x_m$ is that the value of a maximum standardized variance of estimation error is precisely $q+1$. The maximum variance is reached at the accumulation points—this is a necessary condition of optimality of the G-optimum designs. The number $q+1$ is the number of coefficients of the model of the regression function. Also, because of the above conditions, the optimum D- and G-type designs are identical. This equivalence of the designs is applied to determine a particular design for the models in the form of an algebraic or trigonometric polynomial, where the domain is a generalized interval or a simplex. For example when the model is a qth order polynomial (6.5), the optimum location of the points is determined by the roots of the polynomial falling in the interval $\langle -1, 1 \rangle$

$$(1-x^2)P_q'(x), \tag{6.42}$$

where P_q' is the derivative of the qth Legendre polynomial [19]. The Legendre polynomials are listed in Table 3.2 for $k = (q+1)/n$, where $n = 1, 2, 3, \ldots$ (over the interval $\langle 0, 1 \rangle$).

The above properties are not valid for discrete designs, in particular when the quotient of the number of points and the number of model coefficients is not a natural number

$$\frac{k}{q+1} \neq n = 1, 2, \ldots \tag{6.43}$$

The G-optimum discrete designs are composed in such a manner that the required optimality condition (6.36) is fulfilled and the maximum value of variance of the estimation error is greater than $q+1$. This fact is illustrated in Fig. 6.3. For a number $q+1$ of measurement points and their multiples, the variance is $q+1$, and for other numbers it is greater by several percent.

A theoretical discussion has been presented [20] of the case of distribution of calibration points according to G-optimum design with $q+1$ accumulation points. When the estimation of the model coefficients is performed with the WLSM, the form of the matrix of weights is

$$\boldsymbol{W} = \begin{bmatrix} p_1 w_1 & 0 & \ldots & 0 \\ 0 & p_2 w_2 & \ldots & 0 \\ \ldots & \ldots & \ldots & \ldots \\ 0 & 0 & \ldots & p_m w_m \end{bmatrix}, \tag{6.44}$$

where $w_1, \ldots, w_m$ are the weights, and $p_1, \ldots, p_m$ are the frequencies of realization of the individual points. In this case, $m = q+1$, $p_j = 1/m$. Applying notation (6.26) instead of (6.28), we have

$$\hat{\boldsymbol{b}} = (\boldsymbol{X}^T \boldsymbol{W} \boldsymbol{X})^{-1} \boldsymbol{X}^T \boldsymbol{W}^T \boldsymbol{\Delta}_2 . \tag{6.45}$$

As the matrices $\boldsymbol{X}$ and $\boldsymbol{W}$ are square and not singular, after the transformation

$$(\boldsymbol{X}^T \boldsymbol{W} \boldsymbol{X})^{-1} = \boldsymbol{X}^{-1} \boldsymbol{W}^{-1} (\boldsymbol{X}^T)^{-1} \tag{6.46}$$

formula (6.45) is reduced to the form

$$\hat{\boldsymbol{b}} = \boldsymbol{X}^{-1} \boldsymbol{\Delta}_2 . \tag{6.47}$$

This result has the following interpretation:

(1) the problem is reduced to solving a system of algebraic equations and is not a regression method;

(2) the coefficients are determined independently of the weights if the WLSM is applied and the values of Δ_{2j} are the mean values (6.39);

(3) the design may be modified with [17, 20]:

(a) $\sigma_\eta^2 = \sigma_\eta^2(x) \neq$ constant,

(b) $d(x) = V(x) \neq$ constant,

(c) both conditions (a) and (b) simultaneously.

The modifications show the possibilities of decreasing the variance of the estimation error.

Because of the fact that the problem is reduced to the condition $m = q+1$, the results cannot be applied in metrology and they just show the limitations of the abilities of G-optimum designs.

6.1.4 The Design Principles of Calibration Processes

Calibration is a technological operation and is a part of the manufacturing process of a measuring instrument. Calibration influences the metrological properties of an instrument as strongly as the quality of the elements, the assembling process, adjustment, etc. The imperfection of the elements and this assembly show up in the errors $h_2(x)$ and $g_2(x)$, more or less reproducible in the technological series. Calibration also introduces errors, which may be adequately controlled.

The choice to the methodology of calibration is an economic problem, with some limitations of a technical nature. The relation of the price of a standard and its accuracy is

$$K = K_0 \delta^{-A}, \tag{6.48}$$

where K_0, A are constants and δ is a relative error. Increasing the number of standards or a multiple repetition of calibration at some points makes the costs rise because of faster wearing and longer time spent or greater effort taken in the calibration process.

Let us denote the tolerances of the construction parameters, the parameters describing the quality of assembling and the degrees of freedom of a calibration procedure as $u_1, \ldots, u_l$. The following relationship may be found:

$$h_2 = f(u_1, \ldots, u_l). \tag{6.49}$$

Each of the above variables involves a respective cost $K_1, \ldots, K_l$; this fact may be written as

$$u_i = c_i K_i, \quad i = 1, \ldots, l, \tag{6.50}$$

After substituting (6.50) in (6.49) and after calculating the total differential we obtain

$$dh_2 = \sum_{i=1}^{l} c_i \frac{\partial f}{\partial K_i} \, dK_i. \tag{6.51}$$

Therefore, a way of decreasing the error may be chosen for a value of $i = 1, \ldots, l$, for which

$$c_i \frac{\partial f}{\partial K_i} = \text{maximum}. \tag{6.52}$$

A calibrated instrument is burdened with the three components of errors h_2, g_2 and Δh.

1. The component h_2, the systematic error, is the effect of the dispersion of values of the construction parameters (see Subsection 4.5.2) and the result of selection of a specific fixed characteristic $\hat{f}_0(x)$ (6.33). The systematic error is eliminated by substituting for the characteristic $\hat{f}_0(x)$ the characteristic $\hat{f}(x) = \hat{f}_0(x) + \hat{h}_2(x)$. This involves the costs of individual fixing of the characteristic, as well as of determining $\hat{h}_2(x)$. When the principal requirement is the linearity of the characteristic, the linearization systems increase the costs as well. The methodology of determining h_2 was presented in Subsection 6.1.1.

2. The component g_2, the random error, cannot be eliminated in the calibration process, althought the value of the estimator may be determined with more accuracy. In some instruments the random error may be reduced by applying the proper calibration procedure and building it into the instrument. For example, application of the averaging procedure (4.22) causes the random error to decrease, at the cost of the measurement time. Solutions like that are known [48] and used; usually up to one thousand measurements are needed to obtain the result [123].

3. The component Δh_2 is the error of estimation of the systematic error; the main source of this error is the calibration process. The formulae derived in Subsection 6.1.2 make possible an assessment of this component and the application of a method of calibration which ensures that the contribution of this error to the total error will be sufficiently small.

The task of selecting the standards for calibration of measuring instruments is usually formulated as follows.

1. The limiting values $h_{\text{lim}}(x)$ and $g_{\text{lim}}(x)$ are assumed for the calibrated instrument, either separately or together, with respect to a proper class of a metrological characteristic, as in Section 4.4. The assumptions are based on the theoretical analysis (Subsection 4.5.2) or on the experimental results (Subsection 4.5.3). The model of the systematic error, e.g. $W_q(x)$, is determined together with the value of q.

2. The distribution of total allowable error (or separate distributions of systematic and random errors) are assumed with respect to the classes of the metrological characteristic (see Section 4.4). The allowable error is here a sum of limiting error, estimation error and error margin, supposed to cover the error

changes during the service time. Examples of distributions of allowable error were shown in Fig. 4.10.

3. The difference of the allowable error and the limiting error is the estimation error $\Delta h(x)$ which cannot be exceeded at the stage of calibration design. The degrees of freedom in the calibration design process are: m—the number of standards, k—the number of calibration points, $k \geqslant m$, X—the matrix of experimental design, expressed by the values of the measures of standards (with a determined number of standards and number of repetitions and with the determined model $W_q(x)$), h_1, g_1—the uncertainties of the measures of the standards.

On the basis of the standards, derived as in Subsections 6.1.2 and 6.1.3, the values of the degrees of freedom are calculated by a trial-and-error method. Condition 2 (not exceeding the allowable error value) should be formulated by means of statistics, i.e., with regard to the measuring range, as (4.142), and to the set of instruments. Here are some comments on some of the practical aspects of the procedure.

The methodology of determination of the systematic error of the calibrated instrument presented in Subsection 6.1.2 applies also to cases when the systematic error is negligibly small. This fact, however, can be proved in the way shown, by proving that the regression function $\hat{h}_2(x)$ is insignificant (see Subsection 3.5.5, formula (3.257)) and the distribution of the systematic error is not $\hat{h}_2(x)$ = constant $\neq 0$. When the regression function is not significant, the uncertainty arising from the estimation error Δh remains a component of the error. In this case, the random errors will be estimated properly.

When selecting a standard or a standard instrument, practical recommendations are often formulated concerning the ratio of the errors of the standard and the calibrated instrument, for which the error of the standard instrument is negligibly small, e.g. $\delta_1 \leqslant 0.1\delta_2$ [105], $\delta_1 \leqslant 0.2\delta_2$, the standard instrument should be 2 clases better [91], etc. The recommendation may come from about 10% accuracy of error estimation and is the negligibility limit, with regard to the metric (5.20). Some other criteria may also be formulated—the details may be found in the bibliography, e.g. [46]. Analysis of the error components described in Fig. 6.1 shows that the error described by the formula (6.29), with variance σ_η^2 (6.41), is significant. Thus, the variance of the errors h_1, g_1 of the standards should be added to the random error g_2 of the calibrated instrument. Instead of giving unconvincing explanations of the fact that the error is negligible, or that it is not, a better idea is to add the random variables. Other factors in formula (6.29) do not influence the relations of the errors discussed.

The course of reading checking, which is similar to calibration when following an adjustment, is somewhat different. In this case, the components $\hat{h}_2(x)$ and $\hat{g}_2$ are the measure of error of the instrument under test and the component

Δh corresponds to the uncertainty of the obtained result of the reading checking. To reduce this uncertainty, one should select appropriately the metrological properties of the standards, the number of standards, the number of repetitions, the experimental design, etc. The formulae derived in this chapter allow proper selection of the parameters and of the course of calibration.

6.2 Selection of the Error Sampling Interval

The *sampling interval* is chosen with regard to the properties of the measured object, in order to compress the information. During measurement, the errors are sampled at intervals whose value is imposed by the properties of the object and by the purpose of measurement. It should be checked then, by the interpretation of the measurement results and by applying the methods of mathematical statistics, whether the realizations of errors are independent. If this condition is not fulfilled, the dependence of the errors should be considered in calculations.

When the task is to determine the errors, the problem is regarded differently. The sampling interval should be chosen so that the task is fulfilled in a rational way. When discussing the sampling interval, both time and spatial coordinates of the space in which a multidimensional random process is mapped should be considered equally.

6.2.1 Error Estimation in a Correlated-Error Measurement Series

The purpose of carrying out a measurement series is to decrease the random errors when the value of the measured quantity is known to be constant, with probability $P = 1$. According to the measurement procedure (see Subsection 5.5.2), the mean value of the measurement result is calculated and the limiting error of the mean value is determined.

The conditions of the analysis of errors are not yet stated; for the time being, let us consider the following estimators of subsequent dependent realizations of errors:

$$\hat{\Delta}_{av} = \frac{1}{N} \sum_{i=1}^{N} \Delta_i, \tag{6.53}$$

$$\hat{s}_N^2 = \mathrm{var}(\Delta_{av}) = \frac{1}{N^2} \sum_{i=1}^{N} \sum_{j=1}^{N} \lambda_{ij}, \tag{6.54}$$

where λ_{ij} are the covariance coefficients, forming the covariance matrix for the errors of the series $\Delta_1, \ldots, \Delta_N$:

$$\Sigma = \begin{bmatrix} \lambda_{11} & \lambda_{12} & \ldots & \lambda_{1N} \\ \lambda_{21} & \lambda_{22} & \ldots & \lambda_{2N} \\ \ldots & \ldots & \ldots & \ldots \\ \lambda_{N1} & \lambda_{N2} & \ldots & \lambda_{NN} \end{bmatrix}. \tag{6.55}$$

The covariance matrix is symmetric, as $\lambda_{ij} = \lambda_{ji}$, and positive semi-definite. The diagonal elements are the variances of particular measurements $\lambda_{ii} = s_i^2$. By applying the model of error (3.136)

$$\{\Delta\} = g(x, T)\{\varepsilon\} + h(x, T),$$

we may describe the correlation coefficients of two errors at time separation $\tau = t_2 - t_1$ and spaced apart by $\xi = x_2 - x_1$, by formula (3.164). Here the formula may be written as

$$\varrho = \frac{1}{\sigma_{\Delta 1}\sigma_{\Delta 2}} K_\Delta(\Delta_1, \Delta_2)_{t_1, t_2, x_1, x_2} = \frac{\mu_{2g}(\xi)\varrho(\tau) + K_h(\xi)}{\sqrt{(g_1^2 + \operatorname{var} h_1)(g_2^2 + \operatorname{var} h_2)}}. \tag{6.56}$$

When the results in the series are corrected, then $h = 0$ and the case becomes simpler. The correlation of the subsequent results is described by the correlation coefficient $\varrho(\tau)$, and we have $\mu_{2g}(\xi) = g^2$. For a series of N measurements with sampling interval Δt, the covariance matrix is

$$\Sigma = g_2 \begin{bmatrix} 1 & \varrho(\Delta t) & \varrho(2\Delta t) & \ldots & \varrho[(N-1)\Delta t] \\ \varrho(\Delta t) & 1 & \varrho(\Delta t) & \ldots & \varrho[(N-2)\Delta t] \\ \varrho(2\Delta t) & \varrho(\Delta t) & 1 & \ldots & \varrho[(N-3)\Delta t] \\ \ldots\ldots & \ldots\ldots & \ldots\ldots & \ldots & \ldots\ldots \\ \varrho[(N-1)\Delta t] & \ldots & \ldots & \ldots & 1 \end{bmatrix}. \tag{6.57}$$

For a series consisting of two measurements, the variance estimator (6.54) is

$$\hat{s}_2^2 = \tfrac{1}{4}[2g^2 + 2g^2\varrho(\Delta t)] = \tfrac{1}{2}[1 + \varrho_1]g^2. \tag{6.58a}$$

For $N = 3$

$$\hat{s}_3^2 = \tfrac{1}{3}[1 + \tfrac{4}{3}\varrho_1 + \tfrac{2}{3}\varrho_2]g^2, \tag{6.58b}$$

and for N

$$\hat{s}_N^2 = \frac{1}{N}\left[1 + 2\frac{N-1}{N}\varrho_1 + 2\frac{N-2}{N}\varrho_2 + \ldots + 2\frac{1}{N}\varrho_{N-1}\right] g^2$$

$$= \frac{1}{N}\left[1 + \sum_{k=1}^{N-1}\left(2\frac{N-k}{N}\varrho_k\right)\right] g^2, \tag{6.58c}$$

where $\varrho_1 = \varrho(\Delta t)$, $\varrho_2 = \varrho(2\Delta t)$, etc.

To show the influence of correction on the variance of the mean value, the variance has been calculated for selected correlation coefficients

$$\varrho_k = \varrho = \text{constant}, \qquad s_N^2 = \frac{1}{N} g^2(1-\varrho)+\varrho g^2, \tag{a}$$

$$\varrho_k = \frac{N-k}{N}, \qquad s_N^2 = \frac{1}{3N^2} g^2 + \frac{2}{3} g^2, \tag{b}$$

$$\varrho_k = \left(\frac{N-k}{N}\right)^2, \qquad s_N^2 = \frac{1}{2N^2} g^2 + \frac{1}{2} g^2. \tag{c}$$

$$\varrho_k = \left(\frac{N-k}{N}\right)^3, \qquad s_N^2 = \left(\frac{2}{3}\frac{1}{N^2}+\frac{1}{15}\frac{1}{N^4}\right) g^2 + \frac{2}{5} g^2. \tag{d}$$

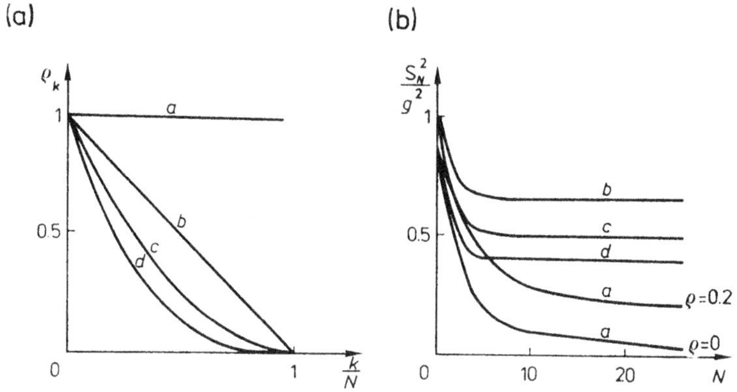

Fig. 6.5. The effect of correlation of errors: (a) properties of the selected correlation functions, (b) decreasing variance of an error in a measurement series in cases of different correlation functions (a, b, c, d as in Fig. 6.5a).

In Fig. 6.5, the function ϱ_k is shown together with the relationships for the variance estimator for different N. As may be seen from the formulae and from the figure, when the measurement series is carried out, the estimator of the error variance in the series with correlated realizations is biased and inconsistent. This property has been pointed out by Elyasberg [150], who gave the example a for ϱ = constant. The effect of correlation of subsequent realizations of errors is that the value of the variance of the series error is greater than its value as calculated from the law of large numbers. The stronger the correlation, the greater the variance. The condition

$$\frac{1}{N} g^2 \leqslant \hat{s}_N^2 \leqslant g^2 \tag{6.59}$$

is fulfilled. In the limiting case $\varrho = 1$, the variance of error in a measurement series is equal to the variance of error in a single measurement, and thus there is no reason for performing a series of measurements.

Consideration of the correlation of errors may be practically important in the case of short measurement series. For a long series, only a fraction of realizations of errors may be correlated. If, for example, in an N-element series only K subsequent realizations are correlated, the form of expression (6.58c) will be

$$\hat{s}_N^2 = \frac{1}{N}\left[1+2\sum_{k=1}^{K-1}\left(1-\frac{k}{K}\right)\varrho_k\right]g^2. \tag{6.60}$$

For a constant value of $\varrho_k = \varrho$, the value of the sum in (6.60) is $(K-1)\varrho/2$ and then

$$\hat{s}_N^2 = \frac{K}{N}\left[1+\frac{1-\varrho}{K}\right]g^2. \tag{6.61}$$

This result means that the estimator bias fades, but the fading ratio is more than K times smaller. It is possible to decrease the bias to the required limits, but the measurement series must be long enough. The above properties of the variance estimator in a series with correlated results were discussed by Shannon [181], who refers to Geisler's analysis [154].

Let us present the problem of correlation of errors differently. Let the aim of the series be the determination of the variance of a single-measurement error from a sample of N realizations. The variance estimator is

$$\hat{s}^2 = \frac{1}{N-1}\sum_{k=1}^{N}(\Delta_i-\Delta_{av})^2. \tag{6.62}$$

$$\Delta_i-\Delta_{av} = \Delta_i-\frac{1}{N}\sum_{i=1}^{N}\Delta_i = \frac{1}{N}\sum_{j=1}^{N}(\Delta_i-\Delta_j). \tag{6.63}$$

After substituting (6.63) in (6.62) we may calculate the expected value of the estimator as

$$E(\hat{s}^2) = \frac{1}{N-1}\,\frac{1}{N^2}\,E\left\{\sum_{i=1}^{N}\left[\sum_{j=1}^{N}(\Delta_i-\Delta_j)\right]^2\right\}. \tag{6.64}$$

For $N = 2$

$$E(\hat{s}^2) = \tfrac{1}{4}E\{(\Delta_1-\Delta_2)^2+(\Delta_2-\Delta_1)^2\} = \tfrac{2}{4}g^2(2-2\varrho_1) = g^2(1-\varrho_1), \tag{6.65a}$$

for $N = 3$

$$E(\hat{s}^2) = \frac{1}{2\times 9} E\{[(\Delta_1-\Delta_2)+(\Delta_1-\Delta_3)]^2+[(\Delta_2-\Delta_3)+(\Delta_2-\Delta_1)]^2+$$
$$+[(\Delta_3-\Delta_1)+(\Delta_3-\Delta_2)]^2\} = g^2(1-\tfrac{2}{3}\varrho_1-\tfrac{1}{3}\varrho_2), \tag{6.65b}$$

and for N

$$E(\hat{s}^2) = g^2\left[1-\frac{2}{N-1}\sum_{k=1}^{N-1}\frac{N-k}{N}\varrho_k\right]. \tag{6.65c}$$

For the exemplary correlation coefficients we obtain

$$\varrho_k = \varrho = \text{constant}, \qquad E(\hat{s}^2) = g^2(1-\varrho), \qquad (a)$$

$$\varrho_k = \frac{N-k}{N}, \qquad E(\hat{s}^2) = \frac{1}{3}\frac{N+1}{N}g^2, \qquad (b)$$

$$\varrho_k = \left(\frac{N-k}{N}\right)^2, \qquad E(\hat{s}^2) = \frac{1}{2}\frac{N+1}{N}g^2, \qquad (c)$$

$$\varrho_k = \left(\frac{N-k}{N}\right)^3, \qquad E(\hat{s}^2) = \frac{3}{5}\frac{N+1}{N}g^2\left[1-\frac{1}{gN^2}\right]. \qquad (d)$$

As the examples show, the expected value of the sample variance in the case of correlated realizations is smaller in comparison with the statistically independent result. The variance estimator (6.62) is biased and is not consistent with the estimator calculated from independent results. When, in a series of N realizations, K subsequent realizations are correlated so that $\varrho_k = \varrho =$ constant, the expected value of the variance is

$$E(\hat{s}^2) = g^2\left(1-\frac{K-1}{N-1}\varrho\right). \tag{6.66}$$

In this case the variance estimator is not biased, but the bias fades more slowly; thus the measurement series should be $K\varrho$ times longer.

6.2.2 Total Error Analysis for a Measuring Instrument

In the simplest case of a metrological characteristic (Class 5), the total error of an instrument is limited and the systematic and random errors are not distinguished. Thus, it is natural to determine the error $\{\Delta\}$ for the whole domain x (i.e., for the whole measurement range) and to check that the error does not exceed the allowable value. Let the set

$$\{\Delta_j\} = g(x_j)\{\varepsilon\}+h(x_j) = g_j\{\varepsilon\}+h_j \tag{6.67}$$

be the set of realizations of errors for different values of x_j. Hence

$$\mathop{E}_{x}\{\Delta_j\} = \Delta_{av} = h_{av}, \tag{6.68}$$

$$\operatorname{var}\{\Delta_j\} = \frac{1}{N}\sum_{j=1}^{N}(\Delta_j - \Delta_{av})^2 = E[g_1\{\varepsilon_j\} + h_j - h_{av}]^2 = g_{av} + \operatorname{var} h \tag{6.69}$$

since $E[(h_j - h_{av})(g_j\varepsilon_j)] = 0$.

It is clear from formula (6.69) that the variance of the random and systematic errors has been calculated. The variance is the mean variance of the random error and of the variance of systematic error, calculated with respect to x. The systematic error is expressed by the value h_{av}, i.e., the average value of the systematic error (with respect to x).

For any point of the measuring range, the error falls within the interval

$$\Delta_{av} - t\sqrt{\operatorname{var}\Delta} \leqslant \Delta \leqslant \Delta_{av} + t\sqrt{\operatorname{var}\Delta}, \tag{6.70a}$$

where t is the confidence interval for a given confidence level. The confidence interval may be written as

$$h_{av} - t\sqrt{g_{av}^2 + \operatorname{var} h} \leqslant \Delta \leqslant t\sqrt{g_{av}^2 + \operatorname{var} h}\ . \tag{6.70b}$$

If $g(x)$ = constant, the above formula will express strictly the confidence interval of the error. If $g(x) \neq$ constant, the formula is approximate, since the value of $g(x)$ remains unknown.

Moreover, because of the unknown probability distributions of both random, $p(\varepsilon)$, and systematic, $p[h(x)]$, errors, the confidence interval t in formula (6.70) is hard to determine. The method expressing of and estimating the total error is correct in substance, yet some of its interesting properties should be pointed out.

(1) To determine h_{av} or var$h$, the sample should be extracted from the general set $\{h\}$. The notation for $\{h\}$ is parametric and the value of h remains unknown. Therefore, the condition of acquiring a representative sample is uniform sampling in the domain of error, i.e. $x \in \langle x_1, x_u \rangle$. This principle comes from a property of the distribution (the so-called geometric probability [22]). When the sample distribution is uniform continuous $p(x) = 1/(x_u - x_l)$ or discrete $p(x) = (1/N)\sum_{j=1}^{N} \delta(x - x_j)$, $j = 1, \ldots, N$; $x_{j+1} - x_j = \Delta x$ = constant, then the values of h_{av} obtained from (6.68) and the variance var$h$ are calculated correctly. This, according to the present author, is the implication of the methodology of checking the error at the points of the scale labelled with numbers.

(2) Let us check whether the subsequent realizations of errors are correlated. The correlation function of two realizations of errors is given by formula (3.164), but the properties of the error components, which are necessary to assess the

degree of correlation, remain unknown. To determine the correlation function, the values of μ_{2g}, K_h and $\varrho(\tau)$ must be known, as the values of $\varrho_1, \ldots, \varrho_{N-1}$ must be calculated at known time and space intervals. On the other hand, formula (6.64) says how much the variance is underestimated. The approximate data may be gathered in type tests, and the influence of the error correlation may be limited by increasing the number of points at which the errors are determined. Such a recommendation is reasonable for a type (class) of instruments, but is not appropriate for a given specimen. It must then be supposed that the variance of error is underestimated because of the correlation of errors. More complete data concerning the correlation of errors ex-post (i.e., after the errors are determined) are included in the metrological characteristics, Classes 4 and 3.

(3) The confidence interval of error (6.70) is derived for a given confidence when the probability distribution of error is unknown. In the case discussed, the distribution of a 2-component error is calculated as a convolution of the component distributions. The distribution of a random error $p(\varepsilon)$ arises from its physical nature, and the parameters of the distribution may depend on x. The distribution of the systematic error $p[h(x)]$ is less typical. The example (Fig. 4.15) presented in Chapter 4 demonstrates that there may be a variety of distributions of the systematic errors. There are no grounds for recommendations concerning the distributions of errors. The results of research by Novitskii [78] (see Fig. 4.11) show that the uncertainty of the value of y may be even greater than $\pm 30\%$. As may be seen, the lack of a number of data results in the total error as a measure of errors not being calculated precisely. For this reason, the total error is not a part of the metrological characteristics.

Appendix

The Lebesgue Measure of a Set

Different measures are applied to characterize sets. The place of the Lebesgue measure is quite specific. Let the set of errors Δ be mapped in a k-dimensional R^k space X. The measure of a set is a number $\mu(\Delta)$, determined within the field M. The measure conforms to the following axioms:

$$0 \leqslant \mu(\Delta) \leqslant \infty \quad \text{for every set} \quad \Delta \in M,$$

$$\mu(\emptyset) = 0,$$

$$\mu\Big(\bigcup_{i=1}^{\infty} \Delta_i\Big) = \sum_{i=1}^{\infty} \mu(\Delta_i).$$

To determine the measure of a set, one should:

(1) Create closed intervals $J \in R^k$ included in M, such that:

— each J_n is a set of points $(x_1, x_2, \ldots, x_k) \in R^k$ for which the inequalities

$$a_{1n} \leqslant x_{1n} \leqslant b_{1n};\ a_{2n} \leqslant x_{2n} \leqslant b_{2n};\ \ldots;\ a_{kn} \leqslant x_{kn} \leqslant b_{kn};$$

$$\frac{i_1 - 1}{2^{n-1}} \leqslant x_{1n} \leqslant \frac{i_1}{2^{n-1}};\ \frac{i_2 - 1}{2^{n-1}} \leqslant x_{2n} \leqslant \frac{i_2}{2^{n-1}};\ \ldots;\ \frac{i_k - 1}{2^{n-1}} \leqslant x_{kn} \leqslant \frac{i_k}{2^{n-1}}$$

are satisfied, where $i_1, i_2, \ldots, i_k = 0, \pm 1, \pm 2, \pm 3, \ldots$; $n = 1, 2, 3, \ldots$;

— each J_k is the sum of 2^k intervals J_{n+1};

— the measure of each J_n is the number

$$\mu(J_n) = (b_{1n} - a_{1n})(b_{2n} - a_{2n}) \ldots (b_{kn} - a_{kn}).$$

(2) Create the sets

$$[\Delta]_n = \bigcup_{J_n \subset \Delta} J_n,$$

where $\Delta = \bigcup_{n=1}^{\infty} [\Delta]_n$.

(3) Calculate the measure

$$\mu(\Delta) = \lim_{n \Rightarrow \infty} \mu([\Delta]_n).$$

The Lebesgue measure concerns both open and closed sets. Every open set is measurable in the sense of the Lebesgue measure. Not all closed sets are measurable, although no example of a non-measurable closed set has yet been found. The measurability of a set—a set of errors—is an important property, as it makes expressing the properties of a set by numbers possible as well as allowing comparison of different sets.

The following theorem is particularly important. *If f is a non-negative (L)-measurable function definite in an (L)-measurable set $A \subset R^k$, then the set $Z \subset R^{k+1}$ defined as $Z = \{(x_1, \ldots, x_k, y)\colon [(x_1, \ldots, x_k, y) \in A] \wedge [0 \leqslant y \leqslant f(x_1, \ldots, x_k)]\}$ is (L)-measurable and*

$$\mu_{k+1}(Z) = \iint_A \ldots \iint f(x_1, \ldots, x_k) \mathrm{d}x_1 \ldots \mathrm{d}x_k .$$

References

1. Ajdukiewicz, K.: *Pragmatic Logic* (in Polish), PWN, Warszawa, 1974.
2. Anderson, T.: *The Statistical Analysis of Time Series*, J. Wiley & Sons, New York, 1971.
3. Bendat, J. S., Piersol, A. G.: *Random Data: Analysis and Measurement Procedures*, J. Wiley & Sons, New York, 1971.
4. *BIPM Bulletin de XIV Session de Commité International des Poids et Mesures*, 1976.
5. Bobrowski, D.: *Probability Theory in Technical Applications* (in Polish), WNT, Warszawa, 1980.
6. Bolshev, L. N., Smirnov, N. V.: *Tables of Mathematical Statistics* (in Russian), Wychislitel'nyi Centr AN SSSR, Moskva, 1968.
7. Box, G. E. P., Jenkins, G. M.: *Time Series Analysis, Forecasting and Control*, Holden Day, New Jersey, 1976.
8. Brandt, S.: *Statistical and Computational Methods in Data Analysis*, North-Holland Publishing Co., Amsterdam and New York, 1970.
9. Brownlee, K. A.: *Statistical Theory and Methodology in Science and Engineering*, J. Wiley & Sons, New York–London, 1965.
10. Campbell, N. R.: *Physics: The Elements*, Cambridge University Press, Cambridge, 1920.
11. *Catalogue of Reference Materials From National Laboratories of Pure and Applied Chemistry*, Vol. 48, 1976.
12. Cox, D. R., Hinkley, D. V.: *Theoretical Statistics*, Chapman Hall, London, 1974.
13. Crown, E. L.: Analysis of the accumulated error in a hierarchy of calibration, *IRE Transactions and Instrumentation*, Sept. 1960.
14. Crown, E. L.: Optimum allocation of calibration errors, *Industrial Quality Control*, Nov. 1966.
15. Danzer, K. et al.: *Analytik.: Systematischer Überblick*, Akad. Verl. Geest & Portig. K-G, Leipzig, 1987.
16. Draper, N. R., Smith, H.: *Applied Regression Analysis*, J. Wiley & Sons, New York, 1973.
17. Dziambor, K.: Selection of a Fisher matrix for a metrological characteristic of an instrument for the design of an experiment for determining the errors with different optimality criteria applied (in Polish), M. Sc. Thesis, Institute of Automatic Control, Silesian Technical University, Gliwice, 1982.
18. Eykhoff, P.: *System Identification—Parameter and State Estimation*, J. Wiley & Sons, New York, 1974.
19. Fedorov, V. V.: *Theory of Optimal Experiment* (in Russian), Nauka, Moskva, 1971.
20. Felhorski, W., Stanioch W.: *Trichromatic Colorimetry* (in Polish), WNT, Warszawa, 1973.
21. Finkelstein, L.: Grundgedanken zum Messen, *Messen Steuern Regeln*, H. 6 (1973). 16.

22. Fisz, M.: *Probability Theory and Mathematical Statistics*, John Wiley, New York, London, 1963.
23. Franks, L. E.: *Signal Theory*, Prentice-Hall, Englewod Cliffs, N. J., 1967.
24. Frączek, J., Romer, E., Piotrowski, J.: Thermomagnetic oxygen analyzer KTM-7 (in Polish), *Pomiary. Automatyka. Kontrola*, No 10/11 (1963).
25. Froehlich, M.: *Das Normalelement*, Akademische Verlagsgeselschaft, Wiesbaden, 1978.
26. Fukunaga, K.: *Introduction to Statistical Pattern Recognition*, Academic Press, New York–London, 1972.
27. GOST 8.009-84 *Standardized Metrological Characteristics of Measuring Instruments* (in Russian), Izd. Standardov, Moskva, 1985.
28. Hagel, M.: A mathematical model of an ultrasonic flowmeter (in Polish), *Pomiary. Automatyka. Kontrola*, No 1 (1980).
29. Hagel, M.: Reliability studies of water meters, Problem NB-171/RAu1/81, Institute of Automatic Control, Silesian Technical University, Gliwice, 1981.
30. Hagel, R.: *Dynamic Measurements* (in Polish), WNT, Warszawa, 1975.
31. Hamer, W. J.: Standards cells, their construction, maintenance and characteristics, *NBS Monograph*, **84**, 1975.
32. Hartig, G., Hart H.: Kontinuerliche Fertigungsüberwachung durch statistische Auswertung von Prüfdaten, *Nachrichtentechnik. Elektronik*, **30** (1980).
33. Harvey, J. K., MacFarlane, J. C.: Controle a long terme de l'etalon australien de force elektromotorice au moyen de l'effet Josephson, *Bull. BIPM*, 1976.
34. Hinton, J. R. B.: Organizing an instrument department in a large oil company, *Instrument Practice*, **18** (1963).
35. Hofmann, D.: Heuristische Algoritmen zur Rationalisierung von Messprozessen, *Messen Steuern Regeln*, **14** (1971).
36. Hölder, O.: Die Axiome der Quantität und die Lehre von Mass, *Berichte über die Verhandlungen der König, Sächsischen Gesellschaft der Wissensch. Math-phys. Klasse*, **53** (1901).
37. Jarnot, J.: A photodiode detection circuit for a colorimeter (in Polish), M. Sc. Thesis, Institute of Automatic Control, Silesian Technical University, Gliwice, 1978.
38. Jaworska, I.: Analysis of the influence of the measuring unit on the operating quality of a linear control system with a quadratic figure of merit (in Polish), Ph. D. Thesis, Warsaw Technical University, Warszawa, 1978.
39. Jaworski, B. M., Dietlaf, A. A.: *Physics: A Concise Encyclopaedia* (in Polish), PWN, Warszawa, 1966.
40. Jaworski, J.: Mathematical models of measurement systems and procedures (in Polish), *Prace naukowe. Elektryka*, No 46, Wyd. PW, Warszawa, 1977.
41. Jaworski, J.: *Mathematical Principles of Metrology* (in Polish), WNT, Warszawa, 1979.
42. Jellonek, A. et al.: *Principles of Metrology in Electrical and Electronic Engineering* (Appendix by R. Rymaszewski) (in Polish), PWN, Warszawa, 1980.
43. Jędrzejewski, A.: Analysis of fidelity error of a lever-and-tooth transducer (in Polish), Ph. D. Thesis, Institute of Metrology and Instrumentation, Warsaw Technical University, Warszawa, 1978.
44. Kacprzyński, B.: *Experimental Design. Mathematical Principles* (in Polish), WNT, Warszawa, 1974.
45. Kailath, T.: *Lectures on Wiener and Kalman Filtering*, Springer-Verlag, Berlin, New York, Heidelberg, 1981.
46. Kałuski, J.: An integral accuracy figure for a selected class of measuring instruments (in Polish), Ph. D. Thesis, Silesian Technical University, Gliwice, 1976.

47. Kamke, D., Kramer, K.: *Physikalische Grundlagen der Masseinheiten*, Teubner-Verlag, Stuttgart, 1977.
48. Kavalerov, G. I., Mandelshtam, S. M.: *Introduction to the Theory of Measurement* (in Russian), Nauka, Moskva, 1982.
49. Kiczma, B.: Application of magnetic atomic resonance in current measurements (in Polish), Ph. D. Thesis, Silesian Technical University, Gliwice, 1979.
50. Kiefer, J., Wolfowitz, J.: The equivalence of two extremum problems, *Can. J. Math.*, **12**, (1960).
51. Kondratev, G. M.: *Thermal Measurements* (in Russian), Masgiz, Moskva, 1957.
52. Kopacz, S.: Influence of the asymmetry of dynamic properties on the supplementary errors of a differential transducer (in Polish), Ph. D. Thesis, Institute of Automatic Control, Silesian Technical University, Gliwice, 1975.
53. Kostyrko, K.: Some considerations of the allocation of errors in a hierarchy of calibration, *IMEKO Symposium Proc.*, TC7, Dubrovnik, 1981.
54. Krantz, D. et al.: *Foundations of Measurement*, Vol. 1, Academic Press, New York–London, 1971.
55. Krauss, M., Woschni, E. G.: *Messinformationssysteme*, Verlag Technik, Berlin, 1972.
56. Król-Stan, A.: At a given waveform and interferences, design an algorithm of waveform forecasting and define the conditions of optimum accuracy. Verify the theses in model tests (in Polish), M. Sc. Thesis, Silesian Technical University, Gliwice, 1972.
57. Krzywiecki, J.: Melted steel temperature measurement in a complex control system. Analysis of a measurement channel. Software design (in Polish), *Pomiary. Automatyka. Kontrola*, No 2 (1981).
58. Kukułka, R.: Studies of the limiting distribution of systematic errors for given tolerances of construction parameters (in Polish), M. Sc. Thesis, Institute of Automatic Control, Silesian Technical University, Gliwice, 1982.
59. Kulikowski, R.: *Control in Large Systems* (in Polish), WNT, Warszawa, 1970.
60. Lathi, B. P.: *Signals, Systems and Communication*, J. Wiley & Sons, New York, 1965.
61. Likharev, K. K., Ulrich, B. T.: *Principles of Systems with Josephson Junctions* (in Russian), Izd. Mosk. Univ., Moskva, 1978.
62. Lindenlaub, J. C., Cooper, G. R.: Noise limitations of system identification techniques, *IEEE Trans. Automatic Control*, **8** (1963).
63. Liptak, R. G.: Costs of process instruments, *Chem. Eng.*, **77** (1970).
64. Lisak, L.: A statistical analysis of typical noise sources (in Polish), M. Sc. Thesis, Institute of Automatic Control, Silesian Technical University, Gliwice, 1969.
65. Maćkowiak, R.: A metrological system in electronic industry (in Polish), Ph. D. Thesis, Wrocław Technical University, Wrocław, 1971.
66. Mandelshtam, S. M., Solopchenko, G. N.: Metrological service in industrial measurement systems (in Russian), *Pribory i Sredstva Upravleniya*, No 1 (1977).
67. Mańczak, K.: *Methods of Identification of Multidimensional Objects of Automatic Control* (in Polish), WNT, Warszawa, 1971.
68. Mańczak, K.: *Experimental Design Techniques* (in Polish), WNT, Warszawa, 1976.
69. Mańczak, K., Nahorski, Z.: *Computer Identification of Dynamic Objects* (in Polish), PWN, Warszawa, 1983.
70. Means of Measurement. Standardizable General Metrological Characteristics. III draft of OIML Recommendation SP21 SD1, 1981.

71. *Measurement Systems* (in Polish), ed. J. Piotrowski, Silesian Technical University Handbook 573, Gliwice, 1974.
72. *Measurements, Definitions and Theories*, eds. C. W. Churchman and P. Ratoosh, J. Wiley & Sons, New York, 1959.
73. Mention of reference materials in international standards, *ISO Guide*, **6** (1977).
74. Metrology. Standardized metrological characteristics of measuring instruments, *COMECON Recommendations MS*, 53–78, Berlin, 1978.
75. Mirskii, G. J.: *Characteristics of Stochastic Relationships and their Measurements* (in Russian), Energoizdat, Moskva, 1982.
76. Nalimov, V. V.: *Theory of Experiment* (in Russian), Nauka, Moskva, 1971.
77. Niederliński, A.: *Digital Industrial Automatic Control Systems* (in Polish), Vol. 2, WNT, Warszawa, 1977.
78. Novitskii, P. V.: *Principles of Information Theory of Measuring Instruments* (in Russian), Energia, Moskva–Leningrad, 1968.
79. Nowak, J.: Quad-electrode cell with calculable constant for electrolyte conductivity measurements, *IX Conference of Metrologists* (in Polish), Gliwice, 1974.
80. Olszewska, D.: Algorithms for calibration and control of a measuring instrument applying a mathematical model of instability error (in Polish), Ph. D. Thesis, Silesian Technical University, Gliwice, 1980.
81. Ordyncew, W. N.: *Mathematical Description of the Object of Automatic Control* (in Polish), WNT, Warszawa, 1968.
82. Ostrovskii, L. A.: *Principles of the General Theory of Electrical Measuring Instruments* (in Russian), Energia, Leningrad, 1971.
83. Otnes, R. K., Enochson, L.: *Digital Series Analysis*, J. Wiley & Sons, New York, 1972.
84. Pawłowski, Z.: *Introduction to Mathematical Statistics* (in Polish), 3rd ed., PWN, Warszawa, 1979.
85. Pfanzagl, J.: *Theory of Measurement*, J. Wiley & Sons, New York, 1968.
86. Pierzchała, W.: Definitions and basic concepts in measurement systems (in Polish), *Zesz. Nauk. Pol. Śl. Automatyka*, **22** (1973).
87. Piotrowski, J.: Influence of some of the construction parameters on the metrological properties of a thermomagnetic oxygen analyser (in Polish), Ph. D. Thesis, Silesian Technical University, Gliwice, 1964.
88. Piotrowski, J.: Classification and definitions of thermal errors of gas analysers (in Polish), *Pomiary. Automatyka. Kontrola*, No 8/9 (1966).
89. Piotrowski, J.: Methodology of measurements of thermal errors of gas analysers (in Polish), *Pomiary. Automatyka. Kontrola*, No 1 (1968).
90. Piotrowski, J.: Application of mathematical models in measurement system design (in Polish), *Proc. Symp. of the Metrology Commitee of the Polish Academy of Sciences, Wisla*, 1975 (ed. Gliwice, 1974).
91. Piotrowski, J.: *Principles of Metrology* (in Polish), PWN, Warszawa, 1976.
92. Piotrowski, J.: Statistical interpretation of the concept of accuracy class applying geometric distribution (in Polish), *Normalizacja*, No 5 (1980).
93. Piotrowski, J.: Statistical interpretation of the concept of accuracy class applying order statistics (in Polish), *Normalizacja*, No 6 (1980).
94. Piotrowski, J.: Estimation of the systematic and random errors at a small number of error measurements (in Polish), *Normalizacja*, No 2 (1981).
95. Piotrowski, J.: Classification of measuring tools according to the properties of errors (in Polish), *Normalizacja*, No 3 (1981).

96. Piotrowski, J.: Evaluation of measurement results by analysis of a mathematical model of a measurement system (in Polish), *Pomiary. Automatyka. Kontrola*, No 4 (1981).
97. Piotrowski, J.: Static metrological properties of measurement tools as a multidimensional random process. Basic definitions (in Polish), *Pomiary. Automatyka. Kontrola*, No 2 (1982).
98. Piotrowski, J.: Propagation of standards and errors (in Polish), *Normalizacja*, Nos 2, 3, 4 (1983).
99. Piotrowski, J.: Mathematical modelling of the errors of measurement systems, applying a multidimensional non-stationary random process (in Polish), *Zesz. Nauk. Pol. Śl. Automatyka*, 77 (1984).
100. Plebański, T.: *Physicochemical Standards* (in Polish), Wyd. Normalizacyjne, Warszawa, 1979.
101. Pociask, U.: A methodology of selection of a measurement transducer based on effectiveness analysis of the controlled process (in Polish), Ph. D. Thesis, Szczecin Technical University, Szczecin, 1974.
102. Poleck, H.: Die Sicherheit statischer Fehlergrenzen bei der Fehlergrenzenfortpflanzung, *Archiv für technisches Messen*, J021, 8, 9, 10 (1964).
103. Polish Standard PN-65/M-53950. Flow rate measurement by means of orifice plates, nozzles and Venturi tubes.
104. Polish Standard PN-71/N02050. Metrology.
105. Polish Standard PN-77/E-06509. Standard resistors. General requirements and tests.
106. Popielarz, S.: Tests of the PT-type telemetric system (in Polish), M. Sc. Thesis, Institute of Automatic Control, Silesian Technical University, Gliwice, 1978.
107. Rabinovic, S. T., Reznik, K. A.: Determination of the number of steps in checking systems (in Russian), *Izmeritel'naya Tekhnika*, 9 (1972).
108. Ralston, A.: *A First Course in Numerical Analysis*, McGraw-Hill, Toronto–London–Sydney, 1965.
109. Report of Codata Task Group on fundamental constants, *Codata Bull.* No 11 (1973).
110. Romer, E.: *Industrial Measurements* (in Polish), PWN, Warszawa, 1978.
111. Rozenberg, V. Y.: *Introduction to the Theory of Error in Measurement Systems* (in Russian), Nauka, Moskva, 1979.
112. Rubincev, N. A., Frumkin, W. D.: Optimum structure of a calibration scheme (in Russian), *Izmeritel'naya Tekhnika*, 3 (1970).
113. Sage, A. P., Melsa, J. L.: *Estimation Theory with Application to Communications and Control*, McGraw-Hill, New York, 1971.
114. Sage, A. P., Melsa, J. L.: *System Identification*, Academic Press, New York–London, 1971.
115. Schwartz, L.: *Analyse Mathematique*, Hermann, Paris, 1967.
116. Schweppe, F. C.: *Uncertain Dynamic Systems*, Prentice-Hall, Englewood Cliffs, N. J., 1973.
117. Silvey, S. D.: *Statistical Inference*, J. Wiley & Sons, New York, 1970.
118. Sotskov, B. S. *Reliability of the Elements and Devices of Automatic Control* (in Russian), Vysshaya Shkola, Moskva, 1970.
119. Springer, J.: Methodological aspects and mathematical evaluation of analytical spectral standards, on the example of high-alloy nichrome steel (in Polish), Ph. D. Thesis, Silesian Technical University, Gliwice, 1981.
120. Stakhov, A. P.: *Introduction to an Algorithmic Theory of Measurement* (in Russian), Sov. Radio, Moskva, 1977.

121. Strauch, E.: *Metrology and Environment* (in Polish), PWN, Warszawa, 1975.
122. Sveshnikov, A. A.: *Basic Methods in Random Function Analysis* (in Russian), Sudpromgiz, Leningrad, 1961.
123. Świder, W.: Calculating corrections to the constitution of standard gases (in Polish), *Pomiary. Automatyka. Kontrola*, No 5 (1964).
124. Tabin, S.: Analysis of a method of direct digital measurement of frequency shift in Doppler ultrasonic measurements of a biphase mixture flow rate (in Polish), Ph. D. Thesis, Silesian Technical University, Gliwice, 1982.
125. *Technical Physics* (in Czechoslovakian), SNTL, Praha, 1961.
126. *Theory of Measurement* (in Polish), ed. H. Szydłowski, PWN, 1981, Warszawa.
127. Thompson, A. M.: The cylindrical cross-capacitor as a calculable standard, *Proc. IEEE*, No 2887M (1959).
128. Thompson, A. M.: An absolute determination of resistance based on a calculable standard of capacitance, *Metrologia*, **4** (1968).
129. Tou, J. T., Gonzales, R. C.: *Pattern Recognition Principles*, Addison-Wesley Publ. Co., Massachusetts, 1974.
130. Unbehauen, R.: *Systemtheorie*, Springer-Verlag, München–Wien–Oldenburg, 1971.
131. Urban, A.: *Principles of Measurement*, Part 1 (in Polish), Warsaw Technical Univ. Publ., Warszawa, 1977.
132. Van der Ziel, A.: *Noise, Sources, Characterization, Measurement*, Prentice-Hall, New York, 1970.
133. Vasilenko, G. I.: *Theory of Signal Reproduction* (in Russian), Sov. Radio, Moskva, 1979.
134. Vostroknutov, N. N., Zemelman, A. A., Kaslakov, V, M.: Selection of standards for a periodic calibration applying reliability critetia (in Russian), *Izmeritel'lnaya Tekhnika*, 7 (1977).
135. Waluś, S.: Determination of the metrological properties of ultrasonic flowmeters from a mathematical model (in Polish), Ph. D. Thesis, Silesian Technical University, Gliwice, 1980.
136. Waluś, S., Żelezik, J.: Analysis of a measurement system for decarburization rate measurement in a converter (in Polish), *Proc. of VI Nation. Conf. on Automatic Control*, Vol. III, Poznań, 1974.
137. Woschni, E. G.: *Messdynamik*, Hirzel-Verlag, Leipzig, 1972.
138. Wójcicki, R.: *Formal Methodology in Empirical Sciences* (in Polish), Ossolineum, Wrocław, 1974.
139. Zausznica, A.: *Theory of Colour* (in Polish), PWN, Warszawa, 1959.
140. Żelezik, J.: Model of a decarburization process in a converter with heat applied to final carbon contents control in the bath (in Polish), Ph. D. Thesis, Silesian Technical University, Gliwice, 1979.

Supplementary Literature

141. Aliev, T. M., Seidel, L. R., Tier-Chatchaturov, A. A.: Method of improving accuracy of digital measurements of analog quantities (in Russian), *Avtometria*, No 5 (1969).
142. Alijev, T. M., Sekkhanov, A.: Iterationsmesssystem zur analog-digitalen Signalverarbeitung. *Messen, Prüfen, Automatik*, No 12 (1983).
143. Allisy, A.: Les incertitudes des mesures. Applications pratiques, *Recueil de Travaux du BIPM*, Vol. 9, 1983–84.

144. Bachmair, H. et al.: Determination of the SI unit ohm based on a calculable cross capacitor, *Metrologia*, **22**, No 3 (1986).
145. Bjerhammar, A.: *Theory of Errors and Generalized Matrix Inverses*, Elsevier Scientific Publishing Co., Amsterdam–London–New York, 1973.
146. Bonfig, K. W., Melik-Schachnazarov, A. M., Alijev, T.: Methoden zur Erhöhung der Präzision von Messsystemen, *Messen. Prüfen. Automatik* No 1/2, (1983).
147. Bromberg, E. M., Kulikovskii, K. L.: *Test Methods of Improving the Accuracy of Measurements* (in Russian), Energia, Moskva, 1978.
148. Delahaye, F. et al.: Precise quantized Hall resistance measurements in $GaAs/Al_xGa_{1-x}As$, and in $Al_xGa_{1-x}As/InP$ heterostructures, *Metrologia*, **22**, No 3 (1986).
149. DIN 1319. Grundbegriffe der Messtechnik. Teil 3—Begriffe für die Messunsicherheit und für die Beurteilung von Messgeräten und Messeinrichtungen, August 1983; Teil 4—Behandlung von Unsicherheiten bei der Auswertung von Messungen, Dezember 1985.
150. Elyasberg, P. E.: *Measurement Information: How Much is Needed? How to Process It?* (in Russian), Nauka, Moskva, 1983.
151. *Etalons and Accurate Measurements of Electrical Quantities* (in Polish), ed. J. Dudziewicz, WKiŁ, Warszawa, 1982.
152. Fomin, A. F., Novotselov, O. N., Plushchev, A. V.: *Rejection of Abnormal Measurement Results* (in Russian), Energoatom Izd., Moskva, 1985.
153. Freedman, H. W.: The "little variable factor". A statistical discussion on the reading of seismograms, *Bull. Seismol. Soc. Am.*, **56** (1966).
154. Geisler, M. A.: The sizes of simulation samples required to compute certain inventory characteristics with stated precision and confidence, RM-3242-RP, The Rand Corp., Santa Monica, Calif., Oct. 1962.
155. Godziszewski, J., Mania, R., Pampuch, R.: *Principles of Experiment Design and Processing Measurement Results* (in Polish), AGH, Kraków, 1982.
156. Goyal, R., Brodie, B. T.: Recent advances in precision AC measurement, *IEEE Trans. Instrum. and Measurement*, IM-33, No 3 (1984).
157. *Handbook of Measurement Science*, ed. P. H. Sydenham, J. Wiley & Sons, New York, 1982.
158. Huber, P. J.: Robust estimation of a location parameter, *Ann. Math. Stat.*, **35** (1964), 73–101.
159. Huber, P. J.: *Robust Statistics*, J. Wiley & Sons, New York, 1981.
160. Kavalerov, G. I., Mandelstam, S. M., Solopchenko, G. N., Khrumalo, V. M.: Main problems of estimation of dynamic measurement errors (in: *Instrument Construction and Automatic Control*) (in Russian), Masinostroenie, Moskva (1978).
161. Korn, G. A.: *Random-Process Simulation and Measurements*, McGraw-Hill Inc., New York, 1966.
162. Kose, V., Wöger, W.: Fundamental constants and the units of physics, *Metrologia*, **22**, No 3 (1986).
163. Kurotshkin, E. P., Muminov, N. A., Kopeykin, S. V.: *Adaptive Methods of Processing the Measurement Information* (in Russian), Izd. FAN, Tashkent, 1986.
164. Landwehr, G.: The discovery of the quantum Hall effect, *Metrologia*, **22**, No 3 (1986).
165. *Modern Signal Processing*, ed. Kailath T., Hampshire Publishing Corp., Washington, & Springer-Verlag, Berlin, 1985.
166. Mudrov, J. J., Kushko, V. L.: *Methods of Processing the Measurement Information* (in Russian), Radio i Svaz', Moskva, 1983.

167. Müller, J. W.: The treatment of measurement uncertainties, *Recueil de Travaux du BIPM*, **9** (1983/1984).
168. Müller, S.: Ein hochgenaues Messverfahren für Gruppennormale, dargestellt am Beispiel der Normalelemente. *Elektric* **35**, No 3 (1981).
169. NBS Spec. Publ. 617—*Precision Measurement and Fundamental Constants* II, eds. Taylor B. N., Phillips W. D., US Government Printing Office, Washington D. C., 1984.
170. Novitskii, P. V.: *Gütenkriterien fur Messeinrichtungen*, VEB Verlag Technik, Berlin, 1978.
171. Novitskii P. V., Zograf, I. A.: *Estimation of Measurement Errors* (in Russian), Energoatomizdat, Leningrad, 1985.
172. Osmolovskii, P. F.: *Iterative Multi-Channel Automatic Control Systems* (in Russian), Sov. Radio, Moskva, 1969.
173. Petzold, B., Richter, W.: Störspannung analyse im Tieffrequenzbereich, *Messen. Steuern, Regeln*, No 2 1986.
174. Piotrowski, J. On the need of systematic error—polemic with the CI-1981 CIPM Recommendation, *Proc. XIth IMEKO World Congress* (*Houston* 16–21 *Oct.* 1988), "*Theory, Simulation, Calibration, Testing, Education*". pp. 241–252.
175. Pugachev, V. S.; *Theory of Random Functions and Its Application in Automatic Control* (in Russian), Fizmatgiz, Moskva, 1962.
176. Rao, C. R.: *Linear Models in Mathematical Statistics*, J. Wiley & Sons, New York, 1976.
177. Rau, Z.: Algorithms of intercomparison of a group standard (in Polish), *Referaty problemowe Instytutu Łączności*, Warszawa, 1983, No 62.
178. Report on the BIPM Enquiry on Error Statement. *Proces-Verbaux CIPM*, **48** (1980), Annexe III, p. C 1–36.
179. Rozenberg, V. J.: *Introduction to a Theory of Accuracy of Measurement Systems* (in Russian), Moskva, Sov. Radio, 1975.
180. Shalygin, A. S., Palagin, Yu. I. *Methods of Statistical Modelling* (in Russian), Mashinostroenie, Leningrad, 1986.
181. Shannon, R. E.: *System Simulation. The Art and Science*. Prentice-Hall, Englewood Cliffs, N. J., 1975.
182. Skubis, T.: Inaccuracy reduction method of the true conventional values of standards, *EMISCON Conference*, Gottwaldov, 1987.
183. Srinath, M. D., Rajesekaran, P. K.: *An Introduction to Statistical Signal Processing with Applications*, J. Wiley & Sons, New York, 1979.
184. *Statistical Methods for Digital Computers*, eds. K. Enslein, A. Ralston, M. S. Wilf, J. Wiley & Sons, New York, 1977.
185. Tikhonov, A. N., Goncharskii, A. V., Stepanow, V, V., Yagola, A. G.: *Regularizing Algorithms and a priori Information* (in Russian), Izd. Nauka, Moskva, 1983.
186. Tukey, J. W.: The future of data analysis, *Ann. Math. Stat.*, **33** (1967), 1–67.
187. von Klitzing, K., Dorda, G., Pepper, M.: Realization of a resistance standard based on fundamental constants, *Phys. Rev. Lett.*, **45** (1980).
188. Wagner, S. R.: On the quantitative characterization of the uncertainty of experimental results in metrology, *PTB Mitteilungen*, **8**, No 2 (1979).
189. Weygand, K., Willenberg, G. D.: Realization of the ampere based on high- and low-field NMR experiments, *Metrologia*, **22**, No 3 (1986).
190. Wilbur-Ham, J.: Internal comparison of a large group of standard cells, *IEEE Trans. on Instr. and Measurement*, IM-22, No 3 (1973).
191. Winiecki, W.: Method of processing measurement data by applying piecewise-uniform

functions, and its applications (in Polish), Ph. D. Thesis, Warsaw Technical University, Warszawa, 1986.

192. Yershov, A. A.: Stable methods of parameter estimation (in Russian), *Avtomatika i Telemekhanika*, No 8 (1978).
193. Helmholtz, H. L. F.: Zahlen und Messen erkenntnis—theoretisch betrachtet. *Philosoph. Aufsätze Eduard Zeller... gewidmet*, Leipzig, Fues, 1887.
194. Niemeyer, J., Hinken, J. H., Vollmer, E., Grimm, L., Meier, W.: A new Josephson voltage standard with tunnel-junction series arrays, *Metrologia*, **22**, No 3 (1986).
195. Braun, E., Warnecke, P. Leontiew, H.: Reproduction of the ohm using quantum Hall effect, *Metrologia*, **22**, No 3 (1986).
196. Ivachnienko, A. G., Müller, J. A.: *Selbstorganization von Vorhersagemodellen*, Verlag Technik, Berlin; Verlag Technika, Kiev, 1984.
197. Kochsiek, M.: Fortschritte bei der Darstellung der Messskala, *PTB Mittellungen*, **89**, No 6 (1979).
198. Isayev, L. K., Mechannikov, A. I.: *Time Variability of the Errors of Measuring Instruments. General Problems in Metrology* (in Russian), Izd. Standardov, Moskva, **130** (190), 1972.
199. Pugachev, V. S.: *Reliability and Mathematical Statistics* (in Russian), Nauka, Moskva, 1979.
200. Aliev, T. M., Seidel, L. R., Tier-Chatchaturov, A. A., Sekhukhanov, A. M.: *Iterative Methods of Improving Accuracy of Measurement* (in Russian), Energoatomizdat, Moskva, 1986.
201. Granovskii, V. A.: *Dynamic Measurements—Methods of Metrological Treatment* (in Russian), Energoizdat, Leningrad, 1984.
202. Andryanov, A. V., Shpak, J. J.: *Digital Information Processing in Measurement Systems and Instruments* (in Russian), Vysshaya Shkola, Minsk, 1987.
203. *Robustness in Statistics*, eds. Launer R. L., Wilkinson G. N., Academic Press, New York, San Francisco, London, 1979.
204. Morawski, R. Z.: Model inversion as a fundamental problem of metrology, *1st International IMEKO TC4 Workshop on "Measurement—Oriented Modelling of Electrical Objects"*, Prague, March 27–30, 1988.
205. Morawski, R. Z.: Direct methods of measurement signal reconstruction, *Proc. of XI IMEKO World Congress* (*Houston, 16–21 Oct. 1988*), "*Applications*", pp. 111–120.
206. Quinn, T. J.: New techniques in the manufacture of platinum–iridium mass standards, *Platinum Met. Rev.*, **30** (2) (1986).
207. Preston-Thomas, H., Quinn, T. J., Hudson, R. P.: The International Practical Temperature Scale, *Metrologia*, **21** (1985), 75–79.
208. Kochsiek, M.: The mass unit "kilogram", precision measurement of mass, attainable uncertainties and possibilities of a new definition, in: [169].
209. Cage, M. E. et al.: Status of the NBS-NRL determination of the fine-structure constant using the quantized Hall resistance effect, in: [169].
210. Delahaye, F.: Present state of quantized Hall resistance metrology, *Metrologia*, **25** (1988), 73–79.
211. Niemeyer, J., Grimm, Z.: Present status of Josephson voltage metrology, *Metrologia*, **25** (1988), 135–140.
212. Hampel, F. R.: Robust estimination: a condensed partial survey, *Zeitschr. Wahrscheinlichkeitstheorie und verw. Gebeite*, t. 27, 1973.

Index

accuracy class 199
adjustment 208, 229, 246
ambiquity of mapping 22, 26
approximation of characteristic 73, 268
— of a model 74
averaging 151, 154, 161, 208

balance detector 85, 225

calibration 43, 227, 229, 232
— direct 232
— indirect 232
— procedure parameter 150
category qualitative 60
certified reference material 60
characterictic approximation 73, 268
— fixing 230
— of measurement system 72, 229, 230
— metrological 164, 171, 175
classification of errors 153
— of measurement procedures 151
class of images 33
coefficient of correlation 113, 157, 180, 194
— — model an MS 93, 143
cognition, rational 1
— sensory 1
— subject 1
comparator 46
confidence interval 25
— — for regression function 137, 271
— level (confidence) 25, 138
construction parameter 72, 75, 93, 100, 102
conversion equation 70, 104, 227
correlation coefficient 113, 157, 180, 184, 194, 286
— function 114, 127, 177
— of errors 178, 194, 286
covariance matrix 124, 286
Cramer–Rao inequality 137, 200
criterion of small error 236

decision rule 19
definition of error 144, 150
distance of sets 31, 234
— set's elements 22
distribution of calibration points 273
— — errors 165, 167, 188, 273
domain of image 4

entropy 26, 173
— coefficient 173
error 105, 146, 149
— allowable 170
— classification 153
— correction 163, 246, 260
— correlation 178, 194
— criterion of small 236
— definition 146, 153, 161
— distribution 165, 167, 188
— dynamic 161, 163, 258, 260
— fundamental 155
— instability 155, 161
— limiting distribution 171, 187, 191
— — value 249, 255
— model 104, 111, 154
— of calibration 160, 161
— — comparison 46
— — sensitivity 158, 176, 217, 245
— — zero 158, 176, 217, 245
— random 153
— — of a standard 43, 44, 57
— — supplementary 157

— source 93, 100, 149, 268
— supplementary 156, 158, 161
— systematic 153
— — of a standard 43, 44, 47, 56
— — supplementary 157
environmental infiuence on MS 91
etalon (standard) 58
experimental design 216, 279

figure of merit 150
Fisher matrix 140, 270
fixing characteristic 230
— — algorithmic 230
— — constrictional 230
— — mixed 231
formulation of the MS's model 72, 234
F-test 138, 195
function mapping 3, 19
— inverse mapping 3, 19
— of correlation 114, 177
— — cross-correlation 177
— — regression 137, 267
— — — confidence interval 137, 271

group standard 40

image of matter 60
— — reality 2, 6, 225
— — state 19, 20, 27
information theory 218
interpretation of the accuracy class 199
interval of uncertainty of measure 47

kernel of operator 78
— — — conjugated 162
— — the Volterra operator 161

Lebesque measure of a set 23, 29
LSM (least squares method) 131, 140, 143, 267

mapping of state 2, 19, 69
— operator 20, 70, 263
matrix of covariance 124, 286
maximum likelihood function 137
measurement 1, 220
— direct 227
— equation 225, 226
— indirect 228
— method 221, 225
— — comparison 52, 222, 226
— — compensation 222, 226
— — difference 222, 226
— — iterative 213
— — null 222, 226
— — substitution 222, 226
— — swing 221, 226, 228
— multiple 250
— procedure 222, 225
— — for constant quantities 247
— — — time-dependent quantities 256
— result 221
— — complete 227
— — corrected 147, 226
— — raw 226
— scale 4, 9, 70
— — empirical 7, 9, 14
— — metric 5, 7, 9, 224
— — multudumensional 28
— — vector 30
— series 250, 285
— single 248, 257
— system (MS) 67
— — environment 234
measure of ambiguity of mapping 24
— — a standard 46, 52, 54, 266
— — — nominal 266
— of measurement 221, 225
model mathematical 67, 71
— — digital 76
— — discrete-time 82, 111
— — dynamic 76
— — linear 77
— — nonlinear 74
— — of an MS 68, 72, 148
— — — — coefficients 93, 143
— — — — extended 150
— — — — formulation 72, 234
— — — — prediction 150, 152
— — — — simplified 150
— — — error 104
— — physical 72, 79
— — static 73
— — statistical 72, 76
— regressive 131

neighbourhood of state 20, 147
norm of space 22

object of measurement 1, 220
optimization of calibration procedure 151, 242
optimum design 216, 272
order statistic 202

prediction of MS model 150, 152
procedure 222, 225
— of adjustment 208, 229, 246
— — calibration 71, 148, 150, 160, 230, 266
— — —, averaging 151, 161
— — —, dynamic 151, 162
— — —, natural 156, 162
— — —, optimization 151, 242
— — —, predictive 152
— — —, static 151
— — measurand reproducing 70
— — measurement 222, 256
— — reading checking 229, 242
process random, bidimensional 155
— —, non-stationary 43, 105, 175
— —, stationary 43, 108
— — unidimensional 106
propagation of standards 42, 266
purpose of measurement 234

quantitity influencing 72
— input 72
— metric 7
— output 72
— time-dependent 30, 77, 256
— unmeasurable 15, 17
— value 2, 6, 146
— —, correct 147

range 26, 202
reading checking 229, 242
reference conditions 155
— state 9, 36, 207, 225
regression method (analysis) 131, 267, 278
relation of equivalence 5
— — strict order 5

sampling interval 178, 285
scale measurement 4, 9, 70
— — empirical 7, 9, 14
— — mathematical 4, 18
— — metric 5, 7, 9, 71, 224
— — multidimensional 28
— — of colour, trichromatic 10
— — physical 18
— — theory 8
scheme of measurement method, functional 222
— — — system, structural 222
— — standards comparison 57
series of measurements 250, 285
service time 43, 168, 244
set of states 2
— — values 2, 18
signal 67, 79
standard deviation 24
— of unit of measure, establishing 37, 224
standard primary 42
— secondary 42
standards comparison, single 46
— — multiple 49, 50
state of quality 1, 6, 18, 70, 146

transfer of measure 266

uncertainty of measure of a standard 43, 45, 49, 58, 62

value of error limiting 249
Volterra operator 78, 98, 260